HISTORICAL MANUSCRIPTS

INTELLIGENCE IN PLANTS AND ANIMALS

HISTORICAL MANUSCRIPTS

Additional books and e-books in this series can be found on Nova's website under the Series tab.

HISTORICAL MANUSCRIPTS

INTELLIGENCE IN PLANTS AND ANIMALS

THOMAS G. GENTRY

snova
New York

NOTICE TO THE READER

The Publisher has taken reasonable care in the preparation of this book, but makes no expressed or implied warranty of any kind and assumes no responsibility for any errors or omissions. No liability is assumed for incidental or consequential damages in connection with or arising out of information contained in this book. The Publisher shall not be liable for any special, consequential, or exemplary damages resulting, in whole or in part, from the readers' use of, or reliance upon, this material. Any parts of this book based on government reports are so indicated and copyright is claimed for those parts to the extent applicable to compilations of such works.

Independent verification should be sought for any data, advice or recommendations contained in this book. In addition, no responsibility is assumed by the Publisher for any injury and/or damage to persons or property arising from any methods, products, instructions, ideas or otherwise contained in this publication.

This publication is designed to provide accurate and authoritative information with regard to the subject matter covered herein. It is sold with the clear understanding that the Publisher is not engaged in rendering legal or any other professional services. If legal or any other expert assistance is required, the services of a competent person should be sought. FROM A DECLARATION OF PARTICIPANTS JOINTLY ADOPTED BY A COMMITTEE OF THE AMERICAN BAR ASSOCIATION AND A COMMITTEE OF PUBLISHERS.

Additional color graphics may be available in the e-book version of this book.

Library of Congress Cataloging-in-Publication Data

ISBN: 978-1-53619-476-0

Published by Nova Science Publishers, Inc. † New York

CONTENTS

List of Illustrations

PREFACE*

Nothing is more charming to the mind of man than the study of Nature. Religion, moderation and magnanimity have been made a part of his inner being through her teachings, and the soul has been rescued by her influence from obscurity. No longer doth man grovel in the dust, seeking, animal-like, the gratification of low and base desires, as was his wont, but on the wings of thought is enabled to soar to the very gates of Heaven and hold communion with God.

Though made "a little lower than the angels," yet, through the mighty play of forces that have been at work in the world, which we, in the latter half of this enlightened century, are just beginning to recognize and comprehend, he has been lifted from the mire of degradation and placed upon a higher social, intellectual, moral and spiritual level. Out of the animal, in the scheme of Deity, the spiritual system of things is to be elaborated, and not the animal out of the spiritual. This natural world, so to speak, is the raw material of the spiritual. Therefore, ere man can understand the spiritual, he must understand the natural. Though his knowledge was at first about material things, or such as pertained to natural phenomena, yet from this through the ages has been builded, little by little, that mountain-height of knowledge, intellectual and moral, which, if rightly directed, is to bring him into fellowship with Deity. "As we have borne the image of the earthy, we shall also bear the image of the heavenly," or, Lord from heaven.

When is considered, therefore, the immense good which the study and investigation of nature have accomplished, it is not at all surprising that the literature on the subject should be markedly in the ascendant. Natural science bids fair to be in a preëminent degree the pursuit of the coming man. There is no end to the books that have been written upon the subject during the past few decades, if not by specialists, but by men and women who have

* This is an edited, augmented and reformatted version of "Intelligence in Plants and Animals" by Thomas G. Gendry, Sc.D., originally published by New York, Doubleday, Page & Co., dated 1900. The views, opinions, and nomenclature expressed in this book are those of the authors and do not reflect the views of Nova Science Publishers, Inc.

been well informed and who have made themselves fully capable of contemplating understandingly the world which lies about them.

Our libraries are to-day quite affluent in books that are the handmaids of natural science. Michelet and Hugh Miller, in their day, opened glorious new worlds before a rising generation, and that generation is now doing excellent work under the inspiration of the impetus which it then received. Tait, Balfour Stewart, Dawson, Gray, McCook, Thompson, Scudder, Mrs. Treat, Olive Thorne Miller and others have done much to continue the interest, pleasure and enthusiasm awakened by those earlier writers, and even Darwin and Huxley themselves, in detailing their experiments, have not scorned to bring their thoughts within the range of narrower minds.

But in the popularization of natural science no man has done more than Rev. J. G. Wood in his numerous works. Not only have his writings created in thousands a taste for nature-studies, but they have been no less the means of cultivating the observation, awakening enthusiasm and directing effort in the lines of original research and discovery. Certainly no one, as his many writings so abundantly attest, possessed a larger fund of knowledge concerning the powers and capabilities of the lower animals than this author. Few knew our domestic animals better than he, and none was more capable of judging of the mental and moral *status* which they should occupy in the world of animals. It is true that men and women, eminent in theology, literature and science, had expressed a belief in the idea that the "latent powers and capacities" of the lower animals might be developed in a future life, but no one had felt secure enough in this belief to warrant more than a passing thought or two upon the subject.

Bishop Butler, in his "Analogy of Religion," undoubtedly believed the lower animals capable of a future life. In speaking of them in this connection in the opening of his work, he says: "It is said these observations are equally applicable to brutes; and it is thought an insuperable difficulty that they should be immortal, and by consequence capable of everlasting happiness. And this manner of expression is both invidious and weak; but the thing intended by it is really no difficulty at all, either in the way of natural or moral consideration."Referring then to the undeveloped powers and capacities of the so-called brutes, the Bishop could perceive no reason why they should not attain their development in an existence beyond the earth-life. It was in pursuance of this same train of thought that Rev. J. G. Wood was led to show in a work, entitled "Man and Beast Here and Hereafter," that the lower animals do possess those mental and moral characteristics—the attributes of reason, language, memory, moral responsibility, unselfishness and love—which we admit in man as belonging to the immortal spirit, rather than to the perishable body. Having previously cleared away the difficulties which certain passages in the Old Testament seemingly interposed, and proved that the Scriptures do not deny futurity of life to lower animals, he very naturally concluded that as man expects to retain these qualities in the future life there is every reason to suppose that they may share his immortality in the Hereafter as in the Now they are partakers of his mortal nature.

Few minds, unswayed by thoughts materialistic, can study the living works of God, whether vegetal or animal, and fail to be convinced that they, as living exponents of Divine conceptions, are as needful in the world of spirit as in the world of matter. While many are disposed to believe that man will share the future life with beast, bird, insect and such like, yet but few, if any, can be found who believe that tree and shrub and flower will be there to continue the life begun on earth and reach out to higher and fuller development. In announcing this belief, the author but expresses a conviction as deep as any that could occupy a human mind. The possession of soul and spirit can be predicated no less of plants than of man and the lower animals. They have all one breath or life and one spirit, and as such are living souls, living, breathing frames or bodies of life. From being living, breathing frames, and endowed with the same life and spirit as man and the lower animals, they have all one destiny, for "all go unto one place; all are of the dust, and all turn to dust again." But of the new life which Christ came down to earth to proffer to man that he might inherit the kingdom of God. While to man it was only offered, and had for its purpose the uplifting and improvement of his earth-life by the promise of something higher and better to those who are accounted worthy, yet there can be no doubt that it was equally intended through his uplifting to place all the creatures of the earth over which he was given dominion by God upon a more elevated and nobler plane, so that those which had been profited in the earth-life by his beneficent influence should become partakers with him in the new life, when Christ shall "transfigure the body of our humiliation, that it may become of like form with the body of His glory, by the power of that which enables Him even to subdue all things to Himself." As all existence is a unit, which the author has taken especial pains through the body of this book to impress upon the minds of his readers, it can hardly be conceived that an all-wise God, who is infinite in love, mercy and justice, would look to the preservation in a future state of but a very small part of the life which He has been instrumental in placing upon this earth. It would be more consistent with His attributes, and with the scheme of development of life upon our planet, whereby life has been progressive, the fittest only being allowed to survive, to have provided in the grand plan of redemption, not merely the salvation of the highest of earth-life, but of all life, the purest and the best, that would represent in the heaven-life, in spiritualized form, the highest living exponents of Divine ideas. No other belief accords so well with the teachings of science and philosophy. In its acceptance, for it makes all life related to the Divine life, can there be any hope of escape from materialism, that curse of the age.

Thomas G. Gentry, Sc. D.
Philadelphia, February 28, 1897.

Figure 1. Snapping-turtles fighting. Copyright 1900 by A. R. Dugmore.

Chapter 1

Life and Immortality: Life and Its Conditions

All natural objects, roughly divided, arrange themselves into three groups, constituting the so-called Mineral, Vegetable and Animal kingdoms. Mineral bodies are all devoid of life. They consist of either a single element, or, if combined, occur in nature in the form of simple compounds, composed of more than two or three elements. They are homogeneous in texture, or, when unmixed, formed of similar particles which have no definite relations to one another. In form they are either altogether indefinite, when they are said to be amorphous, or have a definite shape, called crystalline, in which case they are ordinarily bounded by plane surfaces and straight lines. When mineral bodies increase in size, as crystals may do, the increase is produced simply by accretion. They exhibit purely physical and chemical phenomena, and show no tendency to periodic changes of any kind. Fossils or petrifactions, which owe their existence and characters to beings which lived in former periods of the earth's history, cannot, though made up of mineral matter, be properly said to belong to the mineral kingdom.

But objects belonging to the vegetable and animal kingdoms differ markedly from inert, lifeless, mineral matter. Carbon, hydrogen, oxygen and nitrogen are the most important of the few chemical elements which enter into their composition, and these elements are combined into complex organic compounds, which always contain a large percentage of water, are very unstable, and prone to spontaneous decomposition. They are composed of heterogeneous, but related, parts, termed organs, the objects possessing them being called organized bodies. Some of the lowest forms of animals have bodies whose substance is so uniform that they exhibit no definite organs, but this exception does not affect the general value of this distinction. They are always more or less definite in shape, presenting concave and convex surfaces, and being limited by curved lines. When they increase in size, or grow, as we properly term it, it is not by the addition of particles from the outside, but by the reception of foreign matter into their interior and its consequent assimilation. Certain periodic changes, which follow a definite and discoverable order, are invariably passed through by organized bodies. These changes constitute what is known as life. All the objects, then, which fulfil these conditions are said to be alive, and they all appertain either to the vegetable or the animal kingdom. The study of living objects, no matter to which kingdom they belong, is therefore conveniently called by the general name of Biology, which means a discourse on life. And as all living objects may be referred to one or other of these kingdoms, so Biology may be divided into Botany, which treats of plants, and Zoölogy, which treats of animals.

Now that we have divided all organized bodies into plants and animals, it becomes necessary to inquire into the differences which subsist between them, and which will enable us to separate the kindred sciences of Botany and Zoölogy. Nothing was thought so easy by older observers than the determination of the animal or vegetable nature of any given organism, but, in point of fact, no hard-and-fast line can be drawn, in the existing state of our knowledge, between the animal and vegetable kingdoms, and it is sometimes difficult, or even impossible, to decide with positiveness whether we are dealing with a plant or an animal. In the higher orders of the two kingdoms there is no difficulty in reaching a decision, the higher animals being readily separated from the higher plants by the possession of a nervous system, of a locomotive power which can be voluntarily exercised, and of an internal cavity adapted for the reception and digestion of solid food. No so-called nervous system or organs of sense are possessed by the higher plants, although some of them doubtlessly manifest conscious and intelligent action, nor are they capable of voluntary changes of place, nor provided with any definite internal cavity, their food being generally fluid or gaseous.

Descending the scale to the very bottom, we reach a class of animals, the Protozoa, which cannot be separated in many cases from the Protophyta by these distinctions, since many of the former have no digestive cavity, nor the slightest trace of a nervous system, while many of the latter possess the power of active locomotion. As to external configuration, no certain rules can be laid down for separating animals and plants, many of the lower plants, either in their earlier stages, or in their maturity, being exactly similar in form to some of the lower animals. This is the case with some of the Algæ, which resemble very closely in form certain Infusorian animalcules. Again, many undoubted animals, which are rooted to solid objects in their adult state, are so plant-like in appearance as to be popularly regarded as vegetables. The Sea-firs, and the more highly organized Flustras or Sea-mats, which are usually considered as sea-weeds by sea-side visitors, are a few of many examples that might be taken from the so-called Hydroid Zoöphytes. No decided distinction between animals and plants can be drawn as to their minute internal structure, both alike consisting of molecules, of cells, or of fibres. Some decided, though not universal, differences exist in chemical composition. Plants exhibit a decided predominance of ternary compounds, or compounds which, like sugar, starch and cellulose, are made up of the three elements, carbon, hydrogen and oxygen, but are, comparatively speaking, poorly supplied with quaternary compounds, or those which contain an additional element of nitrogen. Animals, on the contrary, are rich in quaternary nitrogenized compounds, such as albumen or fibrin. Still, in both kingdoms we find nitrogenized and non-nitrogenized compounds, and it is only in the proportion which these sustain to each other in the organism that animals differ in any way from plants.

Before the invention of the microscope, no independent voluntary movements, if we except the opening and closure of flowers, and their turning towards the sun, the drooping of the leaves of sensitive plants under irritation, and some other kindred phenomena, were known in plants. Now, however, we know of many plants which are endowed, either when young or throughout life, with the power of effecting voluntary movements apparently as spontaneous and independent as those performed by the lower animals, the movements being brought about by means of little vibrating cilia, or hairs, with which a part or the whole of the surface is furnished. When it is added that many animals are permanently rooted, in their fully-grown condition, to solid objects, it will at once be apparent that no absolute distinction can be made between animals and plants merely because of the presence or absence of independent locomotive power.

There is, however, a test, the most reliable of all that have been discovered, by which an animal may be distinguished from a plant, and that is the nature of the food and the products which are elaborated therefrom in the body. Plants live upon such inorganic substances as water, carbonic acid and ammonia, and they have the power of manufacturing out of these true organic materials, and are therefore the great producers of nature. All plants which contain green coloring matter, technically called chlorophyll, break up carbonic acid in the process of digestion into its two constituents of carbon and oxygen, retaining the former and setting the latter free. And as the atmosphere always contains carbonic acid in small quantities, the result is that plants remove carbonic acid therefrom and give out oxygen. Animals, on the other hand, have no power of living on water, carbonic acid and ammonia, nor of converting these into the complex organic substances of their bodies. That their existence may be maintained animals require to be supplied with ready-made organic compounds, and for these they are all dependent upon plants, either directly or indirectly. In requiring as food complex organic bodies, which they ultimately reduce to very simply inorganic ones, animals are thus found to differ from plants. Whilst plants are the great manufacturers in nature, animals are the great consumers. Another distinction, arising from the nature of their food, is that animals absorb oxygen and throw out carbonic acid, their reaction upon the atmosphere being exactly the reverse of that of plants. There are organisms, it must be understood, which are genuine plants so far as their nutritive processes are concerned, but which, nevertheless, are in the possession of characters which could locate them among the animals. Volvox, so abundant in our streams during the proper seasons, affords a splendid illustration of the truth of this statement. Plants, which are devoid of chlorophyll, as is the case with the Fungi, do not possess the power of decomposing carbonic acid under the influence of sunlight, but are like animals in requiring organic compounds for their food. Two points must therefore be borne in mind in regarding the general distinctions between plants and animals which we have thus briefly outlined, and these are that they cannot often be applied in practice to ambiguous microscopic organisms, and certainly not to plant-forms that are destitute of chlorophyll.

That life should manifest itself certain conditions are essential, but some of which, though generally present, are not absolutely indispensable. One condition, however, seems to be very necessary, and that is that the living body should be composed of a certain material. This material, which forms the essential and fundamental parts of everything living, whether vegetable or animal, is technically called protoplasm. Other substances than it are often found in living bodies, but it is in protoplasm only that vitality appears to be inherent.

But whether it is the same in plants as in animals is a matter of opinion. One thing, however, seems reasonably certain, and that is that it is the medium or vehicle through which vital force is made manifest. Used in its general sense, protoplasm is chemically related in its nature to albumen, and generally has the character of a jelly-like, semi-fluid, transparent material, which, in itself, exhibits no definiteness of structure. When heated to a certain temperature it coagulates, just as the white of an egg does when boiled. Living protoplasm has the power of movement, of increasing in size or of maintaining its existence by the assimilation of fresh and foreign materials, and of detaching portions of itself which may subsequently develop into fresh masses. Though protoplasm be present in the ova of animals and the seeds of plants, yet there is no external and visible manifestation of life. There is in them what is called a dormant vitality, which may remain for a long time unchanged, until altered external circumstances cause the organism to pass into a state of active life.

Generally, certain external conditions must be present before any external vital phenomena can be manifested. The presence of atmospheric air, or rather of free oxygen, is in an ordinary way essential to active life. Life, that is its higher manifestations, is only possible between certain ranges of temperature, varying from near the freezing point to about 120° Fahrenheit. As water is a necessary constituent of protoplasm in its living state, so it becomes an absolutely essential requisite to the carrying on of vital processes of all kinds, for the mere drying of an animal or plant will, in most cases, kill it outright, and will always bring about a suspension of all visible life-phenomena.

While the large majority of living beings are organized, or composed of different parts, called organs, which sustain certain relations with one another, and which discharge different offices, yet it must not therefore be concluded that organization is a necessary accompaniment of vitality, or that all living creatures are organized. Innumerous low forms of life, so low that they occupy the very lowest place in the scale of animated existences, absolutely exhibit no visible structure, and cannot, therefore, be said to be organized, but they, nevertheless, discharge all their vital functions just as well as though they possessed special organs for the purpose. Concluding our theme, we are forced to admit that animals are organized, or possess structure, because they are alive, and not that they live because they are organized. By carefully comparing the morphological and physiological differences between different animals and plants, naturalists have divided the entire animal and vegetable kingdoms into a number of divisions, whose leading characteristics may be found in almost every text-book. All that we promised ourselves when this work was first thought of was a brief treatment of a few of the most interesting life-forms of this planet of ours in the light of their ways and doings, and the direction of human thought to those traits of character and manifestations of conscious intelligence which fit them to become partakers with man of that new life which awaits him beyond the grave.

Chapter 2

PLANTS THAT FEED ON INSECTS

Perhaps it would be difficult to find in the whole range of vegetable creation anything more curious than the carnivorous or flesh-eating plants. That animals eat plants creates in us no emotion of curiosity, for this is the common law of nature. But that plants should devour animals is a marvel to which few minds uninitiated in science would give credence. Though these strange forms of vegetable life have been known for about a century, yet it has been but a few years since the attention of naturalists was first specially called to their habits and character. No one has probably done more to explain the life and operations of the flesh-eating plants than Mr. Darwin.

For centuries strange rumors had been circulated of the existence of huge plants in the more remote and unvisited parts of Asia which would imprison and destroy large animals and men that would venture within reach of their great quivering leaves armed with hooked spines, the flesh of the dead victim being absorbed into their structure, but all these giant flesh-eating trees or plants have so far proved to be mere myths. Science has discovered, however, that there is some foundation for these exciting fictions, and it has not been obliged to go to the distant East to find it, for flesh-eating plants are by no means uncommon in this country and Europe. But these plants confine their destructive propensities to the crawling and flying insects which are beguiled by some tempting reward to rest on their leaves. Such a strange provision of nature is no less interesting than if these plants had the power to destroy the larger animals, for it is the fact itself which startles the attention by its seeming reversal of natural laws.

No better example of carnivorous plants could be taken than *Dionæa muscipula*, or to use the common name, Venus's Fly-trap. It is a species that is indigenous to North Carolina and the adjacent parts of South Carolina, affecting sandy bogs in the pine forests from April to June, and a representative of the *Droscraceæ*, or Sundew Family. One cannot fail after once seeing it of becoming impressed with its peculiar characteristics. It is a smooth perennial herb with tufted radical leaves on broadly-winged, spatulate stems, the limb orbicular, notched at both ends, and fringed on the margins with strong bristles. From the centre of the rosette of leaves proceeds at the proper time a scape or leafless stalk which terminates in an umbel-like cyme of from eight to ten white bracted flowers, each flower being one inch in diameter. The roots are small and consist of two branches each an inch in length springing from a bulbous enlargement. Like an epiphytic orchid, these plants can be grown in well-drained damp moss without any soil, thus showing that the roots probably serve for the absorption of water solely. Three minute pointed processes or filaments, placed triangularly, project from the upper surface of each lobe of the bi-lobed leaf, although cases are observed where four and even ten filaments are found. These filaments are remarkable for their extreme sensitiveness to touch, as shown not only by their own movement, but by that of the lobes also. Sharp, rigid projections, diminutive spikes as it were, stand out from the leaf-margins, each of which being entered by a bundle of spiral vessels. They are so arranged that when the lobes close they interlock like the teeth of an old-fashioned rat-trap. That considerable strength may be had, the mid-rib of the leaf, on the lower side, is quite largely developed.

Minute glands, of a reddish or purplish color, thickly cover the upper surface of the leaf, excepting towards the margins, the rest of the leaf being green. No glands are found upon the spikes or upon the foliaceous footstalk. From twenty to thirty polygonal cells, filled with purple fluid, constitute each gland. They are convex above, somewhat flattened underneath, and stand on very short pedicels, into which spiral vessels do not enter. They have the power of secretion under certain influences, and also that of absorption. Minute octofid projections, of a reddish-brown color, are scattered in considerable numbers over the footstalk, the backs of the leaves and the spikes, with a few on the upper surfaces of the lobes.

The sensitive filaments, which are a little more than one-twentieth of an inch in length, and thin, delicate and tapering to a point, are formed of several rows of elongated cells, filled with a purplish fluid. They are sometimes bifid or even trifid at the apex, and towards the base there is a constriction formed of broader cells, and beneath the constriction an articulation, supported on an enlarged base, consisting of differently shaped polygonal cells. As the filaments project at right angles to the surface of the leaf, they would have been in danger of being broken off whenever the lobes closed together had it not been for the articulation, which allows them to bend flat down. So exquisitely sensitive are these filaments, from their tips to their bases, to a momentary touch, that it is hardly possible to touch them even so lightly or quickly with any hard object without causing the lobes to close, but a piece of delicate human hair, two and a-half inches in length, held dangling over a filament so as to touch it, or pinches of fine wheaten flour, dropped from a height, produce no effect. Though not glandular, and hence incapable of secretion, yet the filaments by their sensitiveness to a momentary touch, which is followed by the rapid closure of the lobes of the leaf, assure to Dionæa the necessary supply of insect food for all its wants.

Inorganic bodies, even of large size, such as bits of stone, glass and such like, or organic bodies not containing nitrogeneous matter in a soluble condition, as bits of cork, wood, moss for examples, or bodies containing soluble nitrogeneous matter, if perfectly dry, such as small pieces of meat, albumen, gelatine, etc., may be long left on the lobes, and no movement is excited. But when nitrogeneous organic bodies, which are all damp, are left on the lobes, the result is widely different, for these then close by a slow and gradual movement and not in a rapid manner as when one of the sensitive filaments is touched by a hard substance. Small purplish, almost sessile glands, as has already been stated, thickly cover the upper surface of the lobes. These have the power both of secretion and absorption, but they do not secrete until excited by the absorption of nitrogeneous matter. No other excitement, as far as experiments show, produces this effect. When the lobes are made to close over a bit of meat or an insect, the glands over the entire surface of the leaf emit a copious discharge, as in this case the glands on both sides are pressed against the meat or insect, the secretion being twice as great as when the one or the other is laid on the surface of a single lobe; and as the two lobes come into almost close contact the secretion, containing dissolved animal matter, diffuses itself by capillary attraction, causing fresh glands on both sides to begin secreting in a continually widening circle. The secretion is almost colorless, slightly mucilaginous, moderately acid, and so copious at times in the furrow over the mid-rib as to trickle down to the earth. But all this secretion is for the purposes of digestion. Be the animal matter which the enclosed object yields ever so little, it serves as a peptogene, and the glands on the surface of the leaf pour forth their acid discharge, which acts like the gastric juice of animals.

Now as to the manner in which insects are caught by the leaves of *Dionæa muscipula*. In its native country they are caught in large numbers, but whether they are attracted in any special way no one seems to know. Both lobes close with astonishing quickness as soon as a filament is touched, and as they stand at less than a right angle to each other, they have an excellent chance of capturing any intruder. The chief seat of the movement is near the mid-rib, but is not restricted to this part. Each lobe, when the lobes come together, curves inwards across its whole breadth, the marginal spikes alone not becoming curved. From the curving inwards of the two lobes, as they advance towards each other, the straight marginal spikes intercross by their apices at first, and ultimately by their bases. The leaf is then completely shut and encloses a shallow cavity. If made to shut merely by the touching of one of the sensitive filaments, or by the inclusion of an object not yielding soluble nitrogeneous matter, the two lobes retain their inwardly concave form until they re-expand. The re-expansion, when no organic matter is enclosed, varies according to circumstances, a leaf in one instance being fully re-expanded in thirty-two hours.

But the lobes, when soluble nitrogeneous matter is included, instead of remaining concave, thus containing within a concavity, slowly press closely together throughout their entire breadth, and as this takes place the margins gradually become a little everted, so that the spikes, which at first intercrossed, at last project in two parallel rows. So firmly do they become pressed together that, if any large insect has been caught, a corresponding projection is clearly visible on the outside of the leaf. When the two lobes are thus completely closed, they resist being opened, as by a thin wedge driven with astonishing force between them, and are generally ruptured rather than yield. If not ruptured, they close again with quite a loud flap. The slow movement spoken of, excited by the absorption of diffused animal matter, suffices for its final purpose, whilst the movement brought on by the touching of one of the sensitive filaments is rapid, and thus indispensable for the capturing of insects.

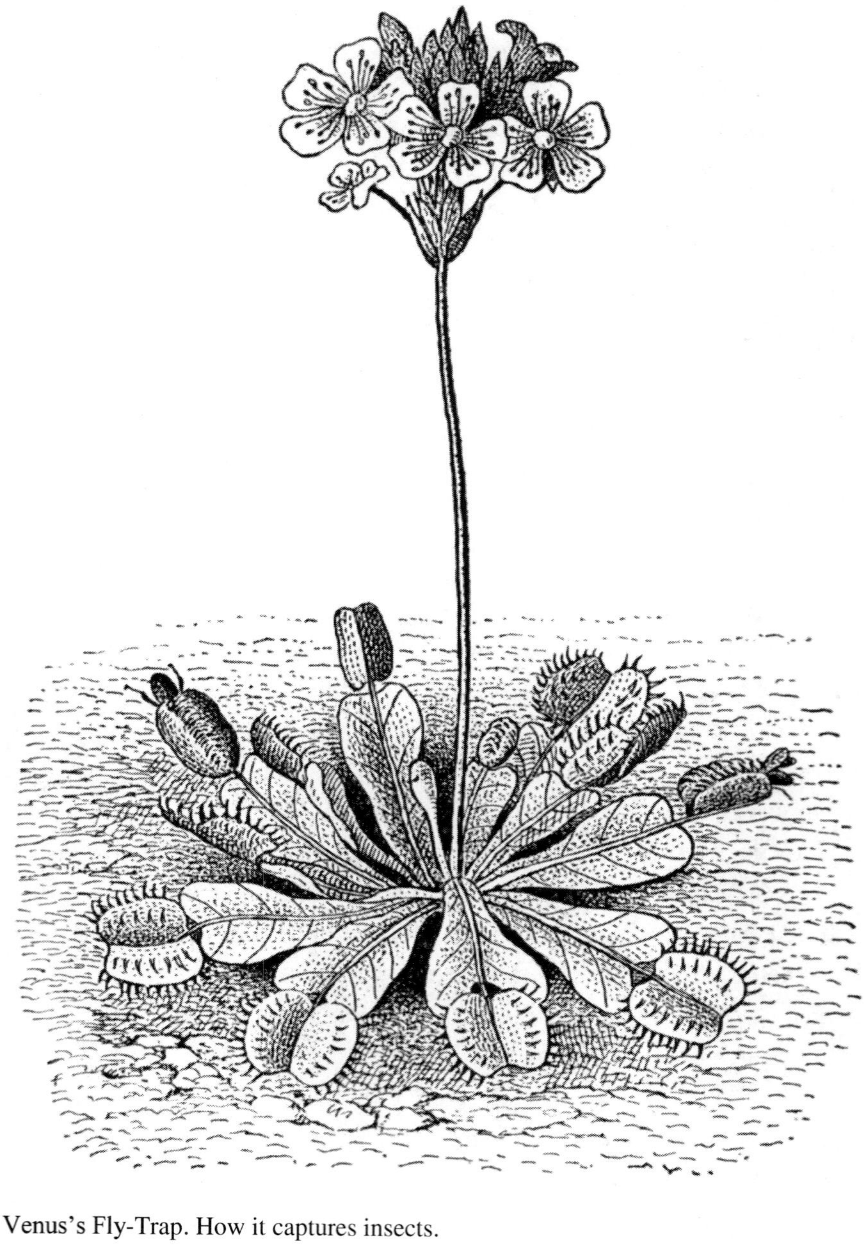

Figure 2. Venus's Fly-Trap. How it captures insects.

Leaves remain shut for a longer time over insects, especially if the latter are large, than over meat. In many instances where they have remained for a long period over insects naturally caught, they were more or less torpid when they reopened, and generally so much so during many succeeding days that no excitement of the filaments caused the least movement. Vigorous leaves will sometimes devour prey several times, but ordinarily twice, or, quite often, once is enough to render them unserviceable.

What purpose the marginal spikes, which form so conspicuous a feature in the appearance of the plant, subserve was unknown until the genius of Darwin solved the mystery. It was he that showed that elongated spaces between the spikes, varying from one-fifteenth to one-tenth of an inch in breadth according to the size of the leaf, are left open for a short time before the edges of the lobes come into contact, consequent upon the intercrossing of the tips of the marginal spikes first, thus enabling an insect whose body is not thicker than these measurements to escape, when disturbed by the closing lobes and the increasing darkness, quite easily between the crossed spikes. Moderately sized insects, if they try to escape between the bars, will be pushed back into the horrid prison with the slowly closing walls, for the spikes continue to close more and more until the lobes are brought into contact. Very strong insects, however, manage to effect their release. It would manifestly be a great disadvantage to the plant to remain many days clasped over a minute insect, and as many additional days or weeks in recovering its sensibility, inasmuch as a very small insect would afford but little nourishment. Far better would it be for the plant to wait until a moderately large insect was captured, and to allow the little ones to escape, and this advantage is gained by the slow intercrossing of the marginal spikes, which, acting like the large meshes of a fishing-net, allow the small and worthless fry to pass through.

Touching any one of the six filaments is sufficient to cause both lobes to close, these becoming at the instant incurved throughout their entire breadth. The stimulus must therefore radiate in all directions from any one filament, and it must also be transmitted with considerable rapidity across the leaf, for in all ordinary cases, as far as the eye can judge, both lobes close at the same time. Physiologists generally believe that in irritable plants the excitement is transmitted along, or in close connection with, the fibro-vascular bundles. Those in Dionæa seem at first sight to favor this belief, for they run up the mid-rib in a great bundle, sending off small bundles almost at right angles on each side, which bifurcate occasionally as they stretch towards the margin, the marginal branches from adjoining branches uniting and entering the marginal spikes. Thus a continuous zigzag line of vessels runs round the whole circumference of the leaf, while in the mid-rib all the vessels are in close contiguity, so that all parts of the leaf seem to be brought into some degree of communication. The presence of vessels, however, is not necessary for the transmission of the motor impulse, for it is transmitted from the apices of the sensitive filaments, which are hardly one-tenth of an inch in length, into which no vessels are seen to enter. Slits made close to the bases of the filaments, parallel to the mid-rib, and thus directly across the course of the vessels, sometimes on the inner and sometimes on the outer sides of the filaments, do not interfere with the transmission of the motor impulse along the vessels, and conclusively show that there is no necessity for a direct line of communication from the filament, which is touched towards the mid-rib and opposite lobe, or towards the outer parts of the same lobe. With respect to the movement of the leaves, the wonderful discovery made by Dr. Burdon Sanderson, and published in 1874, offers an easy explanation. There is, says this distinguished authority, a normal electrical current in the blade and footstalk, which, when the leaves are irritated, is disturbed in the same manner as is the muscle of an animal when contraction takes place.

After contraction has endured for a greater or less time, dependent upon circumstances which we do not well understand, re-expansion of the leaves is effected at an insensibly slow rate, whether or not any object is enclosed, both lobes opening in all ordinary cases at the same time, although each lobe may act to a certain extent independently of the other. The re-expansion is not determined by the sensitive filaments, for these may be cut off close to their bases, or be entirely removed, and re-expansion occur in the usual manner. It is believed that the several layers of cells forming the lower surface of the leaf are always in a state of tension, and that it is owing to this mechanical state, aided probably by fresh fluid being drawn into the cells, that the lobes begin to separate as soon as the contraction of the upper surface diminishes.

Six known genera, Drosophyllum, Roridula, Byblis, Drosera, Dionæa and Aldrovanda comprise the Droseraceæ, all of which capture insects. The first three genera effect this purpose solely by the viscid fluid secreted from their glands, and the last, like Dionæa, which has already been described, through the closing of the blades of the leaf. In these last two genera rapid movement makes up for the loss of viscid secretion. But of all the genera none is more interesting than the typical Sundews.

Growing in poor peaty soil, and sometimes along the borders of ponds where nothing else can grow, certain low herbaceous plants, called Droseras, abound. So small and apparently insignificant are they, that to the ordinary observer they are almost unnoticed. But they have peculiarities of structure and nature that readily distinguish them. Scattered thickly over their leaves are reddish bristles or tentacles, each surmounted by a gland, from which an extremely viscid fluid, sparkling in the sunlight like dew, exudes in transparent drops. Hence the common name of Sundew by which the half-dozen species found in the United States east of the Mississippi River are known. A one-sided raceme, whose flowers open only when the sun shines, crowns a smooth scape, which is devoid of tentacles. *Drosera rotundifolia*, our commonest species, has a wide range, being indigenous to both Europe and America. In the United States it extends from New England to Florida and westward, and is occasionally associated with *Drosera longifolia*, a form with long strap-shaped leaves, but whose distribution is mostly restricted to maritime regions, from Massachusetts to Florida.

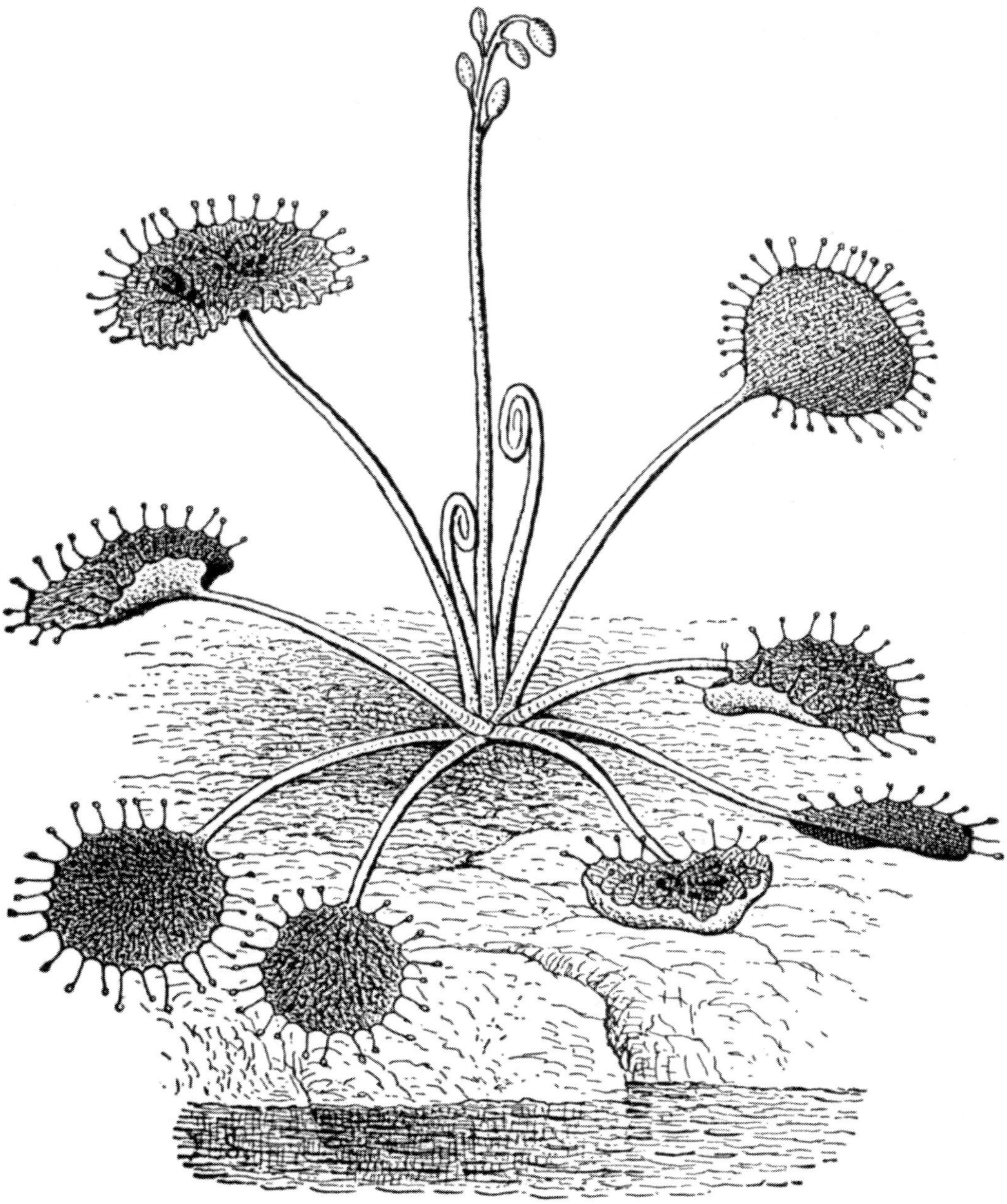

Figure 3. Round-Leaved Sundew. Leaves acting as stomachs.

All of the species are remarkably similar in habits, capturing insects, and digesting and absorbing the soft parts, a circumstance which explains how these plants can flourish in an extremely poor soil where mosses, which depend almost entirely upon the atmosphere for their nourishment, only can live. Although the leaves of the Droseras at a hasty glance do not appear green, owing to the purple color of the tentacles, yet the superior and inferior surfaces of the blade, the stalks of the central tentacles, and the petioles contain chlorophyll, rendering the best of evidence that the plants obtain and assimilate carbon dioxide from the air. But when the poverty of the soil where these plants grow is considered, it is at once apparent that their supply of nitrogen would be exceedingly small, or quite deficient, unless they had the power of obtaining it from some other source. From captured insects this important element is largely obtained, and thus we are prepared to understand how it is that their roots, which consist of only two or three slightly divided branches, from one-half to one inch in length, and furnished with absorbent hairs, are so poorly developed. From what has been stated it would seem that the roots but serve to imbibe water, but there is no doubt that nutritious matters would also be absorbed were they present in the soil.

With the edges of its leaves curled so as to form a temporary stomach, and with the glands of its closely-inflected tentacles pouring forth their truly acid secretion, which dissolves animal matters that are subsequently absorbed, Drosera may be said to feed like an animal. But, unlike an animal, it drinks by means of its roots, and largely, too, for it would not be able to supply its glands with the necessary viscid fluid. The amount needed is by no means an inconsiderable quantity, as two hundred and seventy drops may sometimes be exposed during a whole day to a glaring sun. Such a profuse exudation implies preparations for hosts of insect visitors. In this Drosera has not miscalculated. Its bright pink blossoms and brilliant, glistening dew lure vast numbers of the smaller kinds, and the larger ones, too, to certain death. But the wholesale destruction of life that goes on is much in excess of what the plant requires for food. While the smaller flies remain adherent to the leaves, affording them the needed aliment, the larger insects, after death, fall around the roots, where they decay and fertilize the soil with nitrogen, which doubtless through the proper channels makes its way into the body of the plant, thus helping to give it tone and vigor. There are times when these plants work better than at others, but whether this is caused by the electrical condition of the atmosphere, or the amount of its contained moisture, is a question which science has not positively determined.

Drosera longifolia folds it leaves entirely around its victim, from the apex down to the petiole after the manner of its vernation, but in *Drosera rotundifolia*, whose marginal tentacles are longer, the tentacles simply curve around the object, the glands touching the substance, like so many mouths receiving nourishment. Experimented upon with raw beef, the tentacles of healthy leaves, from within to without, but in periods of time varying from six to eight or nine hours, clasp firmly the beef, almost concealing it from view. Equally vigorous leaves, however, made no move towards clasping a bit of dry chalk, a chip of flint, or a lump of earth. Bits of raw apple cause a curving of the tentacles, but very few of the glands are seen touching them. It would seem, therefore, that these plants are really carnivorous, preferring animal substances, which they, by the aid of some ferment analogous to pepsin, which is secreted by the glands, are able to absorb. A minute quantity of already soluble animal matter is the exciting cause, and this must be taken in by the glands, or there is no secretion of the fermenting material.

In all ordinary cases the glands alone are susceptible to excitement. When excited, they do not themselves move or change form, but transmit a motor impulse to the bending part of their own and adjoining tentacles, and are thus carried towards the centre of the leaf. Stimulants applied to the glands of the short tentacles on the disc indirectly excite movement of the exterior tentacles, for the stimulus of the glands of the disc acts on the bending part of the latter tentacles, near their bases, and does not first travel up the pedicels to the glands, to be then reflected back to the bending place. Some influence, however, does travel up to the glands, causing them to secrete most copiously, and the secretion to become acid, just such an influence as that which in animals is transmitted along the nerves to glands, modifying their power of secretion, independently of the condition of the blood-vessels. Over organic substances that yield soluble matter the tentacles remain clasped for a much longer time than over those not acted upon by the secretion, or over inorganic objects. That they have the power of rendering organic substances soluble, that is, that they have the power of digestion, is no longer a question of dispute. They certainly have this power, acting on albuminous compounds in exactly the same manner as does the gastric juice of mammals, the digested matter being afterwards absorbed. In animals the digestion of albuminous compounds is effected by means of a ferment, pepsin, together with weak hydrochloric acid, though almost any acid will serve, yet neither pepsin nor an acid by itself has any such power. It has been observed that when the glands of the disc are excited by the contact of any object, especially of one containing nitrogeneous matter, the outer tentacles and often the blade become inflected, the leaf thereby becoming converted into a temporary cup or stomach. The discal glands then secrete more copiously, the secretion becoming acid, and, moreover, some influence being transmitted by them to the glands of the exterior tentacles, causing them to emit a more abundant secretion, which also becomes acid. This secretion is to a certain extent antiseptic, as it checks the appearance of mould and infusoria, and in this particular acts like the gastric juice of the higher animals, which is known to arrest putrefaction by destroying the microzymes.

With animals, according to Schiff, mechanical irritation excites the glands of the stomach to secrete an acid, but not pepsin. There is strong reason to believe, too, that the glands of Drosera, which are continually secreting viscid fluid to replace the losses by evaporation, do not secrete the ferment proper for digestion when mechanically irritated, but only after absorbing certain matters of a nitrogeneous nature. The glands of the stomachs of animals secrete pepsin only after they have absorbed certain soluble substances designated peptogenes, showing a remarkable parallelism between the glands of Drosera and those of the stomach in the secretion of their appropriate acid and ferment.

Not only animal matter, but also the albumen of living seeds, which are injured or killed by the secretion, are acted upon by the glands of Drosera. Matter is likewise absorbed from pollen, and from fresh leaves. The stomachs of vegetable-feeding animals, as is only too well known, possess a similar power of extracting nourishment from such articles. Though properly an insectivorous plant, but as pollen, as well as the seeds and leaves of surrounding plants, cannot fail to be often or occasionally blown upon the glands of Drosera, yet it must be credited with being to a certain extent a vegetable feeder.

That a plant and an animal should secrete the same, or nearly the same, complex digestive fluid, adapted for a similar purpose, is a wonderful fact in physiology, but not more remarkable than the movements of a tentacle consequent upon an impulse received from its own gland, the movement at the bending place of the tentacle being always towards the centre of the leaf, and so it is with all the tentacles when their glands are excited by immersion in a suitable fluid. The short tentacles in the middle part of the disc, however, must be excepted, as these do not bend at all when thus excited. But when the motor impulse comes from one side of the disc, the surrounding tentacles, and even the short ones in the middle of the disc, all bend with precision towards the point of excitement, no matter where it may be located. This is in every way a remarkable phenomenon, for the leaf appears as if endowed with animal sense and intelligence. It is all the more remarkable when the motor impulse strikes the base of a tentacle obliquely to its flattened surface, for then the contraction of the cells must be restricted to one, two or a very few rows at one end, and different sides of the surrounding tentacles must be acted on that all may bend with precision to the point of excitement. The motor impulse, as it spreads from one or more glands across the disc, enters the bases of the surrounding tentacles, and instantly acts on the bending place, but does not first proceed up the tentacles to the glands, causing them to reflect back an impulse to their bases, although some influence is sent up to the glands, whereby their secretion is soon increased and rendered acid. The glands, being thus excited, send back some other influence, dependent neither on increased secretion nor on the inflection of the tentacles, which causes the protoplasm to aggregate in cell beneath cell. This maybe called a reflex action. How it differs from that which proceeds from the nerve-ganglion of an animal, if it differ at all, no one can say. It is probably the only known case of reflex action in the vegetable kingdom.

Concerning the mechanism of the movements and the character of the motor impulse little is known. During the act of inflection fluid surely passes from one part to another of the tentacles. In explanation of the fact it is claimed that the motor impulse is allied in nature to the aggregating process, and that this causes the molecules of the cell-walls to approach each other, as do the molecules of the protoplasm within the cells, thereby causing the cells in all to contract. This is probably the hypothesis that best accords with the observed facts, although some strong objections may be urged against this view. The elasticity of their outer cells, which comes into activity as soon as those on the inner side cease contracting with prepotent force, leads largely to the re-expansion of the tentacles, but there is reason to suspect that fluid is continually and slowly attracted into the outer cells during the act of re-expansion, thus augmenting their tension.

With respect to the structure, movements, constitution and habits of *Dionæa muscipula* and *Drosera rotundifolia*, as well as kindred species, little has been made out by patient study and investigation in comparison with what remains unexplained and unknown. Many of their movements, especially of Dionæa and Drosera, seem so sensible and intelligent that the reflecting mind of man can hardly hesitate to assign them high positions in organic nature and the possession, even though in a very small degree, of that consciousness with which animal life is endowed. That man is psychically related to all life is the belief of millions in the old world, and the hope of millions in the new. In this thought is the escape from materialism, that threat of the ignorant and unbelieving. Higher conceptions of beauty and greatness are now being entertained by the multitudes, and we begin to feel that the next great step is being taken when we shall become, instead of poor trembling denizens of a perishable world, proud and conscious citizens of an imperishable universe. That we of the upper ranks of God's creation alone possess an inner life which shall transcend all change is no longer a general belief, but there is a growing hope that all nature shares it, and that love is its expression and its method. All existence is a unit. Life, law and love are divine. Man, looking calmly about him, cannot set himself apart as something essentially different from nature, but must recognize himself as a part, and include love in the universal scheme of development. All other expressions of life must share with him in the divine love and progress. His dogmas, founded on mistaken traditions, have given way to science, and he cannot but believe that love is in and of the soul, and that all life has some sort of development of soul. Because plant-life has no brain, and therefore has no intelligence, no mind, no soul, is preposterous to contemplate. Who can positively affirm that brain alone is the seat of conscious intelligence? None but He alone, the Giver of all life, who sits enthroned and exalted in the everlasting heavens.

Chapter 3

SLIME-ANIMALS

Possibly the simplest of life's children are the singularly unique and structureless little Finger Slimes, which live not only in the sea but also in puddles and pools, and in the gutters of our streets and of our house-tops. Anywhere that stagnant water abounds these tiny drops of slime will grow up and make it their home. Sometimes few and far between, and sometimes in such immense crowds that the entire pond would seem, if they could be seen with the unaided vision, literally alive with them, they live, and multiply and die under our very feet.

Nothing can be less animal-like than one of these shapeless masses of pure protoplasm, yet under a microscope of strong power it may be seen moving lazily along by pulling out a thick finger of slime and then letting all the rest of its body flow after it. When coming into contact with food it may be said to flow over it, dissolving the soft parts and sending out the hard, indigestible refuse anywhere, no matter where, for its body is devoid of skin, being merely one general mass of homogeneous slime.

But what can these little slime specks tell us about the wonderful powers of life? Nothing at all, it would seem, for in these tiny creatures life has nothing better to work with than a mere drop of living matter, which is all alike throughout, so that if broken into a hundred pieces every piece would be as much a living being as the whole. And yet by means of the wonderful gift of life, with which the all-wise Omnipotence has endowed it, this slime-drop lives, and breathes, and eats, and increases, shrinks away when you touch it, feels for its food, and moves from place to place, changing its shape to form limbs and feeling-threads, which are let into the general organism when they have served the purpose of their existing, only to be succeeded by others as short-lived as themselves when necessity requires their development.

So small are these creatures that the largest specimen will be found to be smaller than the smallest pin's head. Examine how we will, there will be found no mouth, no stomach, no muscles, no nerves, no parts of any kind. The animal looks merely like a minute drop of gum with fine grains diffused throughout, floating in the water, some times with outstretched arms, and at other times as a simple drop. An analysis of the matter of which it is composed shows it to be much the same as a speck of white-of-egg. Yet it is alive, for it breathes. Kept in a drop of water, it uses up the oxygen it contains, and renders the water foul by the carbonic acid it breathes out. The arms, so necessary in the procurement of food, can be drawn in and thrown out when and where the animal chooses, showing that some option is undoubtedly exercised in the matter. Minute jelly-plants, that live in the water, and even higher animals than itself, constitute its food. The presence of an animal with a shell does not deter it from attack, for it is just as able to deal with it as with the softer, shell-less kinds, sucking their jelly-like contents, and discarding the empty, innutritious shells.

Quite as interesting among the Moners, to which the Finger Slime belongs, is the *Protomyxa aurantiaca*, a shapeless bit of transparent matter, containing merely circulating granules. Locomotion is effected by extending the body into pseudopodia, or false feet, and contracting them. Its movement is slow and gliding. When at rest it appears as a mere lump of jelly, but its whole demeanor changes when in the presence of a living animal suited for food. Fine threads immediately begin to shoot out from all sides, which fuse about the unsuspecting prey, while all the little grains in the slime course to and fro. For five or six hours the little fellow hugs closely round the prey until it has become thoroughly absorbed, at least the nutritious parts, into its body-mass, when it draws itself away, or back into its original place, leaving by its side the skeleton of its late victim. Without eyes or ears or parts of any kind it knows how to find its food; without muscles or limbs it is able to seize it; without a mouth it can suck out its living body, and without a stomach it can digest the food in the midst of its own slime, and cast out the parts for which it has no use.

When Protomyxa has become a burden to itself it divides itself by a simple process of fission, each part being complete in itself, or it assumes a thick covering, becoming encysted, as it is termed. In a little while the enclosed mass divides into spheres, the cell-wall bursts, and the little spheres, which have now taken on a sort of tadpole shape, float out upon the water, where they soon assume the parent-form.

Like all living things, these Moners have a desire for food, which their protoplasm first appropriates, then converts into available material. They thus grow and increase in size, but when they become too large to be comfortable they usually split into two, in obedience to the law of their being, and each half goes its own way as a living animal. This is the earliest form of parentage, the simplest form of reproduction. Thus yielding to this necessity of a separation of one into more than one, these Moners live on forever, or as long as the earth continues to support life, thus becoming immortal in the scientific sense in which the term is used to devote a continuance of the physical life on earth. They only and their nearest relatives, as simple in structure as themselves, achieve this stupendous result, for in such a division of their entire substance they know no loss, no death of any part, violence only being able to sunder them from life. They resolve themselves into their own offspring, and nothing perishes.

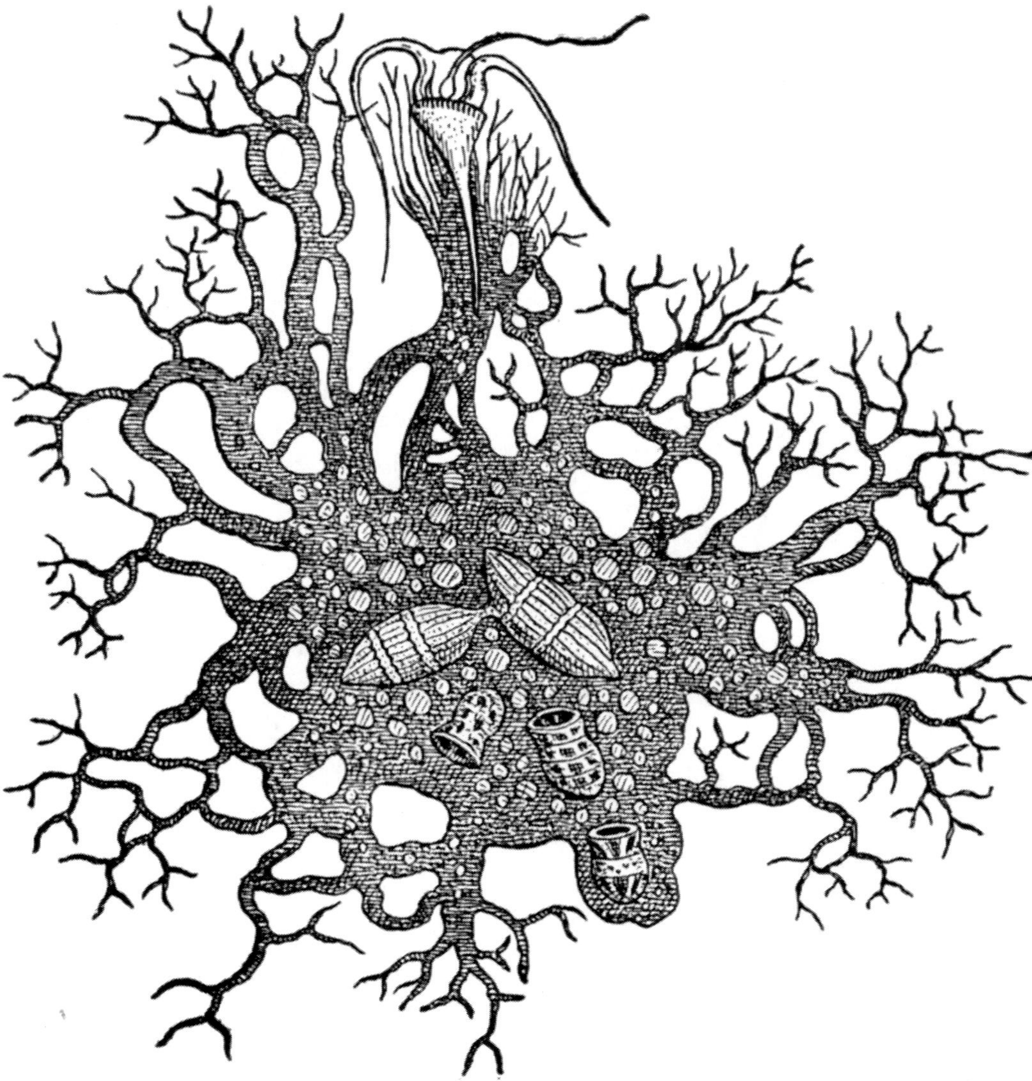

Figure 4. Protomyxa Feeding.

Chapter 4

PRIMITIVE LASSO-THROWERS

Everyone knows that the long cord or thong, called the lasso, is the peculiar weapon of the South American hunter. Almost from his earliest childhood the young Gaucho learns to amuse himself with it, and as soon as he is able to walk takes great pleasure in catching young birds and other animals around his father's hut, hurling the long lash with such dexterity that the noose drops over their bodies and brings them to his feet. Did we wish to select from among all the denizens of life the most brilliant, graceful, and sylph-like, whose very life-histories read more like the romance of poetry than sober reality, we would choose those which might be appropriately designated the lasso-throwers.

Now among animals, as is only too well known, any weapons which they could be called upon to use must develop in their own bodies, and therefore it could hardly be suspected that a simple jelly-animal could be provided with a lasso ready grown in its own flesh. Yet it is so, for in that class of animals, which ranks just above the sponges, we discover a weapon of this kind as simple and as deadly, and far more wonderful in its action than any used by man.

In fresh-water ponds, attached by its base to the under surfaces of aquatic plants, may be found a very small animal, just large enough to be seen without the aid of a lens, usually pale green, but sometimes of a brown color. This is our common hydra, technically called *Hydra fusca*. It is nothing more than a tube or sac, with a sucker at one end to hold on with, and a mouth at the other, surrounded with from five to eight hollow tentacles or feelers, which opens into a central cavity or stomach. Firm and muscular are the walls of the sac, so that the little creature, which is not fixed permanently to whatever it is found clinging to, may stretch itself out or draw back as its own volition dictates, or move slowly along by means of its sucker, or float easily or contentedly upon the water. But the most remarkable, as well as the most interesting thing about this odd creature is the power which it possesses of overcoming animals more powerful and active than itself.

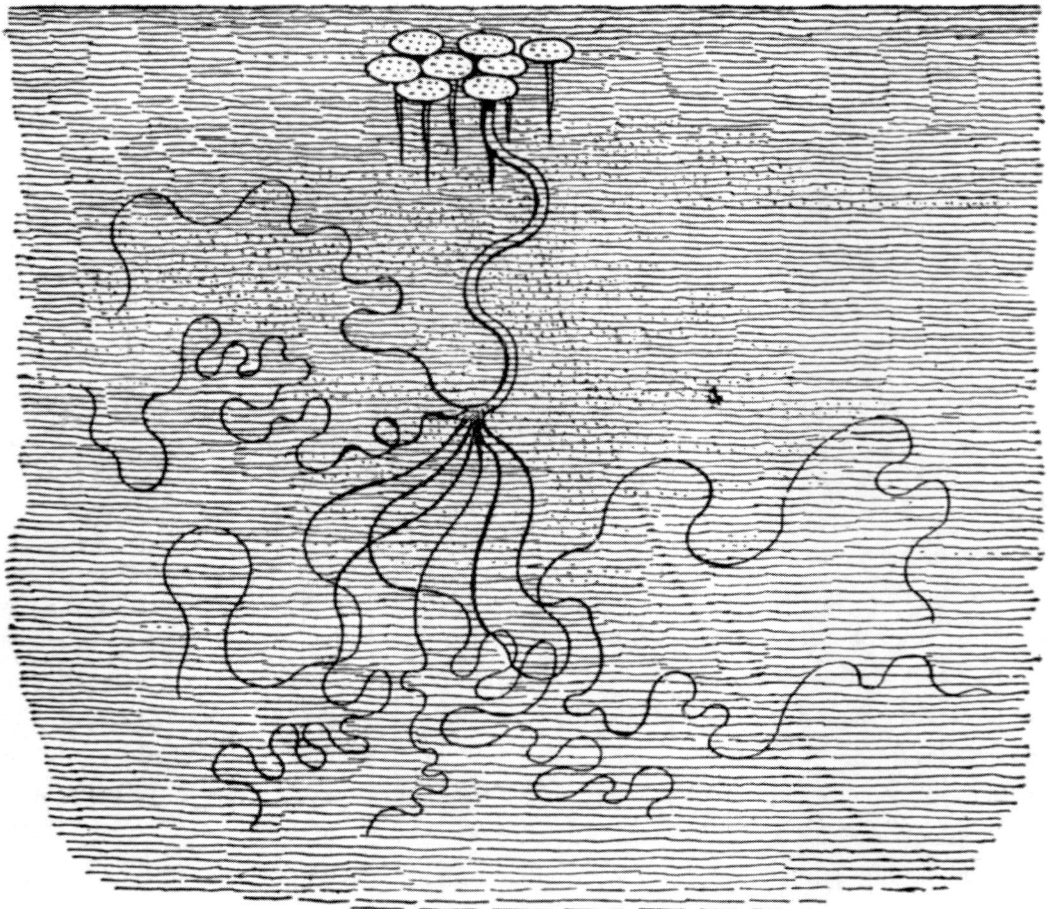

Figure 5. Fresh-Water Hydra moored and searching for prey.

Groping about with its flexible arms, which are closely invested with fine jelly-hairs, with which it seemingly feels, or attached to some leaf or bit of floating stick, its tentacles reaching out in all directions, the Hydra instantly paralyzes any minute insects, young snail or infusorian that touches its feelers, and complacently closing its arms over the helpless victim, carefully tucks it away, so to speak, into its stomach, where it is speedily digested. This power of paralyzing and thus readily capturing active living creatures is due to the presence in the skin of the tentacles and body of what are called lasso-cells, or nettling-organs, which are minute, transparent cells, so small that two hundred of the largest would occupy but the distance of an inch, each being armed with a long barbed thread coiled up within its walls. This delicate thread, which is often from twenty to forty times the length of the cell, lies bathed in a poisonous fluid, and only waits for the cell-walls to burst, which they do when the Hydra touches an animal swimming near it, when thousands of these little barbed cords dart into the victim, quickly paralyzing it and rendering it an easy prey to its captor. All Cœlenterates, such as jelly-fishes and coral polyps, possess these nettling-organs.

Thus we see where the Hydra's strength lies. He has no need to struggle, for his victim, penetrated by a multitude of darts, and made powerless by the poison instilled, becomes as manageable as an equal bulk of inert matter. It behooves the little creature to take things quietly, for a cell once burst cannot be used again, and he is therefore compelled to wait until a new one is grown to take the place of the one that has become exhausted. So he patiently bides his time till his victim is half-conquered, when he draws him gently into his body. He lives and catches his food, as must be apparent, without the necessity of moving very far from the place where he had his birth.

All the summer through the Hydra puts out buds from its side, which, when their tentacles have grown, drop from the parent-body, and settle down in life for themselves. But when winter comes, and before all life has become extinct, an egg appears near the base of the tubes of those that are living, and these eggs lie dormant till the next spring, when they are hatched, and a new generation of Hydras is produced. Budding, which is but a process of natural self-division, is carried on to a large extent, more individuals being produced in this way than from eggs. These buds are at first a simple bulging out of the body-walls, the bud enveloping a portion of the stomach, until it becomes constricted and drops off, the tentacles meanwhile budding out from the distal end, and a mouth-opening arising between them. In the Hydra, the Actinia, and other polyps, and in truth in all the lower animals, budding is simply due to an increase in the growth and multiplication of cells at a special place on the outside of the body. As in the vertebrates, man included, the Hydra arises from an egg which, after fertilization, passes through two stages, the germ consisting at first of two cell-layers, but the sexes are not separate as in the marine Hydroids, which grow in colonies that may be either male or female.

Like some other animals of simple structure, the Hydra is capable of reproducing to a most wonderful degree when cut into pieces. Divided in two, each becomes a perfect Hydra, and even when sliced into any number of thin rings each ring will grow out a crown of tentacles. You may split them into longitudinal strips and each strip will eventually become a well-shaped Hydra. Two individuals may be fastened together by a horse-hair and in a short time they will have become like Siamese twins, but there will never arise the slightest disagreement between them. A Hydra turned inside out will readily adapt itself to the change, and in a few days will be able to swallow and digest bits of meat, its former stomach-lining having now taken upon itself the condition of skin.

Hydra fusca is our simplest lasso-thrower, and the only one to be found in fresh waters in this country. Such a wonderful and deadly weapon is his, that it is easy to understand how his numerous relatives in the wide ocean have made good use of the weapon with which nature has provided them, and secured, under all kinds of shapes and forms, homes and resting-places throughout the vast waste of waters. From the Arctic to the Tropics, and from the shallow seaside pools at low tide to the fathomless abysses of the ocean, we meet the lasso-throwers. Now in the form of huge jelly-fishes, covering the sea for miles and miles, transparent domes by day and phosphorescing lights by night, and now as tiny balls of jelly, glistening by millions in some quiet bay and splintering into light upon the beach; or in the form of living animal-trees waving their graceful arms over rocks in waters deep, or creeping like delicate threads over shells and stones and seaweed on the shore, where they often lose their identity and are mistaken for plants. There is scarcely a nook or cranny in the bed of ocean where these tree-like forms, associated with the beautiful sea-anemone, whose brilliant crimson, green and purple are unmatched in color by gem and flower, are not to be found.

All these beautiful creatures, as well as the living coral that nestles in the bosom of the warm Mediterranean or the sea that lashes our Southern shores, or that struggles boldly against Pacific's waves, are lasso-throwers. *Cœlenterata*, the "hollow-bodied animals," because of the large cavity within their bodies, is the name by which they are known to science. They naturally fall into two families, the *Hydrozoa*, or Water Animals, and the *Actinizoa*, or Ray-like Animals, our little Hydra, about which so much has been written, being representative of the former and the Anemones of the latter division.

Chapter 5

FIVE-FINGERED JACK ON THE OYSTER

Quite as infinite in number, variety and form is the life of the sea as that of the land. But of all marine animals, however, there is none more curious than the echinoderm, a name derived by science from two Greek words, indicating an animal bristling with spines like the hedgehog. These creatures are sometimes free, but quite as often attached by a stem, flexible or otherwise, and radiate after the fashion of a circle or star, or are of the form of a star, with more or less elongated arms. They are covered with shell-like plates, which they secrete for themselves, and are still further protected by spines or scales.

Perhaps the most common of the echinoderms is the Star-fish, or Five-fingered Jack, as it is called by sailors. Whoever has spent any time on the seashore has doubtless made the acquaintance of this animal, for it is readily distinguishable by its shape, its upper surface being rough and tuberculous, and armed with spine-like projections, while the under portion is soft, containing the essential organs of life and locomotion.

When first seen stranded on the shore the Star-fish, by the uninitiated, is thought to be a creature incapable of movement of any kind. But this is far from being the case, for in its native element it moves along the bottom of the sea with the greatest ease, being provided with an apparatus specially adapted for the purpose. Ordinarily its arms are kept upon the same level, but in passing over obstacles that lay in its path, the animal has the power of raising any one of its several arms. Elevations are ascended with the same ease and facility as progression on plane surfaces is effected. Perforating the arms, or rays, and issuing from apertures, will be found large numbers of membranous tubes, which prove to be the feet of the animal. Upon careful examination the latter will be found to consist of two parts, a bladder-like portion, resident within the body, and a tubular outlying projection, ending in a disk-shaped sucker, thus showing the feet to be muscular cylinders, hollow in the centre, and very extensible. In progression the animal extends a few of its feet, attaches its suckers to the rocks or stones and then, by retracting its feet, draws the body forward. Like that of the tortoise, its pace is slow and sure. But the most singular thing about this singular animal is its manner of overcoming obstructions, which it must certainly perceive, judging from the preparations to surmount them which it makes at the opportune moment.

In addition to organs of locomotion Star-fishes possess blood-vessels, digestive and respiratory apparatus, and a nervous system of a very low order, an inference to which its seeming capacity of enduring vivisection without pain unmistakably leads.

Interesting as its manner of progression, even under the most trying circumstances, must be, yet there is nothing in the life of this lowly-organized animal that has half the charm to the true lover and student of nature than the mother Star's devotion to her young. Her eggs she carries in little pouches placed at the base of the rays. When emitted through an opening, which occasionally and unintentionally occurs, the mother does not abandon them to the cruel charities of the ocean world, but gathers them together, forming a kind of protecting cover of them, very much like a hen brooding over her chickens. Her actions bespeak an anxiety which could only be born of an affection, as real and sympathetic as that which a human mother feels for the loss of any of her offspring. No matter how often the eggs become accidentally scattered, the mother does not grow weary of her charges and leave them to themselves, but gathers them to the maternal fold with the same tender, patient solicitude as characterized her first efforts. Confined to a tank, when with ova, the mother Star has been known to traverse the entire length of the vessel until she has found and recovered her scattered treasures.

Reproduction by eggs is not the only means of generation in vogue. In common with other sea animals the Star-fish has the strange capacity of detaching one or more of its arms, each of the cast-off members becoming in time a perfect creature of its own kind, while a new arm, fully equipped to perform all necessary functions, will grow out in place of the lost member. From twelve to fifteen weeks are required to reproduce a lost ray, the animal meanwhile seeming not the least discontented, but acting as utterly unconscious of any changes in its anatomy.

As found upon the shore, Star-fishes appear dead when really they are alive. Put one of these perfectly still creatures into fresh sea-water, and in a short time it will probably be disporting itself as freely as ever it did. But as the dead and the living, when stranded by the tide, present nearly the same appearance, some certain test seems necessary to distinguish them apart. If a Star-fish hangs loose and limp, it is dead; but, however dead it may look, if on touching it there are manifest a firmness and consistency in its substance, one may feel reasonably sure that it is playing the 'possum and will revive when placed in the water. Quite as certain a mode of ascertaining whether your starry friend is living or dead, is to lay it upon its back, when, if alive, a number of semi-transparent globular objects will be seen to move, reaching this way and that, as though feeling for something to lay hold of wherewith to restore it to its normal position. These globular appendages are the ambulacra, or locomotory organs, seeking to acquire this end. If, however, no movement is manifested, you can wisely conclude that your animal is dead.

The Star-fish, not unlike all other animals of the sea, has an appetite that is never satisfied. Dinner is always welcome. The procurement of food seems its chief concern in life. It is a scavenger of no mean importance, keeping up an incessant chase after all kinds of dead animal matter, and thus largely contributing, it is probable, towards the maintaining of the waters of the ocean in a state of purity. But its feeding is not exclusively restricted to decaying matters. Any species of mollusk, from the humble whelk, not more than five-eighths of an inch in length, to the lordly oyster, so esteemed by epicures, constitutes a dainty tidbit. No more inveterate ravager and brigand, not even excepting man himself, have the oyster-beds to disturb the equanimity and serenity of their existence than the audacious, insinuating Star-fish.

With its five arms, and apparently without any other organ, this comparatively insignificant little being accomplishes a work which man, without the aid of extraneous appliances, is quite unable to execute. It opens an oyster as deftly and effectually as an expert oysterman would do, and that, too, without the habitual oyster-knife, and swallows the slimy bivalve in the same manner as the lords of creation do. Man, with all his genius and skill, were he deprived of all other means of subsistence than the oyster, and having no implement with which to open it, would be severely puzzled to get at the savory morsel shut up in its obstinate valves, yet the Star-fish performs the task seemingly without the least difficulty.

How the Star-fish manages the problem was at first a matter of guess-work. For a long time it was confidently believed that the animal waited for the moment when the oyster opened its shell to introduce one of its arms into the opening. This much gained, the other four arms were got in without much trouble, and the whole business ended with the devouring of the inmate. This belief is no longer tenable. Careful observation has revealed to us the true inwardness of the proceeding. The oyster is seized between the arms of the Star-fish and held under its mouth by the aid of its suckers. Thus secured, the Asterias, or Star-fish, everts its stomach, and envelops the whole oyster in its interior recesses, distilling a poisonous fluid, a secretion from its mouth, which causes the oyster to open its shell, when the robber, as it were, crawls in and takes its dessert. Incredible numbers of oysters are destroyed by Star-fishes, but the oystermen fail to see that their own barbaric ignorance is largely to blame. Star-fishes drawn up in nets, rakes and dredges in immense quantities are tied into bundles, but the cords are made so tight that the pile is cut in twain, the result being that all the pieces, when afterwards thrown overboard, become new and perfect Star-fishes.

Not often has one the pleasure of meeting with these animals on the New Jersey coast, but yet they are occasionally seen, more frequently, perhaps, in the North. *Asterias berylinus*, the commoner form, is a fairly large species, of a more or less greenish color, sometimes waning to brown, and roughly covered with tubercles. Its five arms, at the extremity of each of which is situated a single red-eye speck, are somewhat irregularly arranged, and not rarely one is stumpy through breakage or unequal development.

the least disposition to bite. Different species quickly learn, when treated with kindness, to regard man as their friend. I have seen Agalena take food from the hand out of a pair of forceps, or water from a brush, and even to reach on tiptoe after it from the mouth of a bottle placed for her accommodation. Though naturally timid and shy, and prone to flee to her funnel on man's approach, yet she has been known to permit the most unexpected familiarities without fear or resentment. Many a female has taken from my hand the proffered fly, and submitted to the gentle caresses of my finger down the back and abdomen with the most pleasurable satisfaction. They have come at the sound of my voice, dancing upon their sheeted web like one gone mad, so perfectly carried away with delight. An interesting experience of last summer during a brief stay in the country seems apropos at this time. While sauntering carelessly along a forest-road I came unexpectedly upon a rustic bridge, with a railing on one side, which overspanned a small water-course. Leaning for rest and support against the railing, soon my attention was arrested by a huge female spider, which I recognized as *Epeira domiciliorum*. She was evidently in quest of something, as I was led to suspect from her seemingly thoughtful and deliberate movements. I watched her closely and criticisingly for a long while, and in one of her contemplative moods, when she stood perfectly motionless and fixed as it were to the railing, I reached out my finger rather impulsively and began stroking her along the abdomen, a familiarity which she did not resent, and which seemed to give her the most intense delight. When the caressing had ceased, she would turn round and confront her newly-made acquaintance, but the lifting of the finger was always the signal for her to assume an attitude of the most perfect quiescence. That she enjoyed these little attentions there cannot be a shadow of doubt, or actions are no use in the interpretation of feeling. Had they been painful, she would have sought relief in flight, or in the manifestation of an untoward disposition towards her unintentional persecutor.

Figure 6. Star-Fish opening an oyster.

When a Star-fish is alarmed, or finds itself in strange quarters, it will be seen to curl up the tips of its rays, and there under the point of each ray will be found a thick red spot seated on the extremity of a nerve, and having in it as many as from one hundred to two hundred crystal lenses surrounded by red cells. With such a highly-developed eye, which is far better than the jelly-fish enjoys, it is no wonder that the Star-fish is so quick in discerning food, or enrages the fisherman by the discovery of the bait which he had intended for other animals, for it turns out that this stupid-looking animal is more wide-awake than it is given credit for. Sometimes, as in the beautifully delicate Star-fish, called the "Lingthorn," a soft lid, or feeler, hangs over the eye-spot, which gives to the creature a curiously intelligent look, but in the case of our common form this lid is notably absent.

From all that has been written it must be evident that our first walking animal is by no means a poor or feeble creature. He has a chain armor woven into his leathery skin, with sharp, pointed spines, and snapping, beak-like claws to protect him; an excellent digestion and a capacious mouth to feed his greedy stomach, and a fine array of nerves, quick feeling and eyesight, and a wonderful apparatus for moving over the ground. When it is added to all these possessions the ability to close over the wound in the case of a lost ray and the growing of a new one, we see that his powers of living satisfactorily are by no means insignificant. But this curious walking apparatus of the Star-fish is far from being perfect in all his relations. They do not all walk by means of suckers any more than all sponge-animals build toilet sponge, or all slime-animals make chambered shells. Sure, the Rosy Feather-stars, for example, have no use for feet-tubes, as their lives are generally spent upon the rocks or nestled in bunches of sea-weed. Brittle-stars, as these are called, though closely related to the Star-fishes, are not easily confounded with them, for their arms are found to radiate from a clearly defined central disk, and there is no prolongation of their stomachs and ovaries into their interiors. The tube-feet pass out from the plates along the sides of the arms, instead of from the under surface as in the Star-fishes proper, and probably serve merely as a help for breathing, locomotion over the sands being effected by their long flexible arms. Their home is chiefly among the tangle and eel-grass, where their protecting covering affords them security from their many enemies.

Chapter 6

EARTH-WORMS IN HISTORY

Earth-worms are found throughout the world. Though few in genera, and not many in species, yet they make up in individual numbers, for it has been estimated that they average about one hundred thousand to the acre. Our American species have never been monographed, which renders it impossible to judge of their probable number. Their castings may be seen on commons, so as to cover almost entirely their surface, where the soil is poor and the grass short and thin, and they are almost as numerous in some of our parks where the grass grows well and the soil appears rich. Even on the same piece of ground worms are much more frequent in some places than in others, although no visible difference in the nature of the soil is manifest. They abound in paved court-yards contiguous to houses, and on the sidewalks in country towns, and instances have been reported where they have burrowed through the floors of very damp cellars.

Beneath large trees few castings can be found during certain parts of the year, and this is apparently due to the moisture having been sucked out of the ground by the innumerable roots of the trees, an explanation which seems to be confirmed by the fact that such places may be observed covered with castings after the heavy autumnal rains. Although most coppices and woods support large numbers of worms, yet in forests of certain kinds of tree-growths, where the ground beneath is destitute of vegetation, not a casting is seen over wide reaches of ground, even during the autumn. In mountainous districts worms are mostly rare, it would seem, a circumstance which is perhaps owing to the close proximity of the subjacent rocks, into which it is impossible for them to burrow during the winter, so as to escape being frozen. But there are some exceptions to this rule, for they have been found at great altitudes in certain parts of the world, and especially is this so in India, where they have been observed to be quite numerous upon the mountains.

Though in one sense semi-aquatic animals, like the other members of the great class of Annelids to which they belong, yet it cannot be denied that earth-worms are terrestrial creatures. Their exposure to the dry air of a room for a single night proves fatal to them, while on the other hand they have been kept alive for nearly four months completely submerged in water. During the summer, when the ground is dry, they penetrate to a great depth and cease to work, just as they do in winter when the ground is frozen. They are nocturnal in their habits, and may be seen crawling about in large numbers at night, but generally with their tails still inserted in their burrows. By the expansion of this part of the body, and with the aid of the short reflexed bristles with which they are armed inferiorly, they hold so securely that they can seldom be withdrawn from the ground without being torn in pieces. But during the day, except at the time of pairing, when those which inhabit adjoining burrows expose the greater part of their bodies for an hour or two in the early morning, they remain in their burrows. Sick individuals, whose illness is caused by the parasitic larvæ of a fly, must also be excepted, as they wander about during the day and die on the surface. Astonishing numbers of dead worms may sometimes be seen lying on the ground after a heavy rain succeeding dry weather, no less than a half-hundred in a space of a few square yards, but these are doubtless worms that were already sick, whose deaths were merely hastened by the ground being flooded, for if they had been drowned it is probable, from the facts already given, that they would have perished in their burrows.

After there has been a heavy rain the film of mud or of very fine sand to be seen over gravel-walks in the morning is often distinctly marked with the tracks of worms. From May to August, inclusive, this has been noticed when the months have been wet. Very few dead worms are anywhere to be seen on these occasions, although the walks are marked with innumerable tracks, five tracks often being counted crossing a space of only an inch square, which could be traced either to or from the mouths of the burrows in the gravel-walks for distances varying from three to fifteen yards, but no two tracks being seen to lead to the same burrow. It is not likely, from what is known of the sense-organs of these animals, that a worm could find its way back to its burrow after having once left it. They leave their burrows, it would seem, on a voyage of discovery, and thus they find new sites for the exercise of their powers. For hours together they may often be seen lying almost motionless beneath the mouths of their burrows. But let the ejected earth or rubbish over their burrows be suddenly removed and the end of the worm's body may be seen rapidly retreating.

This habit of lying near the surface leads to their destruction to an immense extent, for, at certain seasons of the year, the robins and blackbirds that visit our lawns in the country may be observed drawing out of their holes an astonishing number of worms, which could not be done unless they lay close to the surface. But what brings the worms to the surface? This is a question whose answer cannot be positively asserted. It is not probable that they behave in this manner for the purpose of breathing fresh air, for it has been seen that they can live a long time under water. That they are there for the sake of warmth, especially in the morning, is a more reasonable supposition, which seems to be confirmed by the fact that they often coat the mouths of their burrows with leaves, apparently to prevent their bodies from coming into contact with the cold, damp earth, and by the still other fact that they completely close their burrows during the winter.

Some remarks about the structure of the earth-worm now appear apropos. Its body consists of from one hundred to two hundred almost cylindrical rings, each provided with minute bristles. The muscular system is well developed, thus enabling these animals to crawl backwards as well as forwards, and to retreat by the help of their affixed tails into their burrows with extraordinary rapidity. Situated at the anterior end of the body is the mouth. It is furnished with a little projection, variously called the lobe or lip, which is used for prehension. Behind the mouth, internally located, is a strong pharynx, which is pushed forwards when the animal eats, corresponding, it is said, with the protrudable trunk of other Annelids. The pharynx conducts to the œsophagus, on each side of the lower part of which are placed three pairs of large glands, called calciferous glands, whose function is the secretion of carbonate of lime. These glands are very remarkable organs, and their like is not to be found in any other animal. Their use is connected in some way with the process of digestion. The œsophagus, in most of the species, is enlarged into a crop in front of the gizzard. This latter organ is lined with a smooth, thick chitinous membrane, and is surrounded by weak, longitudinal, but powerful transverse muscles, whose energetic action is most effectual in the trituration of the food, for these worms possess no jaws, or teeth of any kind. Grains of sand and small stones, from the one-twentieth to the one-tenth of an inch in size, are found in their gizzards and intestines, and these little stones, independently of those swallowed while excavating their burrows, most probably serve, like millstones, to triturate their food. The gizzard opens into the intestine—a most remarkable structure, an intestine within an intestine—which runs in a straight line to the vent at the posterior end of the body. But this curious structure, as shown by Claparède, merely consists of a deep longitudinal involution of the walls of the intestine, by which means an extensive absorbent surface is secured.

Worms have a well-developed circulating system. Their breathing is effected by the skin, and so they do not possess any special respiratory apparatus. Each individual unites the two sexes in its own body, but two individuals pair together. The nervous system is fairly well developed, the two nearly confluent cerebral ganglia being situated very close to the anterior extremity of the body.

Being destitute of eyes, we would naturally conclude that worms were quite insensible to light; but from many experiments that have been made by Darwin, Hofmeister and others, it is evident that light affects them, but only by its intensity and duration. It is the anterior extremity of the body, where the cerebral ganglia lie, that is affected, for if this part is shaded and other parts of the body are illuminated no effect will be produced. As these animals have no eyes, it is probable that the light passes through their skins and excites in some manner their cerebral ganglia. When worms are employed in dragging leaves into their burrows or in eating them, and even during the brief intervals of rest from their labors, they either do not perceive the light or are regardless of it, and this is even the case when the light is concentrated upon them through a large lens. Paired individuals will remain for an hour or two together out of their burrows, fully exposed to the morning light, but it appears, from what some writers have said, that a light will occasionally cause paired individuals to separate. When a worm is suddenly illuminated and dashes into its burrow, one is led to look at the action as a reflex one, the irritation of the cerebral ganglia apparently causing certain muscles to contract in an inevitable manner, without the exercise of the will or consciousness of the animal, as though it was an automaton. But the different effect which a light produces on different occasions, and especially the fact that a worm when in any way occupied, no matter what set of muscles and ganglia may be brought into play, is often regardless of light, are antagonistic to the view of the sudden withdrawal being a simple reflex action. With the higher animals, when close attention to some object leads to the disregard of the impressions which other objects must be producing upon them, we ascribe this to their attention being then absorbed, and attention necessarily implies the presence of mind. Although worms cannot be said to possess the power of vision, yet their sensitiveness to light enables them to discriminate between day and night, and thus they escape the attacks of the many diurnal animals that would prey upon them. They are less sensitive to a moderate radiant heat than to a bright light, as repeated experiments have conclusively shown; and their disinclination to leave their burrows during a frost proves that they are sensitive to a low temperature.

Investigation fails to locate in worms any organ of hearing, from which must be concluded that they are insensible to sounds. The shrill notes of a metallic whistle sounded near them, and the deepest and loudest tones of a bassoon, failed to awaken the least notice. Although indifferent to modulations in the air, audible to human ears, yet they are extremely sensitive to vibrations in any solid object. Even the light and delicate tread of a robin affrights and sends them deep into their burrows. It has been said that if the ground is beaten, or otherwise made to tremble, that worms believe they are pursued by a mole and leave their burrows, but this does not stand the test of experiment, for the writer has frequently beaten the ground in many places where these creatures abounded, but not one emerged. A worm's entire body is sensitive to contact, the slightest puff of air from the mouth causing an instant retreat. When a worm first comes out of its burrow it generally moves the much-extended anterior extremity of its body from side to side in all directions, apparently as an object of touch, and there is good reason to believe that they are thus enabled to gain a general knowledge of the form of an object. Touch, including in this term the perception of a vibration, seems much the most highly developed of all their senses. The sense of smell is quite feeble, and is apparently confined to the perception of certain odors. They are quite indifferent to the human breath, even when tainted by tobacco, or to a pellet of cotton-wool with a few drops of Millefleur's perfume when held by pincers and moved about within a few inches of them. The perception of such an unnatural odor would be of no service to them. Now, as such timid creatures would almost certainly exhibit some signs of any new impression, we may reasonably conclude that they did not perceive these odors. But when cabbage leaves and pieces of onion were employed, both of which are devoured with much relish by worms, the result was different. These, with bits of fresh raw meat, have been buried in pots beneath one-fourth of an inch of common garden soil, or sometimes laid on pieces of tin foil in the earth, the ground being pressed down slightly, so as not to prevent the emission of any odor, and yet they were always discovered by the worms that were placed in the pots, and removed after varying periods of time. These facts indicate that worms possess some power of smell, and that they discover by this means odoriferous and much-coveted kinds of food.

That all animals which feed on various substances possess the sense of taste, is a wise presumption. This is certainly the case with worms. Cabbage leaves are much liked by worms, and it would seem that they are able to distinguish between the different varieties, but this may perhaps be owing to differences in their texture. When leaves of the cabbage, horse-radish and onion were given together, they manifestly preferred the last to the others. Celery is preferred to the leaves of the cabbage, lime-tree, ampelopsis and parsnip, and the leaves of the wild cherry and carrots, especially the latter, to all the others. That the worms have a preference for one taste over another, is still further shown from what follows. Pieces of the leaves of cabbage, turnip, horse-radish and onion have been fed to the worms, mingled with the leaves of an Artemisia and of the culinary sage, thyme and mint, differing in no material degree in texture from the foregoing four, yet quite as strong in taste, but the latter were quite neglected excepting those of the mint, which were slightly nibbled, but the others were all attacked and had to be renewed.

There is little to be noted about the mental qualities of worms. They have been seen to be timid creatures. Their eagerness for certain kinds of food manifestly shows that they must enjoy the pleasure of eating. So strong is their sexual passion that they overcome for a time their dread of light. They seem to have a trace of social feeling, for they are not disturbed by crawling over each other's bodies, and they sometimes lie in contact. Although remarkably deficient in the several sense-organs, yet this does not necessarily preclude intelligence, for it has been shown that when their attention is engaged they neglect impressions to which they would otherwise have attended, and attention, as is well known, indicates the presence of a mind of some kind. A few actions are performed instinctively, that is, all the individuals, including the young, perform each action in nearly the same manner. The various species of Perichæta eject their castings so as to construct towers, and the burrows of the Common Earth-worm—*Lumbricus terrestris*—are smoothly lined with fine earth and often with little stones, and the mouth with leaves. One of their strongest instincts is the plugging up of the mouths of their burrows with various objects, the very young worms acting in a similar manner. But some degree of intelligence is manifested, as will subsequently appear.

Almost everything is eaten by worms. They swallow enormous quantities of earth, from which they extract any digestible matter it may contain. Large numbers of half-decayed leaves of all kinds, excepting a few that are too tough and unpleasant to the taste, and likewise petioles, peduncles, and decayed flowers. Fresh leaves are consumed as well. Particles of sugar, licorice and starch, and bits of raw and roasted meat, and preferably raw fat, are eaten when they come into their possession, but the last article with a better relish than any other substance given to them. They are cannibals to a certain extent, and have been known to eat the dead bodies of their own companions.

The digestive fluid of worms, according to León Frédéricq, is analogous in nature to the pancreatic secretion of the higher animals, and this conclusion agrees perfectly with the kinds of food which they consume. Pancreatic juice emulsifies fat, dissolves fibrin, and worms greedily devour fat and eat raw meat. It converts starch into grape-sugar with wonderful rapidity, and the digestive fluid of worms acts upon the starch of leaves. But worms live chiefly on half-decayed leaves, and these would be useless to them unless they could digest the cellulose forming the cell-walls, for all other nutritious substances, as is well known, are almost completely withdrawn from leaves shortly before they fall off. It has been ascertained that cellulose, though very little or not at all attacked by the gastric juice of the higher animals, is acted on by that from the pancreas, and so worms eat the leaves as much for the cellulose as for the starch they contain. The half-decayed or fresh leaves which are intended for food are dragged into the mouths of their burrows to a depth of from one to three inches, and are then moistened with a secreted fluid, which has been assumed to hasten their decay, but which, from its alkaline nature, and from its acting both on the starch-granules and on the protoplasmic contents of the cells, is not of the nature of saliva, but a pancreatic secretion, and of the same kind as is found in the intestines of worms. As the leaves which are dragged into the burrows are often dry and shrivelled, it is indispensable for the unarmed mouths of worms that they should first be moistened and softened, their disintegration being thereby the more readily effected. Fresh leaves, however soft and tender they may be, are similarly treated, probably from habit. Thus the leaves are partially digested before they are taken into the alimentary canal, an instance of extra-stomachal digestion, whose nearest analogy is to be found in such plants as Dionæa and Drosera, for in them animal matter is digested and converted into peptone, not within a stomach, but on the surfaces of the leaves.

But no portion of the economy of worms has been more the subject of speculation than the calciferous glands. About as many theories have been advanced on their utility as there have been observers. Judging from their size and from their rich supply of blood-vessels, they must be of vast importance to these animals. They consist of three pairs, which in the Common Earth-worm debouch into the alimentary canal in front of the gizzard, but posteriorly to it, in some genera. The two posterior pairs are formed by lamellæ, diverticula from the œsophagus, which are coated with a pulpy cellular layer, with the outer cells lying free in infinite numbers. If one of these glands is punctured and squeezed, a quantity of white, pulpy matter exudes, consisting of these free cells, which are minute bodies, varying in diameter from two to six millimetres. They contain in their centres a small quantity of excessively fine granular matter, that looks so like oil globules that many scientists are deceived by its appearance. When treated with acetic acid they quickly dissolve with effervescence. An addition of oxalate of ammonia to the solution throws down a white precipitate, showing that the cells contain carbonate of lime. The two anterior glands differ a little in shape from the four posterior ones by being more oval, and also conspicuously in generally containing several small, or two or three larger, or a single very large concretion of carbonate of lime, as much as one and one-half millimetres in diameter. With respect to the function of the calciferous glands, it is likely that they primarily serve as organs of excretion, and secondarily as an aid to digestion. Worms consume many fallen leaves. It is known that lime goes on accumulating in leaves until they drop off the parent-plant, instead of being re-absorbed into the stem or roots, like various other organic and inorganic substances, and worms would therefore be liable to become charged with this earth, unless there was some special apparatus for its excretion, and for this purpose the calciferous glands are ably adapted. On the other hand, the carbonate of lime, which is excreted by the glands, aids the digestive process under ordinary circumstances. Leaves during their decay generate an abundance of various kinds of acids, which have been grouped together under the term of humus acids. These half-decayed leaves, which are swallowed by worms in large quantities, would, therefore, after having been moistened and triturated in the alimentary canal, be apt to produce such acids, and in the case of several worms, whose alimentary canals were examined, their contents were plainly shown by litmus paper to be decidedly acid. This acidity cannot be attributed to the nature of the digestive fluid, for pancreatic juice is alkaline, and so also is the secretion which is poured out of the mouths of worms for the preparation of the leaves for consumption. With worms not only the contents of the intestines, but their ejected matter or the castings are generally acid. The digestive fluid of worms resembles in its action, as already stated, the pancreatic secretion of the higher animals, and in these latter pancreatic digestion is necessarily alkaline, and the action will not take place unless some alkali be present; and the activity of an alkaline juice is arrested by acidification, and hindered by neutralization. Therefore is seems probable that innumerable calciferous cells, which are emptied from the four posterior glands in the alimentary canal, serve to neutralize more or less completely the acids generated there by the half-decayed leaves. These cells, as has been seen, are instantly dissolved by a small quantity of acetic acid, and as they do not always suffice to render of no effect the contents of the upper part of the alimentary canal, it is probable that the lime is aggregated into concretions, in the anterior pair of glands, in order that some may be conveyed to the posterior parts of the intestine, where these concretions would be rolled about among the acid contents. The concretions found in the intestines and in the castings often present a worn appearance, but whether due to attrition or chemical corrosion it is impossible to say. That they are formed for the sake of acting as mill stones, as Claparède believed, and of thus assisting in the trituration of food, is not at all likely, as this object is already attained by the stones that are present in the gizzards and intestines.

In dragging leaves into their burrows worms generally seize the thin edge of a leaf with their mouths, between the projecting upper and lower lip, the thick and strong pharynx at the same time being pushed forwards within their bodies, so as to afford a point de resistance for the upper lip; but in the case of broad and flat objects the pointed anterior extremity of the body, after being brought into contact with an object of this kind, is drawn within the adjoining rings, so that it becomes truncated and as thick as the rest of the body. This part is then seen to swell a little, seemingly from the pharynx being pushed a little forwards. By a slight withdrawal of the pharynx, or by its expansion, a vacuum is produced beneath the truncated, slimy end of the body whilst in contact with the object, and by this means the two adhere firmly together. Worms can attach themselves to an object in the same manner under the water.

Figure 7. Common Earth-Worms. Out on a foraging excursion.

As worms have no teeth, and their mouths consist of very soft tissue, it may be presumed that they consume by means of suction of the edges and parenchyma of fresh leaves after they have been softened by the digestive fluid. They cannot attack such strong leaves as those of sea-kale or large and thick leaves of ivy. They not only seize leaves and other objects for purposes of food, but for plugging up the mouths of their burrows. Flower-peduncles, decayed twigs of trees, bits of paper, feathers, tufts of wool and horse-hair are some of the many things other than leaves that are dragged into their burrows for this purpose. Many hundred leaves of the pine-tree have been found drawn by their bases into burrows. Where fallen leaves are abundant, especially ordinary dicotyledonous leaves, many more than can be used are collected over the mouth of a burrow, so that a small pile of unused leaves is left like a roof over those which have been partly dragged in. A leaf in being dragged a little way into a cylindrical burrow necessarily becomes much folded or crumpled, and when another is drawn in, this is done exteriorly to the first, and so on with succeeding leaves, till finally they all become closely folded and pressed together. Sometimes the mouth of a burrow is enlarged, or a fresh one is made close by, so that a larger number of leaves may be drawn in. Generally the interstices between the drawn-in leaves are filled with moist, viscid earth ejected from their bodies, thus rendering them doubly secure. Hundreds of such plugged burrows may be seen during the autumnal and early winter months.

When leaves, petioles, sticks, etc., cannot be obtained for the mouths of their burrows, heaps of stones, smooth, rounded pebbles, are utilized for protection. When the stones are removed and the surface of the ground is cleared for some inches round the burrow, the worms may be seen with their tails fixed in their burrows dragging the stones inward by the aid of their mouths, stones weighing as much as two ounces often being found in the little heaps, which goes to show how strong these apparently weak creatures are. Work of this kind is usually performed during the night, although objects have been occasionally known to be drawn into the burrows during the day. What advantage worms derive from plugging up the mouths of their burrows, or from piling stones over them, cannot be satisfactorily answered. They do not act in this manner when they eject much earth from their burrows, for then their castings serve to cover the mouth. Perhaps the plugs serve to protect them from the attacks of scolopenders, their most inveterate enemies, or to enable them to remain with safety with their heads close to the mouths of their burrows, which they like so well to do, but which, unless protected, costs many a fellow its life. Besides, may not the plugs check the free ingress of the lowest stratum of air, when chilled by radiation at night, from the surrounding ground and herbage? The last view of the matter seems especially well taken, because worms kept in pots where there is fire, having no cold air with which to contend, plug up their burrows in a slovenly manner, and because they often coat the upper part of their burrows with leaves, apparently to prevent their bodies from coming into contact with the cold, damp earth. But the plugging-up process may undoubtedly serve for all these purposes. Whatever the motive may be, it seems that worms much dislike leaving the mouths of their burrows open, yet, nevertheless, they will reopen them at night, whether or not they are able afterwards to close them.

Considerable intelligence is shown by worms in their manner of plugging up their burrows. If man had to plug up a cylindrical hole with such objects as leaves, petioles or twigs, he would push them in by their pointed ends, but if these were thin relatively to the size of the hole, he would probably insert some by their broader ends. Intelligence would certainly be his guide in such a case. But how worms would drag leaves into their burrows, whether by their tips, bases, or middle parts, has been a matter of interest to many. Darwin, who experimented upon the subject, found it especially desirable to experiment with plants not native to his country, for he conceived that although the habit of dragging leaves into their burrows is undoubtedly instinctive with worms, yet instinct could not teach them how to act in the case of leaves about which their progenitors knew nothing. Did they act solely through instinct, or an unvarying inherited impulse, they would draw all kinds of leaves into their burrows in the same manner. Having no such definite instinct, chance might be expected to determine whether the tip, base, or middle might be seized. If the worm in each case first tries many different methods, and follows that alone which proves possible or the most easy, then both instinct and chance are ruled out of the solution of the question. But to act in this manner, and to try different methods, makes what in man would be called intelligent action.

Three species of pine-leaves are mentioned by Darwin as being regularly drawn into the mouths of worm-burrows on the gravel-walk in his garden. These leaves consist of two needles, which are united to a common base, and it is by this point that they are almost invariably drawn into the burrows. As the sharply-pointed needles diverge somewhat, and as several are drawn into the same burrow, each tuft forms a perfect chevaux-de-frise. Many tufts were pulled up in the evening, but by the ensuing morning fresh leaves had taken their places, and the burrows again well protected. Impossible it would be to drag these leaves to any depth into the burrows, except by their bases, as a worm cannot seize hold of the two leaves at the same time, and if one alone were seized by the apex, the other would be pressed against the ground and resist the entry of the one that was seized. That the worms should do their work well, it was very essential that they drag the pine-leaves into their burrows by their bases, that is, where the two needles are conjoined. But how they are guided in this work was at first perplexing. The difficulty, however, was soon settled. With the assistance of his son Francis, the elder Darwin set to work to observe worms in confinement during several nights by the aid of a dim light, while they dragged the leaves of the aforementioned kinds into their burrows. They were seen to move the anterior extremities of their bodies about the leaves, and on several occasions when they touched the sharp end of the needle they suddenly withdrew as though they had been pricked, but it is doubtful that they were hurt, for they are indifferent to sharp objects, being known to swallow rose-thorns and small splinters of glass. It may be doubted whether the sharp end of the needle serves to tell them that is the wrong end to seize, for the points of many were cut off for the length of an inch, and these leaves were always drawn in by their bases and not by the cut-off ends. The worms, it seemed, almost instantly perceived as soon as they had seized a leaf in the proper manner. Many leaves were cemented together at the top, or tied together by fine thread, and these in the majority of instances were dragged in by their bases, which leads to the conclusion that there must be something attractive to worms in the base of pine-leaves, notwithstanding that few ordinary leaves are drawn in by their base or footstalk. Leaves of other plants, and also the petioles of some compound plants, as well as triangular bits of paper, dry and damp, were experimented with, and the manner of seizing the objects and bearing them into their burrows were as amusing as they were novel and interesting. The leaves and stems used were such as the worms had not been accustomed to in their respective haunts.

When the several cases experimented on are considered, one can hardly escape from the conclusion that some degree of intelligence is shown by worms in plugging up their burrows. Each particular object is seized in too uniform a manner, and from causes which we can generally understand, for the result to be attributed to mere chance. That every object has not been drawn in by its pointed end may be accounted for by labor having been saved by some being carried in by their broader ends. There is no doubt that worms are governed by instinct in plugging up their burrows, and it might be expected that they would have been taught in every particular instance how to act independently of intelligence. It is very difficult to judge when intelligence comes into play. The actions of animals, appearing due to intelligence, may be performed through inherited habit without any intelligence, although aboriginally acquired, or the habit may be acquired through the preservation and inheritance of some other action, and in the latter case the new habit will have been acquired independently of intelligence throughout the entire course of its development. There is no à priori improbability in worms having acquired special instincts through either of these two latter means. Nevertheless it is incredible that instincts should have been developed in reference to objects, such as the leaves and petioles of foreign plants, wholly unknown to the progenitors of the worms which have acted in the manner just described. Nor are their actions so unvarying or inevitable as are most true instincts.

As worms are not controlled by special instincts in each particular case, though possessing a general instinct to plug up their burrows, and as chance is excluded, the next most probable conclusion is that they try in many ways to draw in objects and finally succeed in some one way. It is surprising, however, that an animal so low in the scale as a worm should have the capacity to act in this way, as many higher animals have no such capacity, the instincts of the latter often being followed in a senseless or purposeless manner.

We can safely infer intelligence, as Mr. Romanes, who has specially studied animals, says, only when we see an individual profiting by his own experiences. That worms are able to judge either before or after having drawn an object close to the mouths of their burrows how best to drag it in, shows that they must have acquired some notion of its general shape. This they probably acquire by touching it in many places with the anterior extremity of their bodies, which serves them as a tactile organ. Man, even when born blind and deaf, shows how perfect the sense of touch may become, and if worms, which also come into being in the same condition, have the power of acquiring some notion, however rude, of the shape of an object and their burrows, they deserve, it must seem to every sensible mind, to be called intelligent creatures, for they act in such a case in nearly the same manner as a man would under similar circumstances. That worms, which stand so low in the scale of organization, should possess some degree of intelligence, will doubtless strike everyone as very improbable. It may be doubted, however, whether we know enough about the nervous system of the lower animals to justify our natural distrust of such a conclusion. With regard to the small size of the cerebral ganglia, we would do well to remember what a mass of inherited knowledge, with some power of adapting means to an end, is crowded into the minute brain of a worker ant.

Two ways are adopted by worms in excavating their burrows. Either the earth is pushed away on all sides or it is swallowed by the animal. In the former case the worm inserts the stretched-out and attenuated anterior extremity of its body into any little crevice or hole, and the pharynx is pushed forward into this part, which consequently swells and pushes away the earth on all sides, the anterior extremity thus acting as a wedge. When placed in loose mould a worm will bury itself in between two and three minutes, but in earth that is moderately pressed down it often requires as many as fifteen minutes for its disappearance. But whenever a worm burrows to a depth of several feet in undisturbed compact ground, it must form its passage by swallowing the earth, for it is impossible that the ground could yield on all sides to the pressure of the pharynx when pushed forward within the worm's body. Great depths are reached only during continued dry weather and severe cold, the burrows sometimes attaining to a depth of from seven to eight feet. The burrows run down perpendicularly, or, more commonly, obliquely, and are sometimes said to branch. Generally, or invariably as I think, they are lined with fine, dark-colored earth voided by the worm, so that at first they must be made a little wider than their ultimate diameter. Little globular pellets of voided earth, still soft and viscid, often dot the walls of fresh burrows, and these are spread out on all sides by the worm as it travels up or down its burrow, the lining thus formed becoming very compact and smooth when nearly dry and closely fitting the worm's body. Excellent points of support are thus afforded for the minute reflexed bristles which project in rows on all sides from the body, thus rendering the burrow well adapted for the rapid movement of the animal. The lining appears also to strengthen the walls, and perhaps saves the worm's body from being scratched, which would assuredly be the case when the burrows, as is occasionally observed, pass through a layer of sifted coal cinders. The burrows are thus seen to be not mere excavations, but may be compared with tunnels lined with cement. Those which run far down into the ground generally, or at least frequently, terminate in little chambers, where one or several worms pass the winter rolled up into a ball. Small pebbles and seeds as large as grains of mustard are carried down from the surface by being swallowed or within the mouths of worms, as well as bits of glass and tile, whose only use in their winter-quarters seems to be the prevention of their closely coiled-up bodies from coming into contiguity with the surrounding cold soil, for such contact would perhaps interfere with their respiration, which is effected by the skin alone.

After swallowing earth, whether for making its burrow or for food, the earth-worm soon comes to the surface to empty its body. The rejected matter is thoroughly mixed with the intestinal secretions, and is thus rendered viscid. After becoming dried, it sets hard. When in a very liquid state the earth is thrown out in little spurts, and when not so liquid by a slow peristaltic movement of the intestine. It is not cast indifferently on any side, but first on one and then on another, the tail being used almost like a trowel. The little heap being formed the worm seemingly avoids, for the sake of safety, the use of its tail, the earthy matter being forced up through the previously deposited soft mass. The mouth of the same burrow is used for this purpose for a considerable time. When a worm comes to the surface to eject earth, the tail protrudes, but when it collects leaves its head must protrude, and thus worms must have the power of performing the difficult feat, as it seems to us, of turning round in their closely-fitting burrows. Worms do not always eject their castings upon the surface of the ground, for when burrowing in newly turned-up earth, or between the stems of banked-up plants, they deposit their castings in such places, and even hollows beneath large stems lying on the surface of the ground are filled up with their ejections. Old burrows collapse in time. The fine earth voided by worms, if spread out uniformly, would form in many places a layer of one-fifth of an inch in thickness. But this large amount is not deposited within the old unused burrows. If the burrows did not collapse, the whole ground would be first thickly riddled with holes to the depth of ten inches or more, which in fifty years would grow into a hollow, unsupported place ten inches deep.

Hardly any animal is more universally distributed than worms. The earth-worm is found in all parts of the world, and some of the genera have an enormous range. They inhabit the most isolated islands, abounding in Iceland, and also being known to exist in the West Indies, St. Helena, Madagascar, New Caledonia and Tahiti. Worms from Kergulen Land in the Antarctic regions have been described by Ray Lankester, and Darwin has reported them as being found in the Falkland Islands. How they reach such isolated islands is quite unknown. They are easily killed by salt water, and it does not seem likely that young worms or their egg-capsules could be carried in earth adhering to the feet or beaks of land-birds, especially to Kergulen Land, for it is not now inhabited by any terrestrial bird.

We have seen that worms are found in nearly every part of the globe, that they are very numerous, as many as 348,480 having been found in an acre of rich ground in New Zealand, and that by the peculiar economy of their nature they are fitted to accomplish a great deal of good in the earth. They have played a more important part in the history of the world than most persons would at first suppose. In many parts of England, according to Darwin, a weight of more than ten tons of dry earth annually passes through their bodies and is brought to the surface in each acre of land, so that the entire superficial bed of vegetable mould passes through their bodies in the course of every few years; and in most parts of the forests and pasture-lands of Southern Brazil, where several species of earth-worms abound, the whole soil to a depth of a quarter of a metre looks as though it had passed through the intestines of worms, even where scarcely any castings are to be observed upon the surface. The upper crust is continually being eaten and ejected by them, thus aiding the fertility of the soil, as well as conveying water and air to the interior by the myriads of burrows which they drill. The vast quantities of leaves that they drag into their holes tend also to enrich the ground. Nor does their good end here. They cover up seeds, undermine rocks, burying them up, and to their labors is due the preservation of many ruins and ancient works of art. Numerous old-time Roman villas have been discovered beneath the ground in England, whose entombments were undoubtedly caused by the worms that undermined them and deposited their castings upon the floors, till finally, aided by other causes, they disappeared from sight.

When a wide, turf-covered expanse of earth is beheld, we would do well to remember that its smoothness, upon which so much of its beauty depends, is largely due to all the inequalities having been slowly levelled by worms. That all the surface-mould of any such expanse has passed, and will again pass, every few years through the bodies of worms is a marvellous reflection, and one which should not be lightly dismissed from the mind. The most ancient, as well as one of the most valuable of man's inventions, is the plough. But long before man existed the land was in fact regularly ploughed, and still continues to be ploughed, by earth-worms. No other animal has played such a part in history as have these lowly-organized creatures. True it is that corals, which are still lower in the scale of animals, have performed more conspicuous work in the innumerable reefs and islands they have built in the great oceans, but their work is confined to the tropical zones, while that of the earth-worm is well-nigh universal. Verily it is by the little things in life that the Creator has erected the most stupendous monuments to show forth His infinite power and wisdom.

Chapter 7

FIDDLER- AND HERMIT-CRABS

Among our first acquaintances of the sea-shore are sure to be a number of merry little sprites which do not seem to have yet mastered the lesson of walking straight ahead. Their movements will be seen to be in a direction at right angles to that towards which the head points. It is a very interesting sight to watch these apparently one-sided creatures hurrying off in their lateral progression towards their burrows in the sand or mud, or in quest of food. Pass them, and you will be surprised to see how quickly some of them will reverse their motion, seemingly without so much as pausing to glance at their pursuer, their machinery appearing to have given out at one end, thus compelling them to reverse and travel back over their old courses.

These little Fiddler- or Calling-crabs, as they are termed, are the most pronounced offenders against the commonly accepted rule of proper walking. Scattered all over the salt marshes and mud-flats, at about high-water mark, may be noted their burrows, which are about as large as a thrust made by an umbrella point, and from which can be frequently seen the little animal peeping forth, preparatory to making a sally. At another part of the flat, where the noise of your footsteps has not given signals of danger, hundreds of crabblings are busy with their out-door occupations. Draw near to them, and away they scamper to their dwellings, males and females intermingled promiscuously, the former recognizable by the undue development of one of the claws, which is carried transversely in front of the head. When the animal is provoked, this claw is brandished in a somewhat menacing manner, which has been likened by some to the pulling of a violin bow, and by others to the action of beckoning or calling, and hence the names which have been applied to these eccentric creatures.

Have you a desire for a more intimate knowledge of the animal, take him up by the big claw, and you can now examine him without the least fear of incurring the proofs of his displeasure. Two bead-like, compound eyes, supported on long stalks, which can be readily withdrawn into the protecting shield of the carapace, will be observed. From the manner of this support, which allows of vision in almost every direction, the name of stalk-eyed crustaceans has been given to the group in which this structure is found. The two pairs of feelers, which you see in front of the eyes, are known as antennæ and antennules. They are of peculiar interest, for, aside from acting as feelers, they subserve the functions of smelling and hearing, the auditory apparatus being lodged in the base of the smaller pair. There are ten feet, and this is a character of importance, as it is a feature distinctive of the ten-footed, or decapod, crustaceans. At first sight it appears that the animal is devoid of a tail, but if you turn him over upon his back you will find a very short one tucked safely under the body. A comparison of our study of this crab with that of the lobster or cray-fish will show that the tail, or, more properly, the abdomen, is stretched out beyond the body proper, and that the elongation is in proportion to the length of the animal. Two distinct groups of ten-legged, stalk-eyed crustaceans are thus recognized, namely: the short-tailed forms, or crabs, and the opposite, or long-tailed forms, to which the lobster and shrimp belong, the hermit-crabs constituting an intermediate type.

Two species of the Fiddler, considerably resembling each other in color and ornamentation, are to be found upon our Atlantic Coast. The more common form, *Gelasimus vocator*, has a smooth, shining carapace, while that of *Gelasimus minax* is finely granulated and in part tuberculated, the back of both appearing impressed with a figure very similar to the letter H. The latter, which appears to be a vegetable feeder, is the larger, its burrows not infrequently measuring one and a half inches in diameter. Estuarine regions, in close proximity to fresh water, rather than the tidal flats, are its habitat, and, in truth, it seems to be able to get along for weeks, and even months, without any absolute need of salt water.

In the excavation of their homes the Fiddlers throw up the pellets of moist earth by means of their anterior walking legs, depositing their burden usually at some little distance from the mouth of the burrow. As winter approaches, the domiciliary apertures are closed up, and the famine of winter is spent in a state of torpidity.

With the advent of spring they come forth from their brumal retreats, and soon concern themselves with the duties incident to the propagation of their kind. Two males are often observed contending in the fiercest manner for the possession of a female. They strike with the formidable claw most powerful blows, and I have often seen an opponent so completely claw-locked as to be unutterly unable to make any determined resistance. These contests last a long while, and finally conclude with the complete vanquishment of one or the other of the fighting parties, one or both sustaining at times some severe injury as the loss of an eye-peduncle or the joint of a limb. All the while the battle is waging, the female is a silent, passive spectator, and generally allies herself with the successful competitor for her affections. Even during the summer season, when the cares of brood-raising no longer command and enslave the attention of the female, these combats are still indulged in by the males, growing out of, as it would seem, the lingering smarts of old animosities festering in the memory. While these carcinological lords of the sea-side are eminently fitted for the sparring business, the whole physiognomy of their smaller, weaker partners bespeaks a life in which broils can have no part, a life devoted to peaceful and domestic pursuits.

Figure 8. Fiddler-Crabs. Two males fighting for a female.

Differing widely in structure and habits from the Calling-crabs, and affecting watery situations near the shore, are to be found the Hermit-crabs. These sprightly little animals, which are usually of small size, and have truly habits of their own, that stamp them at once as being original and distinctive, are a source of never-failing delight to the student of nature. They derive their name, as is well known, from the seclusion into which they cast themselves as the inhabitants of the shells of other animals, but it is probably not generally known, however, that the rights of tenantry are oftentimes exercised in the most arbitrary manner. Not always satisfied with a dead shell, the Hermit-crab has been seen to raid upon a living possessor and attempt to drag him from his home, in which operation the assailant is often assisted by a number of his fellows, each bearing with him his castle as defensive armor. True, the attack is probably made in many instances for the purpose of getting possession of the enemy as well as his belongings, and, however this may be, forcible possession is by them considered no misdemeanor.

The body of the Hermit-crab, in the greater number of species, is unprovided with a carapace, and, being soft and liable to injury, the animal is compelled to seek shelter usually in a snail-shell, winding himself about the coils, to the inner extremity of which he attaches himself by his modified posterior feet. So securely is he now intrenched that it is only with difficulty he can be withdrawn, retracting himself as he does further and further within cover of the shell. A sudden fracture of the apex of the shell, under which appears to be the most delicate part of the animal's body, will generally effect a speedy dislodgment, the frightened Crab dropping from the aperture.

With his progressive development in size the Hermit requires frequent changes of abode. His methods in securing a new habitation are among the most interesting of his life. He is very circumspect in his movements, and will make several reconnoissances before he is fully satisfied with the conditions of his prospective home, retiring after each visit to the old shell.

Figure 9. Crab waiting for food under a rock. From a photograph taken through water. Copyright 1900 by A. R. Dugmore.

Like many bipeds, he has his first of May, and so he goes house-hunting. He finds a shell. Will it do? He examines it within, feelingly if not courteously, to see whether it is to let. Satisfied on this point, he turns it over, then turns it round, to know if it will suit, the weight of the house being quite an item in the reckoning to one who is to carry it upon his back. All things being right, his mind is made up to move, and quickly, too, at that, lest he miss his chance through some more active fellow house-hunter who is on the alert. Out comes the body from the old house, and pop it goes into the new. The resolution to move, the surrender of the old house, and the occupancy of the new, were all effected within a fraction of a second of time.

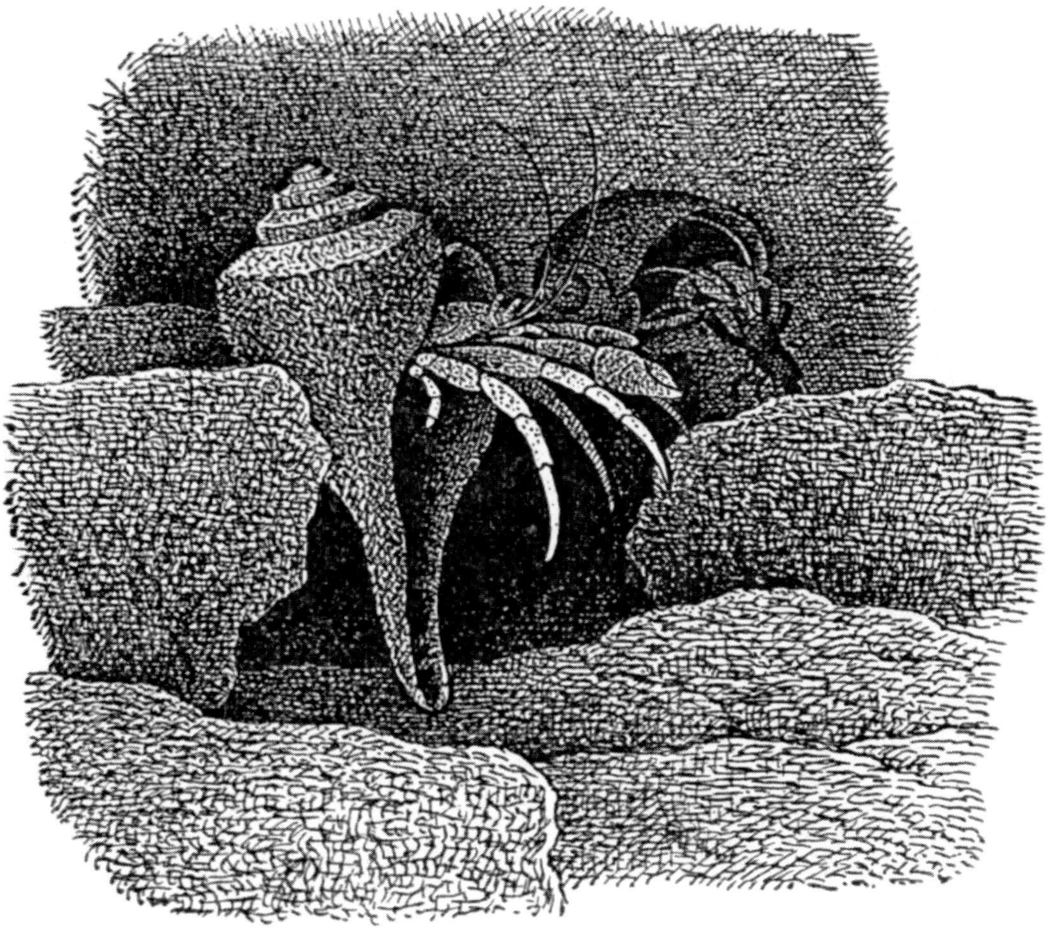

Figure 10. Warty Hermit-Crabs. One at home, the other house-hunting.

But the matter does not always go on pleasantly. Two house-hunters may find the same tenement. Should they both desire it, then comes the tug of war. Dwell together they neither can nor will. Recourse is had to battle, in which the stronger proves his claim right by the rule of might. In these encounters terrible mutilations quite often occur.

As an offset to all this bad feeling and bloodshed, it is a sad sight to see the little Hermit when his time comes to die. However droll his career may have been, he is now very grave, for he knows he must part with life and all its joys and pleasures. Who can explain the strange fact? The poor little fellow comes out of his house to die. Yes, to die. To us humans home is the only fit place to die in, but to Eupagurus it has no attractions at this solemn time. Poor fellow! With a sad look and a melancholy movement he quits of his own will the house for which he fought so well. Those feelers that often stood out so provokingly, and that were quite as often poked into everybody's business, now lie prone and harmless; the eyes have lost their pertness, and dead, stone dead, the houseless Hermit lies upon that moss-covered rock.

There are two species of Hermit-crab occurring on our coast, which are readily distinguishable from each other by their size and the difference in the shape of the big claw. *Eupagurus pollicaris*, the Warty Hermit, is the larger species. He inhabits the shells of the big Naticas and the Fulgurs, and can be easily recognized by his coarse, broad claws, which close up in great part the aperture of the shell which he occupies. In the more common form, *Eupagurus longicarpus*, which seldom attains a length exceeding an inch, the legs are all much elongated, giving the animal a very slender appearance.

Chapter 8

FUNNEL-WEB BUILDER

Simple nests and tubes are all the majority of spiders construct for their homes. The larger and better known webs for catching insects are made by comparatively few species. He who is astir in the grass-fields on damp summer mornings, will everywhere see innumerous flat webs, from an inch or two to a foot in diameter, which weather-wise folks consider prognostic of a fair day. These webs may always be found upon the grass at the proper season, but only become visible from a distance when the dew is upon them, making the earth appear as covered by an almost continuous carpet of silk.

By far the greater number of these nests is of the form which is termed funnel-webs, which consist of a concave sheet of silk, constituted of strong threads, crossed by finer ones, which the author spins with the long hind-spinnerets, swinging them from side to side, and laying down a band of threads at each stroke, the many hundred threads extending in all directions to the supporting spears of grass. The web is so close and tight that the footsteps of the spider can be distinctly heard by the attentive, listening ear as she runs hither and thither over its scarcely bending surface. At one side of the web is a tube, leading down among the grass-stems, which serves as a hiding-place for the owner of the web. Here, at the top, and just out of sight, the spider ordinarily stands, waiting for something to light upon the web, when she eagerly rushes out, seizing the prey unluckily caught and carrying it into her tube to eat. If too formidable an insect comes upon the web, she turns herself round, beating a precipitate retreat out of the lower end of her funnel and soon is lost beneath the mesh of enveloping and interlacing grasses.

Where favorably located, these webs remain through the entire season, and are enlarged, as the spider grows, by additions on the outer edges, and are supported by threads running up into the neighboring plants. Sometimes the webs are built in close proximity to a stone partially imbedded in the earth, the bottom of the funnel opening slightly underneath the stone, which secures to the spider a convenient harbor in case of threatening danger.

Agalenidæ, as our funnel-web weavers are called, are long-legged, brown spiders, in which the head part of the cephalo-thorax is higher than the thoracic part, and distinctly separated from it by grooves or marks at the sides. The eyes are usually in two rows, but in Agalena the middle eyes of both rows are much higher than the others. The feet have three claws, and the posterior pairs of spinnerets are two-jointed and usually longer than the others. *Agalena nævia*, the technical name of our Common Grass Spider, abounds in all parts of the United States, but its very commonness is the principal reason why it is so little known except by the trained naturalist, its very familiarity leading the average man and woman to look upon it with contempt.

Figure 11. Agalena and Her Funnel-Web. House-fly, caught in the toils, becomes a victim.

Persons unfamiliar with spiders find it difficult to distinguish the young from the old, and male from female. This is caused, in part, by the great differences between different ages and sexes of the same spider, on account of which they are supposed to belong to distinct species. The adult males and females, however, are easily distinguished from each other, and from the young, by the complete development of organs peculiar to each sex, the palpal organs on the ends of the palpi in the males, and the epigynum, a hard swollen place just in front of the opening of the ovaries in the females. Usually the males are smaller than their partners, and have, in proportion to their size, smaller abdomens and longer legs. They are generally darker colored, especially on the head and front part of the body, and markings which are distinct in the female coalesce and become darker in the male. In most species these differences are not very great, but in some, Argiope and Nephila for examples, where the males are about one-tenth as large as the females, one would hardly suppose, without other evidence, that the males and females had any relationship to each other. The palpal organs and the epigynum are sexual characters which do not attain their functional value until after the last moult has been effected.

Spiders are naturally very selfish creatures. Their chief concern in life seems to be the gratification of their desires for food. They are eminently unsocial, the sexes preferring to live solitary lives. It is only when actuated by amatory influences that the females will tolerate their weaker lords, and in some instances it is only by stratagem and agility that the latter are able to accomplish the fulfilment of the law of their being, the females by their ugly, vicious tempers resisting to the utmost. In the case of Agalena the male is the stronger of the two. He, at the proper time, when the reproductive cells are matured, takes the female in his powerful mandibles, lays her gently on one side, and inserts one of his palpi, whose little sacs had previously been filled with the fecundating discharge, into the epigynum underneath. After a time, necessarily brief, he rises on tiptoe, turns her around and over, so that she comfortably lies on the other side, her head being in the opposite direction, and inserts the other palpus. All through the operation the female lies as though she was dead. The ends of nature being served, the sexes separate, the male returning to the solitary life he previously led, while the female busies herself in providing for the duties of maternity.

The eggs becoming mature, the latter proceeds to make a little web and lays them in it, practising the utmost care. She now covers them over with silk, which she weaves into a cocoon, where the young remain some time after they are hatched. Seldom is the laying seen, for it generally happens in the night-time, or in retired places. Often, in confinement, the spider refuses to lay at all. An egg of a spider, like that of any other animal, is a cell which separates from the body of the female, and subsequently unites with one or more cells that have separated from the body of the male. This process of union, termed fertilization, doubtless takes place when the eggs have attained their full size and are about to be laid. After being laid and hardened it is a very easy matter to watch their development. All that is necessary to be done is to cover the egg to be examined with oil, alcohol or any liquid that will wet it, for this tends to make the shell transparent. Eggs laid in summer are ready to hatch in a fortnight, while those laid in autumn develop slowly all through the winter. A day or two are occupied in hatching. When the time has arrived the shell, or more properly the skin, cracks along the lines between the legs, and comes off in rags, and the spider slowly stretches itself and creeps about. Pale and soft it appears, and devoid of hairs or spines, but its feet are armed with small claws. In two or three days it gets rid of another skin, and begins to assume a spider-like appearance, the eyes becoming dark-colored, the thoracic marks growing more distinct, and a dark stripe appearing across the edge of each segment of the abdomen. The hairs are now long, but few in number, and arranged in rows across the abdomen and along the middle of the thorax. Before the next moult they usually forsake the cocoon, and live together for a short time in a web spun in common. Where larger broods of young spiders live together, they soon show cannibal-like qualities, and if kept in confinement one or two out of a cocoon-full may be raised without recourse to any other food.

As spiders grow larger, they must moult from time to time. This is an interesting process. The spider hangs herself by a thread from the spinnerets to the centre of the web. In a short time the skin cracks around the thorax, just over the first joints of the legs, and the top part falls forward, being held only at the front edge. The skin of the abdomen now breaks irregularly along the sides and back, and shrinks together in a bunch, leaving the spider suspended only by a short thread from the spinnerets, her legs still being trammelled by the old skin. Fifteen minutes of violent exertion releases her from the encumbrance, when she drops down, hanging by her spinnerets like a wet rag. She can do nothing in this condition, not even draw her legs away from an approaching hand. In ten or twelve minutes the legs show signs of strengthening, and she is able to draw them gradually towards her. A few up-and-down movements, and she manages to get into the web again.

That which, more than anything else, discriminates spiders from other animals is their habit of spinning webs. Some of the mites spin irregular threads upon plants, or cocoons for their eggs, and many insects cocoons in which to undergo their changes from larva to imago, but in the spiders the spinning-organs are much more complicated, and used for a greater variety of purposes, for making egg-cocoons, silk linings to their nests, and nets for catching insects. The spider's thread differs from that of insects, in being constituted of a great number of finer threads laid together, while soft enough to coalesce into one. Each spinneret is provided with a number of little tubes, which convey the viscid liquid that forms the thread from glands in the spider's body. In Agalena the two hinder spinnerets are long, and have spinning-tubes along the under side of the last joint.

When about to produce a thread the spider presses the spinnerets against some object and forces out from each tube enough of the secretion to adhere to it, when the spinnerets are moved away, drawing the viscid liquid out, which hardens at once into threads for each tube. A band of threads is formed when the spinnerets are kept apart, but when closed together the fine threads unite into one or more large ones. Commonly the spinning is aided by the hinder feet, which guide the thread, keeping it clear of surrounding objects, and even pulling it from the spinnerets.

Spiders are best known and hated as animals that bite. Their biting-apparatus, the mandibles, are located in front of the head. Partly in the basal joints of these organs and partly in the head, the poison-glands are seated, from which is discharged through a tube the venom, which makes spiders so much to be feared. This tube opens at the point of the claw of the mandible. When the apparatus is not in use the claws are closed up against the parts between the rows of teeth; but when the jaws are opened to bite the claws are turned outward, so that their points can be made to penetrate anything that comes between the jaws. The ordinary function of the mandibles is the killing and crushing of insects, so that the soft parts can be eaten by the spider, and in this preparation they are substantially aided by the maxillæ. Spiders will sometimes chew an insect for hours, until it becomes a mere ball of skin, only swallowing such bits as may happen to be sucked in with the blood. Let alone and unmolested, they bite nothing except insects that are useful for food. But when attacked and cornered, all species open their jaws and bite if they can, their ability to do so depending upon their size and the strength of their jaws. Notwithstanding the large number of pimples and stings ascribed to spiders, undoubted cases of their biting the human skin are exceedingly rare, and the stories of death, insanity and lameness from spider-bites are probably all untrue. Many experiments have been made to test the effect of the bites of spiders on animals. Insects succumb most readily to their bites, some sooner than others, but birds, except when bitten by the larger Mygale, recover after the lapse of a few hours. The effect upon man, even when the bite is deep enough to draw blood, is like the pricks of a needle, attended by little or no inflammation or pain. Even in cases where death among insects and birds ensues it is claimed by the authorities, men as eminent as Blackwall, Moggridge and Dufour, that the secretion from spiders' jaws is not poisonous, but that the animals die, when bitten, from loss of blood and mechanical injury.

Such is the prejudice against the spider, that its presence, no matter where found, whether in the open field or in a corner of the house, is an inducement for its inveterate enemy, man, to sweep it to the ground or floor and crush its frail life out with one blow of the foot. Few know, or care to know, it would seem, the good it does for man. He owes to it, in a large measure, the protection of his crops, and no little of the comfort he enjoys in life. Spiders are carnivorous creatures, and destroy vast number of insects, many of which are man's worst enemies. They merit, and deservingly, too, his kindness and protection for the benefits they confer.

Tarantulas have been supposed to produce epilepsy by their bites, which could only be relieved by music of certain kinds. Such stories, and they have been widely circulated and believed, are the veriest nonsense, for tarantula-bites produce no such effects nowadays. These spiders, which live in holes in sand, out of which they reach after passing insects, are no more savage in their habits than other spiders, for Dufour, a celebrated French naturalist, once kept one that soon learned to take flies from his fingers without manifesting

Chapter 9

Book-Lovers

Living in chinks and crannies of ranges in our homes, and occasionally in bookcases and closets where glutinous and sugary matters abound, but which has probably not been met with elsewhere, is a strange but beautiful little creature which, as far as can be determined, goes through the brief round of its existence without a name to distinguish it from its fellows.

Few entomologists have given any special attention to its family relationships. The possession of certain bristle-like appendages which terminate the abdomen, and which are no doubt comparable with the abdominal legs of the Myriopods, or Thousand Legs, classes it with the Bristle-tails, or Lepismas. In general form, a likeness to the larva of Perla, a net-veined neuropterous insect, is manifest, or to the narrow-bodied species of Blattariæ, or Cockroaches, when divested of wings.

Lepisma saccharina, of Europe, which is indistinguishable from our ordinary American form, is far from uncommon in old, damp houses. Its structure is less complicated than the heat-loving species to which I have alluded, and there are likewise differences of habits which show themselves to the close investigator of natural phenomena.

Not unlike the cockroaches, which our little denizen of the hearth somewhat vaguely resembles in form, it affects hot, dry localities, and is always astir at nights in quest of its fare, for it disdains the light of the day and the consequent publicity of its deeds of shame and plunder.

Many a housewife in the discharge of duty has unearthed, so to speak, the miscreant from its hidden retreat, and sought by foot or hand to crush the life that dares obtrude its uncleanly presence in her larder, but the cunning, swift-footed Lepisma darts off, like a streak of light, to some near-by crack or breach, where it manages to hide from threatening danger. The bodies of these nimble, silent-moving creatures being coated in a suit of shining mail, which the arrangement of the scales so very much resembles, they have a weird and ghostly look. This appearance, and the swiftness of their movement, which the eye can hardly trace, have led the vivid mind of man, in country town and village, to dub them "silver witches."

So fleet of foot are they, and so like a wave of blurred light they cross the vision, that it is vain to try to figure what they are in shape and look. In death they yield their all of earth to prying science. Their body's form is narrow, flattened; their legs in pairs of threes, each of six joints consisting, the basal joints broad, flat, triangular, the tarsal large, in number two, and armed at end with pair of claws incurved. The three thoracic segments are very like in size, and eight abdominals, of similar length and width. So weak it seems the rather long abdomen is, that two pairs or six of bristles, simple, unjointed, and freely movable, serve as support, and also, as in other groups of insects, as organs locomotive.

The mode of antenna-insertion—and the same prevails in the entire family—is much like that of the Myriopods, the front of the head being flattened and concealing, as in the Centipedes, the base of the antennæ. Indeed, the head of any of the Bristle-tails, as seen from above, bears a general resemblance in some of its features to that of the Centipede and its allies, and so, in a less degree, does the head of the larvæ of certain beetles and neuropters. The eyes are compound, the individual facets constituting a sort of heap. The mouth-parts are readily compared with those of the larva of Perla, the rather large, stout mandibles being hid at their tips by the upper lip, which moves freely up and down when the creature opens its mouth. In length the mandible is three times its breadth, and furnished with three sharp teeth on the outer edge, and with a broad cutting margin within, and still further inwards with a number of straggling small spines. The lower lip is broad and stout, with a distinct medium suture, which indicates a former separation in embryonic life into a pair of appendages. Its palpi are three-jointed, the joints being broad, and directed backwards in life, and not forwards, as in the higher insecta.

Perhaps not more than a half-dozen species of Lepisma are known to exist in this country. Our commonest form is very abundant in the Middle States under stones and leaves in forests, and northward in damp houses, where it has much of the habits of the cockroach, eating clothes, tapestry, silken trimmings of furniture, and doing great mischief to libraries by devouring the paste and mutilating the leaves and covers of books. Our heat-loving form, which is apparently allied to the *Lepisma thermophila* of Europe, and which may be an imported species, is quite as destructive as its nearest of kin *Lepisma saccharina*. It does not confine its ravages to closets and pantries, and feed upon sugar and cake and pastry, but has latterly taken to bookcases, where it leads an easy, comfortable life, without fear of molestation.

So delicately constructed are the Lepismas, and so seemingly feeble the breath of life which animates their frail houses of clay, that nature has endowed them with qualities of mind and body which eminently fit them for the part they have to play in the world. She has made them lovers of darkness rather than light, endowed them with keenness of vision and hearing truly wonderful, and given them a celerity of movement which enables them to outstrip in speed the fleetest of their insect-enemies, and even to baffle the well-directed efforts of man for their destruction. The silver-coated armor with which they are invested is so glossy and smooth that they can slip into a crevice in the wall or floor with the utmost ease and facility. From their actions it would seem that they were always on the alert, for when peril is imminent they do not run aimlessly about for a place of security, but know just where to find it with the least possible expenditure of time and physical strength. Every nook and cranny of their appropriated domain is as well known to these very humble of God's creatures as some forest-tract of country to one skilled in wood-craft. Never have I studied the behavior of Lepisma that I have not been deeply impressed with the intelligence of its actions. There have always been displayed a purpose and an aim, which showed as plainly as could be that no blind instinct was the cause of a conduct so rational and human-like.

Figure 12. Lepismas at Work. How Books Are Destroyed.

Chapter 10

YOU-EE-UP

Hardly a person living in a sandy country district can be found who has not seen or heard of the queer little insect called You-ee-up, a name which the books do not give, and of which writers on entomological subjects seem to be ignorant. The learned call him Myrmeleon, or Ant-lion, and very appropriately too, because, like the great king of beasts, he never attacks his prey in the open field, but by stratagem while lying in wait in some hidden retreat or secret covert.

Should you chance, on a warm summer day, where sunny slopes abound on the outskirts of a woods, or by the side of a frequented path or road, look carefully about and soon will you descry a small funnel-like opening, scarce two inches in depth and in width, upon a bare patch of sand in the midst of an ocean of verdure. This little cavity is the intentional work of the larva of the Ant-lion. A very close scrutiny will show, by the presence of a pair of fierce jaws, the Ant-lion at home.

Would you know the ingenious builder? Lift him out tenderly from his burrow of sand, and when you have placed him upon the palm of your wide open hand, note with the most careful exactness the peculiar make-up of his structure, so that in the future you may have little difficulty in recognizing him should you again meet.

His short, flat head, armed with powerful mandibles, heavy-set chest, and large, soft, fleshy abdomen, amply protected on the sides with stiff, bristly hairs, added to his compact, robust form, the forward projection of his front and middle legs, and the backward prolongation of the stronger and less movable hind ones, which eminently adapts them to a backward manner of walking, are characters which so deeply impress, that we cannot fail to call up, when occasion demands, the possessor of so wonderful a mechanism.

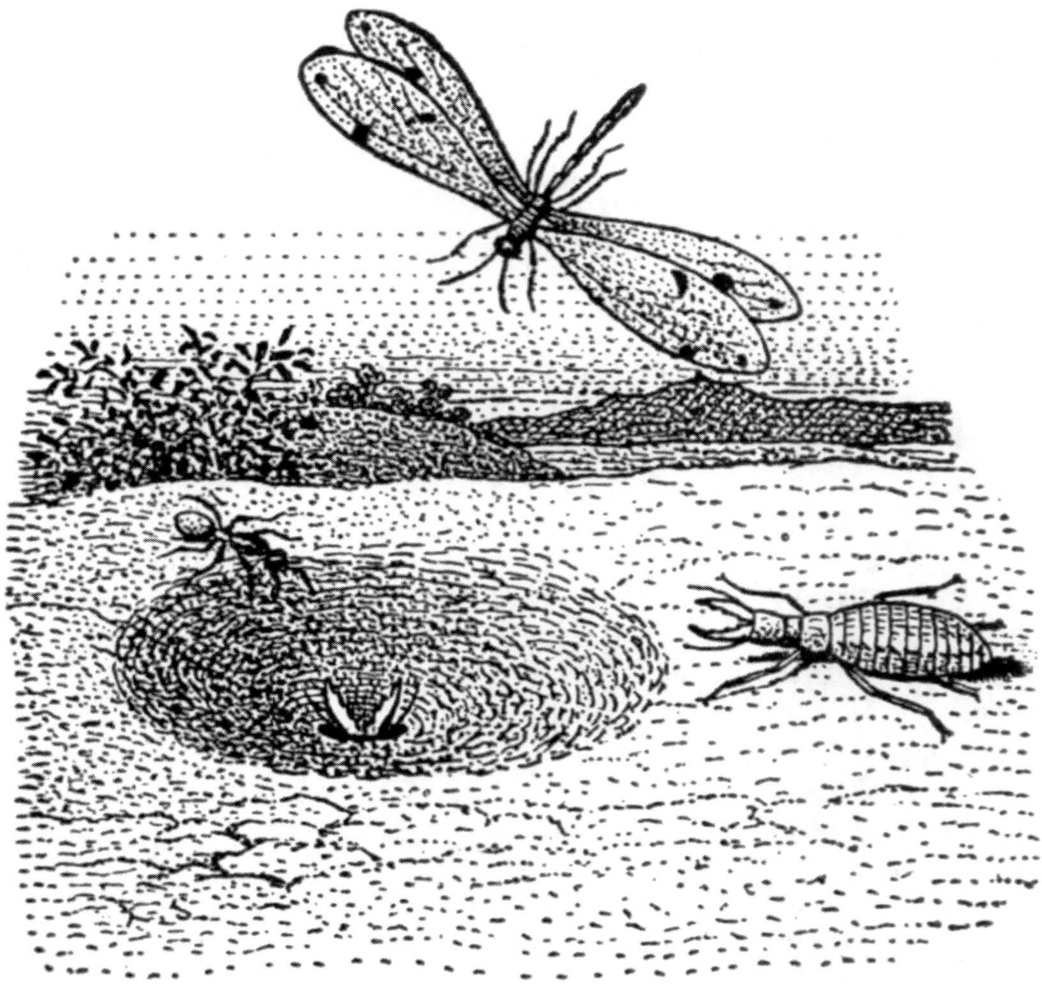

Figure 13. You-Ee-Up in His Den. As he appears in youth and old age.

Now that you have become familiar with the odd creature in form and in mien, set him once more upon his proud realm of sand, and seat yourself on the bank close by to watch and enjoy his curious behavior. In a minute or two his fears will have subsided, and he in control again of his accustomed indifference. See, he moves. Round and round he turns in the loose grey sand, burying himself deeper and deeper, and throwing the grains out from the hole he has made by his twistings, using his short, flat head for a shovel. The sand, as it is thrown over the side of the burrow, forms quite a margin, and when all is completed the Ant-lion sinks himself deep into the bottom of the trap he has digged, leaving only the tips of his mandibles in sight, which are extended and ready to seize any insect that is so luckless as to fall into their reach.

The unfortunate ant that ventures too close to the margin sets the sand off rolling, and it immediately begins to struggle against falling down, but the Ant-lion throws a few shovelfuls of sand against it, and it soon comes tumbling down to the bottom of the funnel, when it is instantly seized between the sharp mandibles in waiting, which, being perforated by slender tubes, enable their blood-thirsty owner to suck out its juices.

Country children, and adults as well, manifest a deep interest in these strange beings. They call them, as has been intimated before, You-ee-ups. How the name originated, and when, I do not pretend to know, nor have I been able upon inquiry to find out from the oldest inhabitants of the regions they affect. Old men and old women in the seventies and eighties knew these insects by this name when they were children, and I have been informed that they were always so spoken of by their fathers and mothers.

Even the insects themselves are believed to know the odd name by which they are designated. So fixed is the belief in the minds of the many that, to contradict it, is sure to subject the person so rash and presumptuous to the grossest abuse from the friends of the strange little creature. They have seen him in his sandy retreat, and have called him by name, and he has never been known to decline a response. "You-ee-up, you-ee-up," cries one, with his mouth just over the opening, and up comes the strange "crittur" as obedient as a lackey. "You-ee-down, you-ee-down," says the same childish voice, and down he goes to his den to await, as is thought, the giving of further orders.

That the Ant-lion does seem to respond when called, cannot be denied, for I have tried the experiment myself, and others have tried it in my presence, and always with the same successful results. But people go through the world not only with their eyes closed and their ears sealed, but also with their minds forever locked against thinking, lest, by thinking, they might do themselves serious injury. Had but a little of thinking been done, or some common sense exercised, the solution of the insect's strange actions could have been reached without any great difficulty.

Let me briefly explain. One cannot talk, as is well known, without some motion being imparted to the outlying air. This moving air impinging upon the loosely arranged sand piled up around the margin of the tiny pitfall, dislodges some particles, and these, falling into the jaws of the hidden Ant-lion, bring him to the surface, for he ascribes the commotion to some ill-fated ant, or other such insect, that has, in its anxious searching for food, tumbled unconsciously into his artfully-laid trap. In a moment the mistake is discovered, and, with all possible dispatch, he backs himself down into his den to await further developments. His appearance on the occasion is greeted by "you-ee-down, you-ee-down," and as he goes down apparently in obedience to the order, but really because it is a matter of business so to do, it is claimed by the unlearned and unwise that his movements are responsive to the command of the person by whom he is addressed.

Two years of larval life, and the subject of our sketch is lost to the sight of the rural folks. A new life, where feeding is no longer necessary, awaits him, but one in which the most radical changes must occur if he is to fulfil the existence which nature designed in her grand scheme of creation. From a silk-gland, which, unlike those of the butterflies and moths, is situated at the end of the body, he spins a cocoon, but there being so little of silk to spare, he needs must supply the deficiency by the utilization of a quantity of sand, which he glues into the walls of his house. Here he dwells a comparatively inactive pupa for three brief weeks, retaining his large, powerful mandibles to the last, which he uses in cutting his way out of the cocoon, when he is ready to emerge as a winged neuropter. In the adult form he resembles the dragon-flies in flight, flapping wildly and irregularly about, as if his muscles were too weak to wield his great stretch of wings. But in repose his alar appendages are folded above each other, forming an acute-angled roof above the long, slender abdomen. The antennæ or feelers are short, stout and club-shaped, and the wings long, narrow and densely veined.

Myrmeleon obsoletus, a name given to this insect by Thomas Say, a naturalist of repute, who lived in Philadelphia in the early half of the present century, is by no means a rare species, if search is made in the proper places. In the cut the larva is found to the right of the burrow, while deep in the bottom, with the jaws only in view, is another, prepared to receive the small ant just above should it lose its foothold and tumble into the trap. On the wing, a little in the background of the picture, may be seen the adult insect, represented in hawking for prey over a meadowy expanse of country.

Chapter 11

TOWER-BUILDING CICADA

Closely allied to the bugs is a group of remarkable insects to which naturalists now apply the name of Cicada, but which are generally, though improperly, designated Locust by the common people. They are readily distinguished by their broad heads, large prominent eyes, with three eyelets triangularly placed between them, and delicately transparent, veined wing-covers and wings. The abdomen is short and pointed, and the legs are short, the anterior femora being much thickened and toothed beneath. The hinder extremity of the body of the female is conical, and the under-side has a longitudinal channel for the reception of the ovipositor, or piercer, which is furthermore protected by four short-grooved pieces which are immovably fixed to the sides of the channel. The piercer itself consists of two outer parts grooved on the inside and slightly enlarged and angular at the tips, which are externally beset with small saw-like teeth, and a central spear-pointed borer which plays between the other two, thus combining the advantages of an awl and a double-edged saw, or rather of two key-hole saws cutting opposite to each other. A hard, horny substance, called chitine, the same as exists in the stings of bees and wasps, is the material of its composition. It would be impossible to conceive of anything more exactly fitted for its required uses than is this beautiful complicated instrument.

But the most peculiar characteristic of this family, however, consists in the structure of the mechanism by which the males make the trilling sound for which they have been so long famous. In the male of the Seventeen-year Cicada the musical instrument consists of two stretched membranes, one on each side of the body, which are plainly to be seen immediately behind the wings. These membranes are gathered into numerous fine plaits, and are played upon by muscles or cords fastened to their under surfaces. When these muscles contract and relax, which they do with great rapidity, the drum-heads, which the membranes resemble, are alternately tightened and loosened, the effect of this alternate tension and relaxation being the production of a rattling sound very much like that caused by a succession of quick pressures upon a slightly complex and elastic piece of tin-plate.

Certain cavities within the body of the insect, which may be seen on raising two large valves beneath the abdomen, and which are separated from each other by thin transparent partitions of the brilliancy of mica or highly polished glass, tend to increase the intensity of the sound.

In the winged state Cicada septendecim, as the subject of our *sketch* was named by the immortal Linnæus, is of a black color, with transparent wings and wing-covers, the thick anterior edge and veins of which being orange-red. Near the tips of the latter there is a dusky zig-zag line which resembles in shape the letter W. The eyes, when living, are also red, while the legs are a dull orange, which color is conspicuous along the edges of the rings of the body. The wings expand from two and a half to three and a quarter inches.

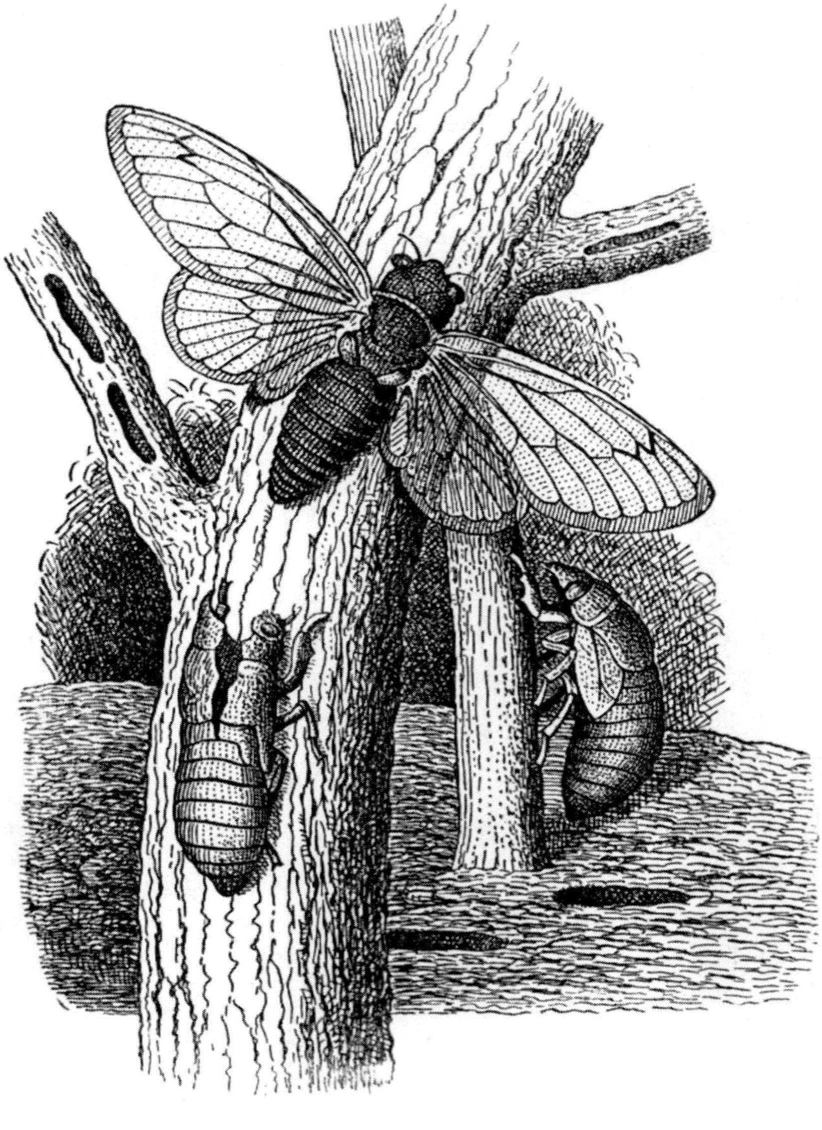

Figure 14. Seventeen-Year Cicada. Adult, Chrysalis-Case, Pupa, entrances to burrows and egg-nests.

About the middle of June the perfect insects make their appearance, and as they generally come in large numbers they do a great deal of damage. In some localities they congregate to such an extent upon the trees as to bend and even to break down the limbs by their weight. The din of their discordant drums resounds in the woods and orchards from morning to evening. As their life is of rather short duration, not lasting for a longer period than a month, they soon begin to pair, and it is not long afterwards that the females may be seen preparing nests for the reception of their eggs. Branches of moderate size are selected for this purpose. Their manner of perforation is curious and interesting. Clasping the branch on both sides with their legs, and bending the ovipositor at an angle of forty-five degrees, they repeatedly thrust it into the bark and wood in the direction of the fibres, at the same time setting the lateral saws at work, thereby detaching little splinters of wood at one end, which are intended to serve as a kind of fibrous cover for the nest. The hole is bored obliquely to the pith, and by a repetition of the same operation is gradually enlarged until is formed a longitudinal fissure of sufficient extent to receive from ten to twenty eggs. The side-pieces of the piercer act as a groove to convey the eggs to the nest, where they are deposited in pairs, but separated from each other by a narrow strip of wood. When two eggs have been thus placed, the piercer is withdrawn for a moment, and then inserted till two more eggs are dropped in a line with the first, and thus the operation is repeated until the fissure has been filled, when the insect removes to a little distance and commences to make another nest to contain two more rows of eggs. It takes about fifteen minutes to prepare a groove and fill it with eggs. As many as twenty grooves are sometimes made in a branch by a single insect, and when the limb has been sufficiently stocked she goes from it to another, or from tree to tree, until she has got rid of her complement of from five hundred to seven hundred eggs. So weak does she at length become, in her continued endeavor to provide for the succession of her race, as to fall, in an attempt to fly, an almost lifeless lump to the earth, where her spirit soon goes out never more to enliven its frail house of clay.

Although Cicadas abound most upon the oaks, yet there seem to be no trees or shrubs that are exempt from their attacks, unless it be the various species of pines and firs. The punctured limbs languish and die soon after the eggs are laid, and as often happens are broken off by the winds; but when this is the case the eggs never hatch, for the moisture of the living branch seems necessary for their proper development.

The eggs are one-twelfth of an inch in length, and one-sixteenth of an inch through the middle, but taper to an obtuse point at each end. They are of a pearl-white color. The shell is so thin and delicate that the form of the inclosed insect can be seen before the egg is hatched. One writer claims that fifty-two days, and others that fourteen days, constitute the period required for the hatching of the egg.

When it bursts the shell the young insect is one-sixteenth of an inch long, and is of a yellowish-white color, excepting the eyes and the claws of the fore-legs, which are reddish. It is clothed with small hairs. In form it is grub-like, larger proportionally than the parent, and provided with six legs, the first pair being very large, shaped like lobster-claws, and armed beneath with strong spines. Little prominences take the place of wings, and under the breast is a long beak for suction. Its movements, after leaving the egg, are very lively, and nearly as quick as some of the ants. But after a few moments their instincts prompt them to reach the ground. They do not attain this end by descending the body of the tree, nor by casting themselves off precipitately, but, running to the side of the limb, deliberately loosen their hold and drop to the ground, making the perilous descent with the utmost safety. This seems almost incredible, but it has been repeatedly observed by scores of honest witnesses.

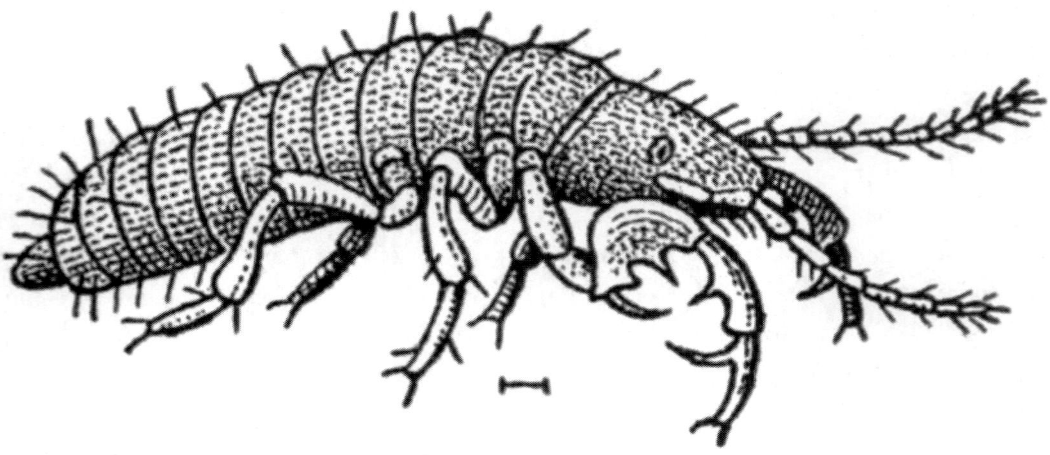

Figure 15. New-Born Cicada. Line below shows natural size.

On reaching the ground the young insects immediately burrow their way into the soil, using their broad and strong fore-feet pretty much after the fashion of the mole. They apparently follow, in their descent, the roots of plants, fastening their beaks into the most tender and succulent, and thus imbibing their juices, which constitute their sole aliment. They do not descend very deeply into the ground, probably not more than ten or twelve inches, although accounts have been published of their discovery at a depth of ten or twelve feet, but their occurrence at such great distances from the top of the ground is doubtless the result of accident.

The only alteration to which the insects are subject during the seventeen years of their subterranean confinement, is an increase in size, and the more complete development of the four small scale-like prominences of the back, which contain their future wings.

When the time of its transformation draws near, the larva, in which stage the insect passes the greater part of its existence, works its way up towards the surface, oftentimes in a very circuitous manner, for local changes make it necessary for it to bore through hard woods and between stones well beaten down. The burrow which it thus produces is cylindrical, about five-eighths of an inch in diameter, and firmly cemented and varnished so as to be water-proof. The upper portion, to the extent of five or six inches, is empty, and serves as a habitation till the period of its exit arrives, while the lower is filled with earthy matter removed by the insect in its progress. In this cell it remains during several days, ascending to the top for the benefit of the sunshine and air when the weather is auspicious, even venturing to peep forth occasionally, but descending on the occurrence of cold or wet weather. But when the favorable moment to leave their subterranean retreats arrives, the Cicada-grubs, or more properly pupæ, for such they are now to be considered, although they still retain something of the grub-like form, issue from the ground in great numbers as evening draws on, crawl up the trunks of trees, the stems of herbaceous plants, or on to whatever is convenient, which they grasp securely with their claws. After resting awhile, their skins, which have become dry and of an amber color, are by repeated exertions rent along the back, and through the slit formed the included Cicada pushes its head and body, and withdraws its wings and legs from their separate cases, and, crawling to a short distance, leaves its empty pupa-case fastened to the tree. At first the wing-covers and wings are small and opaque, but in a few hours they acquire their natural size and shape. It is not, however, for three or four days that the muscles harden sufficiently for them to assume their characteristic flight. The males make their appearance some days in advance of the females, and also disappear sooner. During several successive nights the pupæ continue to issue from the ground, and in some places, as was the case in May of 1868, when these insects appeared in great numbers in the vicinity of Philadelphia, the whole surface of the soil was made by their operations to assume a honey-combed appearance.

In localities where the soil is low and swampy, a remarkable chamber is built up by the larva, where the pupa may be found awaiting the time of its change to the winged state. These chambers were first noticed by S. S. Rathvon, at Lancaster, Pa., and are from four to six inches above the ground, and have a diameter of one inch and a quarter. When ready to emerge the insect backs down to an opening which is left in the side of the structure on a level with the surface of the ground, issues forth and undergoes its transformation in the usual manner. This peculiar habit of nest-building, which is so unlike what is customary with the Cicadidæ, or with Hemiptera in general, points to a high degree of intelligence among these insects, showing a remarkable ability to adapt themselves to environing circumstances. Undue moisture would be prejudicial to the pupa, as the larva seemed to know, through the guidance of the same dumb and unerring instinct which teaches it to cement its underground dwelling, but would that same instinct teach it to construct so wonderful a dome-like house as the one described for the preservation of its after-life, and one so eminently fitted by its position, shape, size and entrance to secure the necessary shelter, warmth and air for its protection and development? I apprehend not. Nothing short of a reason, similar to that in man, but differing in degree, would enable it to grasp the situation in which it found itself to be placed when nearing its final change, and plan with the view of carrying out the ultimate aim of its existence.

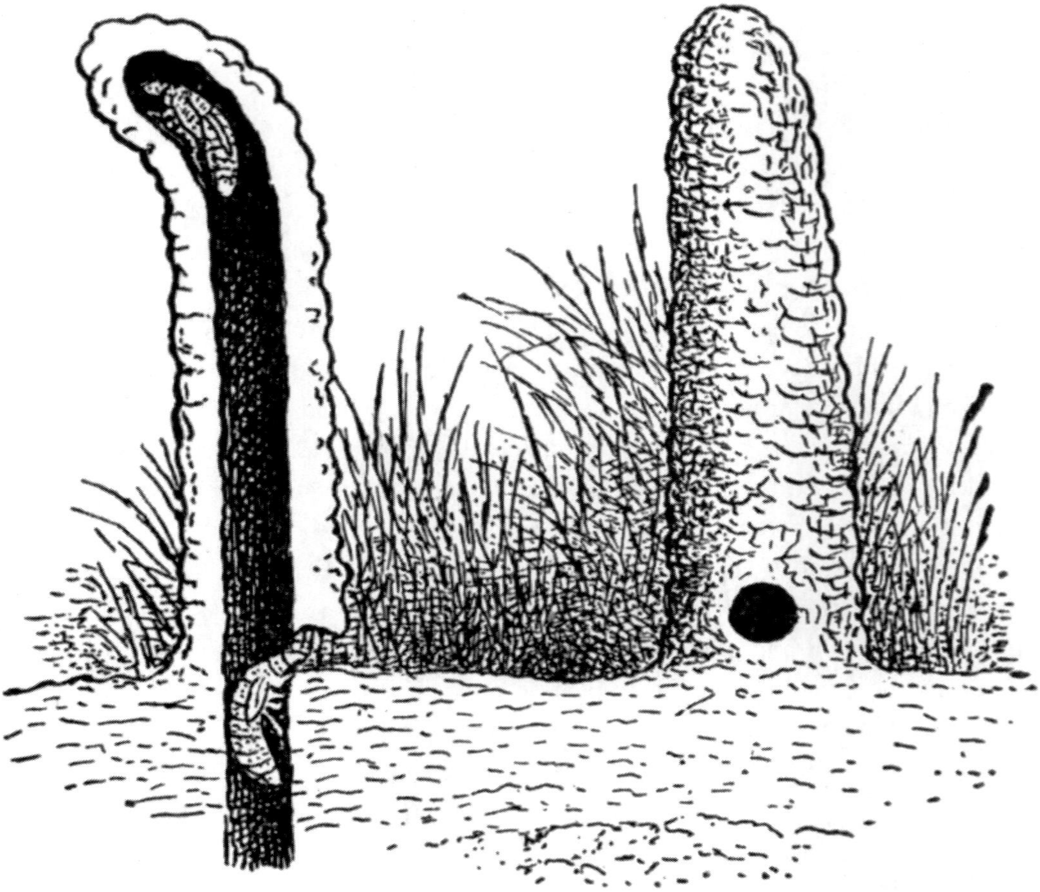

Figure 16. Dome-Like House of Cicada. Longitudinal section showing pupa in two positions.

Fortunately, these insects are appointed to return at periods so distant that vegetation has a chance to recover from the injuries which they inflict. Were they to appear at shorter intervals, our forest- and fruit-trees would be entirely destroyed by them. They are, moreover, subject to many accidents, and have many enemies, which contribute to diminish their numbers. Their eggs are eaten by birds, and the young, when they leave the egg, are preyed upon by ants, who mount the trees for that purpose, or take them upon the ground as they are about to enter upon their protracted larval career. Blackbirds eat them in the spring when turned up by the plough, and hogs, when allowed to run at large in the woods, root them up and devour large numbers, especially just before the arrival of the period of their final transformation, when they are lodged only a few inches below the surface of the soil. Many perish in the egg by the closing up of the bark and wood that constitute the walls of the perforations, thus burying the eggs before they have hatched, and others, no doubt, are killed by their perilous descent from the trees.

As its name implies, this insect generally requires seventeen years to complete its transformations, a fact that was first pointed out many years ago by the botanist Kalm. The late Prof. Riley, who had given this species a great deal of study, was the first to work out the problem of its periodical returns. He found that there are also thirteen-year broods, and that both sometimes occur in the same locality, but that in general terms the thirteen-year brood might be called the southern form, and the seventeen-year the northern form. At the limits of their respective ranges these broods overlap each other. The shorter-lived form he named provisionally Cicada tredecim. It was the existence of this brood that led entomologists to doubt the propriety of Linné's name, because, in calculating each appearance as occurring in any locality at the end of every seventeen years, they could not make the dates of its periodical returns correct. But it was Prof. Riley that cleared up the matter. It happened in the summer of 1868 that one of the largest seventeen-year broods occurred simultaneously with one of the largest thirteen-year broods. Such an event, so far as these two particular broods are concerned, has not taken place since 1647, nor will it take place again till the year 2089. There are absolutely no specific differences between the two broods other than in the time of maturing. There is, however, a dimorphous form that appears with both these broods. It is smaller, of a much darker color, has an entirely different voice, appears a fortnight sooner, and is never known to pair with the ordinary form. Dr. J. C. Fisher, in 1851, described it as Cicada cassinii, but the specific differences are not sufficiently well defined to entitle it to rank as a species.

Chapter 12

Honey-Dew

That aphides secrete, or rather excrete, a saccharine fluid, called honey-dew, which constitutes an important part of the food of ants, is a fact well known to naturalists. It must not be supposed, however, that this was its primitive use. But that it is in some way connected with the preservation of the tender creatures by which it is elaborated, there can exist not the slightest doubt.

Concerning its origin and application, and the benefit which it secures to its authors, various opinions have been hazarded, but they have all been too unsatisfactory to merit more than a passing notice. That it was of some advantage to young aphides was surmised by many, but the proofs necessary to sustain such a surmise were unfortunately wanting. It was left to the latter half of the nineteenth century to throw correct light upon the subject.

Whilst engaged some few years ago in the study of the species that affects the blossoms of one of our gourds—the Cucurbita ovifera of botanists—certain phenomena were observed, which promised an easy and speedy solution of the problem.

Gathered in compact masses, like companies of soldiery preparing for a foray, hundreds of aphides were seen, busily feeding, all over the flowers. There were old and young, not an indiscriminate mingling of ages and sizes, but an orderly arrangement of families, each family preceded by its own appropriate head. First came the very young of each family, only to be followed by those that were older, leaving the oldest of all to lead up the rear.

Here, it was apparent, was a most wonderful manifestation of intelligent design. The newly-born, needing the mother's earliest attention, were in closest proximity, while the almost mature were the farthest removed from her essential presence.

Figure 17. Blossom of Cucurbita. Mother-Aphis and her army of children on tube.

All this seemed to indicate the dearest relationship subsisting between mother and offspring, but judging from outward appearances, little, if any, love existed. It is true that maternal instinct, which is seldom so far gone as to shut its ears to the beseechings of suffering offspring for food, was far from being absent. Instances of its presence were momentarily noted.

But a stimulus seemed, in some cases, quite necessary to its manifestation. There were times when the honey-glands acted without any provocation. It was only, however, when the very tender were a-hungry, that pressure was brought to bear upon the mothers. A few gentle reminders served to arouse them from the apathetic indifference which possessed them. The antennæ of the young were the means employed for this purpose. Two or three caresses almost immediately brought a discharge of honey. Again and again was the process observed, and always with the same invariable result.

Never for a longer period than two days were the very young dependent upon this manner of feeding, for their digestive organs were too weak and delicate to assimilate earlier, without injury, the powerful juices of the food-plant.

But what of the older offspring? That they were far from being disregarded by parental provision, subsequent developments only too plainly showed. The excretion, though less urgent in their case than in that of the very young, was quite as indispensable. Were it not so, what reason can be assigned for their very strict adherence to the course over which the maternal head had already passed in feeding?

From what has been said, there can be no doubt that the newly-born aphis derives material advantage from the excretion. But as the supply is clearly above the requirements, why the excess? It is evident nature does not need it as a kind of compensation for losses sustained through aphides. Then what purpose does it serve? It becomes in part the pabulum of the stronger of the young, and this it accomplishes by mixing with the natural juices of the plant, thereby rendering them fit for use.

To serve as food for the young is then the primary object of aphis-excretion. That a secondary purpose, namely, the preservation of the species, is also subserved, there can be no question. How this is effected, it shall now be my endeavor to show.

Ants, it is well known, are fond of sugar, gums and saccharine solutions, as well as the rich juices and tender tissues of animals. But their appetite for sweets is stronger than for all other diets. To them aphis would prove quite as toothsome a morsel as it is to Coccinella, and would be as eagerly hunted for by them were it not for this matter of sweets.

Way back in the history of time, things were perhaps different from what they are now. Aphis was then a racy tidbit, and shared, no doubt, the murderous assaults of Formica, as it did of other carnivores.

For ages this may have been going on, but how long conjecture only can tell. But there came a time when affairs were changed. A new order of things was initiated. Earth was growing better and impressing new features upon its life. An Ant, wiser than any of its fellows, or any that had ever lived before, doubtless stepped upon the scene, and a new era for Aphis inaugurated.

Finding by accident, or otherwise, the delightful qualities of aphis-excretion, it would not be slow to communicate the information to its companions. And as news travels rapidly, and ants are by no means reticent creatures, but a short time would be necessary to carry it everywhere, till all the families, near and remote, of the great world of the Formicidæ would be made acquainted with the important discovery.

Now, as ants are endowed with a high degree of intelligence, considering the position they occupy in the grand scale of created existences, they would soon perceive that their highest good would be attained by taking under their protection the little creatures which are the authors of this excretion. From this time the ants would begin to abandon their sanguinary propensities and manifest some regard for the aphides. The latter, in return, perceiving the former's friendly disposition, would cease to fear them, and learn to cater to their wants. Thus would be developed, in time, those amicable relations which subsist between the two great, yet widely differentiated, families.

Chapter 13

Milch-Cows of the Ants

While much has been written upon the social relations subsisting between ants and aphides, yet the subject never grows uninteresting or threadbare. New facts are brought to light as observations widen and extend, some tending to confirm, and others to subvert old notions.

That aphides excrete a sweet, viscid, honey-like fluid, which affords food for many species of ants, has been long known to naturalists. Anyone can convince himself of this truth if he will but put himself to the trouble of examining the leaves or branchlets of any plant at the proper season of the year. Scattered upon the foliage and tender twigs thereof will be found millions of aphides, and close beside them countless ants, that ever and anon will be seen to caress, by means of their antennæ, the little creatures for the sweets within their bodies. It has even been asserted that some species of ants keep aphides as human beings do cows, but this by the many has been doubted, or deemed imaginary.

When a young man the writer was disposed to drift with the popular opinion in this particular, but a few facts that fell under his notice whilst searching for carabi and other beetles that live under stones and decayed logs, changed the bias of his mind and established in him the idea that with one species of ant this was at least the case.

It was on an occasion while exploring a neighboring thicket for the objects of his search, that he discovered, underneath a large flat stone which he had raised, a nest of a small red ant, which he took to be the *Lasius flavus* of the books. The ground was covered all over with pits, and divers communicating roads, and round about were hundreds of ants, larvæ in various stages of development, pupæ and eggs, and innumerous flocks of a white aphis, all of which were being tenderly cared for by a large army of thoughtful nurses.

Figure 18. Nest of Lasius. Neuters about their work.

No sooner did the intrusion occur than the colony was a scene of busy activity. Interested in what was before him, the writer seated himself upon a small mound overlooking the nest, where could be clearly observed the minutest details of ant-life. The neuters were everywhere to be noticed, but not a single male or female ant. All the work devolved upon the neuters. These were divided into three sets, each set having a definite part to perform in the unexpected drama before it. Some neuters had the exclusive charge of the mature larvæ, others of the pupæ and very young grubs, and the rest of their aphidian herds.

But it is to those that had the care of the aphides that we shall particularly invite attention. At the time of the disturbance, these specialized neuters were busy milking their cows, which they did by rubbing their long, pliant feelers against the anal nipples of the latter, drawing therefrom, as it seemed, a drop of the coveted fluid with each antennal stroke. No aphis was known to be visited in this business twice in succession, but the ants would go from one to another, and only return to the first when sufficient time had elapsed for the replenishing of its store. So intent were they upon their task, that several minutes must have passed before they took in the danger to which they were exposed.

You should then have seen their anxiety, and the presence of mind they exhibited. Conscious as of attack, and knowing the peril that beset them, they did not flee to their underground galleries, or to the adjoining grasses, for shelter, and thus leave their flocks to the mercy of the invader, but they manifested the deepest concern for the little creatures, so unable to defend themselves, that had so willingly catered to their temporal wants. Not an ant was seen to desert its post, but all remained on duty till the last of their protegés was carried to safe and comfortable apartments in the ground beneath.

What clearer evidence is wanted to show the love these neuters bear the tender objects of their care? It must be plain that man bestows not half the attention upon his flocks than do these ants on theirs. It is true they do not bring them food, but that they build their homes where food, the roots of herbs and grasses, abound, there is no doubt. It may be, too, that they are carried to their pasture-grounds, when that necessity occurs, but this cannot with truth be said. When some would stray, they were returned within the fold, which shows the watch these ants do exercise.

Concluding then, this much may be averred: food, wholesome, sweet, nutritious food, the aphides supply to ants, the neuters and the young, but specially the young. And that they lead most happy, prosperous lives, the ants their masters, must surely be, or looks deceive.

Chapter 14

LIVING ARTILLERY

No more remarkable creature exists, perhaps, than the little Brachinus fumans, which is so very common in the early spring. Damp situations are affected by it, but it is seldom met with except by insect-hunters, for it conceals itself generally under stones, as many as a half-dozen individuals often being found in company in a single locality. Banks of tidal rivers afford excellent hunting-grounds in England for Brachinus, but in America low, dank woods and borders of streams are the places where one must look to discover its presence.

When once you have made the acquaintance of so remarkable a stranger you can never afterwards fail to recognize him in your travels. He is peculiar, but not at all distinguished in looks, as some of his brethren. Picture a yellowish-red beetle, with a bluish frock-coat, which his wing-covers resemble, and possessed of a short, narrow head, a heart-shaped prothorax, as the front of the chest-segments is called, and a long, broad abdomen, three times the size of the rest of his body, and you have a tolerably fair idea of Brachinus.

But it is not so much his odd shape as a most extraordinary property he possesses, which is singularly unique in the animal kingdom, that makes him an object of interest and curiosity. Deep down in his most marvellous body a fluid, highly volatile in its nature, is elaborated, which the little creature can retain or expel at his pleasure. It is only, however, when alarmed that he utilizes this fluid in small quantities in defense, but its effect is wonderful, for in coming into contact with the atmosphere it immediately volatilizes and explodes, looking very much like a discharge of powder from a miniature artillery. In consequence of this phenomenon the insect which produces it is popularly called the Bombardier Beetle.

So small a coleopter, being scarcely one-fourth of an inch in length, and so comparatively weak, is likely to be attacked by the larger Geodephaga, or Earth Devourers, and especially by the Carabi, which inhabit similar retreats. But for this curious defence the smaller insect could have but the barest chance of living in the struggle for existence. Often have I seen a Carabus in hot pursuit of Brachinus. The chase is always an interesting one, and never fails, however frequently it has been observed, of attracting attention and exciting admiration. But the wide-awake, ever watchful Brachinus never loses his head for a second when thus pursued, but like the clever artilleryman that he is, awaits the opportune moment, and then pours a heavy discharge of his fulminating fluid into the very face of the enemy. Baffled, alarmed, Carabus desists from the attack, and backs slowly away from the tiny blue smoke, while Brachinus, in the confusion that ensues, escapes to some place of security for rest and protection.

Figure 19. Brachinus Pursued by an Enemy. His curious and unique method of defence.

Most skillfully has the artist delineated the scene. Carabus serratus, the pursuing beetle, is chasing the Bombardier, and has nearly effected his capture, when, all of a sudden, a discharge of artillery has stopped the pursuit, under cover of which the Bombardier will make off. Meanwhile the Carabus, exchanging his rapid advance for a retreat quite as rapid, throws back his antennæ, a sign of his defeat, and skulks away to recover his wonted self-possession.

The volatile fluid, which produces such curious effects, is secreted in a small sac just within the end of the abdomen. Not only is it capable of repelling the larger beetles by its explosion and cloud of blue vapor, but it is also powerful enough to discolor the human skin, as many who have captured Bombardier Beetles by the hand know only too well. Should the fluid get within the eyelids, the pain and irritation produced are very distressing. Some years ago the writer, while searching for carabi underneath stones and in creviced rocks, met for the first time with Brachinus, but was ignorant as a child of his obnoxious property. Placing a little fellow upon his hand for close examination, he soon experienced a burning and painful sensation of the ball of the eye, but did not for a long while attribute the cause to a discharge from the Beetle. Repeated investigations at very short ranges by means of a microscope were attended with similar results, till eventually an inflammation of the visual organs set in, accompanied by a blurring of the sight, which debarred him from reading and study for nearly a fortnight. One learns wisdom by experience, and the wisdom thus acquired serves for a lifetime.

Even Brachinus has learned by experience, doubtless, to be economical in the use of his resources. The whole of the contents of his tiny magazine are not ejected at one discharge, but there is sufficient to produce a series of explosions, each explosion being perceptibly fainter than its predecessor. By pressing the abdomen of the dead Beetle between finger and thumb these explosions may even be produced. In hot countries, where exceedingly large species abound, the explosions are said to be very loud, and accompanied with quite a cloud of blue vapor.

Chapter 15

Bright and Shining Ones

Probably more than ninety thousand different species of beetles exist in the world, first and foremost among them standing the Cicindelidæ, or Tiger Beetles. From their high position in the coleopterous world they may well demand our attention, but they have other claims upon our consideration. They are beautiful, courageous little creatures, and accomplish a vast amount of good to man. The name Cicindela, by which they are known to scientific people, tells us that they are the "bright and shining ones;" while the cognomen of Tiger Beetle reveals to all English-speaking nations the story of the incessant warfare which they wage upon their fellows.

The Cicindelæ love the merry sunshine. On any bright summer day they may be found running and flying about sunny banks, or revelling in sandy places where the day-god smilingly rejoices. They mostly avoid vegetation, as it checks their easy rapid movements, although some kinds affect grassy spots among the trees. They are the most predaceous of the coleoptera, and behave like the tigers among mammals, the hawks among birds, the crocodiles among reptiles and the sharks among fishes. In the tropics some few genera seek their food on the leaves of trees, but in temperate and sub-tropical regions, where the species are more abundant, they are terrestrial in habits.

Let us now take our instruments of capture and go in quest of some of the dozen or more species that have their home with us. The day is auspicious. Here is a likely spot. See there upon the ground are some specimens of our commonest species—the Cicindela vulgaris of naturalists. Go for that one. He sees you as quickly as you see him, and is off for a few yards, but suddenly drops to the grass from his flight, but always with his head towards the enemy. Again and again you start him, but at last, tiring of the chase, he takes a longer flight that usual. This is a ruse of his, and knowing what it means, you hurry back to where you first saw him in time to see him all unsuspectingly alight, and you easily take him captive in your toils. Now that you have him secure, examine him closely. Watch how savagely he moves his mandibles and tries to pinch. You need not be afraid, for his bite is inoffensive and not very painful. You measure with the eye his size, and you rightly decide that he is not much over an inch in length, and scarcely one-fourth in breadth. His head you will find very large and brainy, his jaws powerful and long and curved, two scimitar-like weapons, which are admirably fitted for cutting and carving the quivering bodies of his prey. His eleven-jointed antennæ are long, slender and graceful. In color his back is dull purple, but beneath he is resplendent in a bright brassy green. Three whitish, irregular bands adorn his wing-covers. His legs, long and slender, are just the things on which to hunt the active insects upon which he feeds.

His next of kin, the Purple Tiger Beetle, is nearly as large as he, and often joins him in company. Beautifully robed in purple he usually is, but sometimes in a greenish garb arrayed. From the outer almost to the inner margin of each wing meanders a reddish line, while lower down a dot, and still another at the farthest tip of the inner border, enhance his beauty. Cold spring days delight him best, and he is often seen when snow is yet upon the ground.

Figure 20. Common Tiger Beetle. Larvæ in Burrows. Two other species in background.

More beautiful by far than either, and no less active, is Cicindela sex guttata, or the Six-spotted Tiger Beetle, whose dress, a brilliant metallic green, flecked with six small silver spots, renders him a pretty sight when you flash the rays of light athwart his burnished armor. Hot, June-like days and dusty road-sides suit him best, and there, what time the sun looks down in all his burning ardor, our little friend is met, his purpose bent on slaughter. Other species might be instanced, for North America contains at least a hundred, but enough have been given for our present object.

Tiger Beetles may well be called beneficial insects. Although they do not, like that brilliant murderess, the dragon-fly, clear the atmosphere of the gnats and flies that torment mankind, but still, with their powerful curved daggers, which serve them for jaws, they accomplish a swift and almost incredible havoc among the smaller insects. We should take care of them, and respect them, for they are an invaluable auxiliary to the farmer.

The ferocity of these insects is remarkable. No sooner have they taken their prey, than they quickly strip it of wings and legs, and proceed at once to suck out the contents of its abdomen. Often when they are disturbed in this agreeable occupation, not wishing to leave their victim, they fly away with it to a place of uninterrupted security, but they are unable to carry a heavy burden to any great distance.

They are true children of the earth. The eggs are laid in the earth, and in the earth the grubs are hatched, and in the earth they spend their days, and in the earth they prepare their shrouds, and, wrapped therein, sleep their pupa-sleep through the long, dreary winter, and with the returning warmth of spring crawl out of their earthy chambers to run and sport on earth, seldom using their new-formed wings to fly away from their beloved mother.

The grubs are hideous hunchbacks, but possessed of brains and stomach. They live in the same localities as their parents, the anxious mother, with wise precision, having carefully deposited her eggs where food would be readily attainable by her children. Have you a desire to examine a larva? There is a hole that has been made by one of these creatures. Place down into it a small straw or a bit of fine twig. The cranky little hermit, who is always wide-awake, resists most fiercely such unprovoked insolence, and instantly seeks, by the aid of his broad, expansive head, to eject the intruding object. Now is your time. When he shows himself, quickly seize him with your fingers. You will find him a perfect Daniel Quilp, with head enormous, flat, metallic in color and armed with long, curved jaws. His legs are six in number, and on the back, half-way between the legs and tail, are two curious, odd-looking tubercles, each terminating in a pair of recurved hooks. The head and first body-division are horny, the rest of the creature being soft and very sensitive.

While the larval Cicindela has all the desire for slaughter which his parents manifest, yet his delicate skin, long body and stubby legs not only prevent him from chasing prey, but also from attempting a struggle with an insect of any size; nevertheless this imperfectly armed creature manages to secure his food without exposing himself to any serious risk. With his short, thick spiny legs he loosens the earth, and with his flat head, which he uses as a shovel, and turning himself into a z-shaped figure, hoists up the clay and upsets it around the mouth of his intended dwelling. With head and legs, and with a perseverance that is truly surprising, he sinks in a very short time a shaft a foot in length and as large in diameter as an ordinary lead-pencil.

Especial pains are taken to see that the tunnel is sufficiently wide, so that the little creature can crawl in with ease. If he wishes to remain set fast, he sticks the back of his body against the sides and rests safely with the aid of his hooks. In this position he can poke his head out of the ground, thus closing the entrance of his burrow, while in patient waiting for some unsuspicious wayfarer to pass over. As soon, however, as the luckless insect touches the top of his head, he relinquishes his hold within the tunnel and descends with great precipitation to the bottom, and thus his victim falls into the hole, where it is seized by the powerful jaws and its juices absorbed in a quiet, leisurely manner. The loose earth around the opening of the tunnel gives way on the approach of an insect, and thus the success of the cunning Cicindela is doubly insured.

Sometimes in the construction of a burrow, after a certain depth has been reached, the young Cicindela meets with a difficulty which he had not expected. A flat stone is encountered, and thus further progress in a vertical direction is prevented. If the obstacle, on account of its size, cannot be gone round, and the shaft is not deep enough for his purpose, it is not unusual for him to desert it and attempt the tunnelling of a home in some more desirable spot. He does not undertake a very long journey, for he knows too well the risk which he runs by so doing, as he is in danger of being assaulted by secret foes in the rear, an attack which the peculiar conformation of his hinder body ill fits him to resist. On land he is timid and cowardly, and well might he be, but within the protecting walls of his underground castle, with a pair of powerful swords with which to defend himself, he is the impersonation of fearlessness and courage.

When fully grown the larva closes up the mouth of its abode, and in quiet and solitude undergoes its metamorphosis, lying dormant during the winter months. But when the breath of warm spring days has melted the icy coldness of the earth, and filled the air with vivifying influences, then comes it forth in all the pomp and splendor of its nature—a winged existence.

It has been seen what a beautiful adaptation of means to an end is shown by the young Cicindela. Even the adult, or mature form, with its long, slender legs, so admirably formed for silence and fleetness of movement, which are alike necessary to pursuit of prey and escape from enemies, displays the wisdom of Him who breathed into all animated nature, no matter how small or how humble, the essence of His being, and endowed one and all with qualities of mind and body which should respond to environing conditions and thus prepare them to survive in the struggle for existence.

Chapter 16

Queen of American Silk-Spinners

No insect affords a better proof of high art in nature, and of the transcendent beauty of the Creator's thoughts, than the Luna moth, which is as preëminent above her fellows as her namesake, the fair empress of the sky, above the lesser lights that dominate the night. Her elegant robes of green, set off with trimmings of purple, and jewelled with diamonds, added to her queenly grace and personal charms, will always distinguish her from the *profanum vulgus* of the articulata.

And now for a short biographical sketch of this remarkable beauty from the cradle to the grave, and beyond, after she has assumed her resurrection-attire, to the day when, her appointed work on earth being ended, she quietly lays her body down to mingle with its native clay.

In her childhood, or caterpillar state, her head is elliptical in shape, of a light pearly color, the rest of the body being a clear bluish-green. A faint yellow band stretches along each side, just below the line of her breathing-organs, from the first to the tenth segment, while the back, between the several body-rings, is crossed by narrow transverse bars, similar in coloration. Each segment, after the fashion of her kith and kin, is adorned with small pearly warts, tinged with purple, some five or six in number, each tipped with a few simple hairs. Three brown spots, bordered above with yellow, ornament the end of the tail. An interesting variety, whose general color is a dull reddish-brown, is sometimes met with, but the lateral and transverse stripes of yellow have disappeared, and the pearl-colored warts with edges of purple have assumed a richer hue and blaze like a coronet of rubies. When at rest, with the rings all bunched and body shortened, the infantile Luna is as thick as a man's thumb, measuring but two inches in linear direction; but when she sets out upon her travels, feeling the dignity of her station in life, she stretches to her full length of three inches.

When have been completed her allotted days of feeding upon the leaves of the hickory, oak, walnut or sweet gum, and she is seriously contemplating the preparing of a shroud and casket in which to await her resurrection-morn, she casts about for leaves, which, when they are found, she securely draws together, and within the hollow space there is soon spun a very close and strong oval cocoon of silk, one and three-fourths inches in length, of chestnut-brown color, thin, and covered with warts and excrescences, but seldom showing the imprints of leaves. Cocoons of Luna so nearly resemble those of polyphemus, that many an experienced collector is greatly chagrined, after getting together a large supply of what he deems Luna cocoons, to find dusky, one-eyed polyphemi to issue from the silken tombs rather than a goodly throng, in delicate bridal attire, of proud empresses of the night. Polyphemus cocoons are, however, somewhat smaller than Lunas, white or dirty-white in color, rounded at each end, and sometimes angular, because of the leaves being unevenly moulded into their surfaces, and generally covered with a whitish meal-like powder.

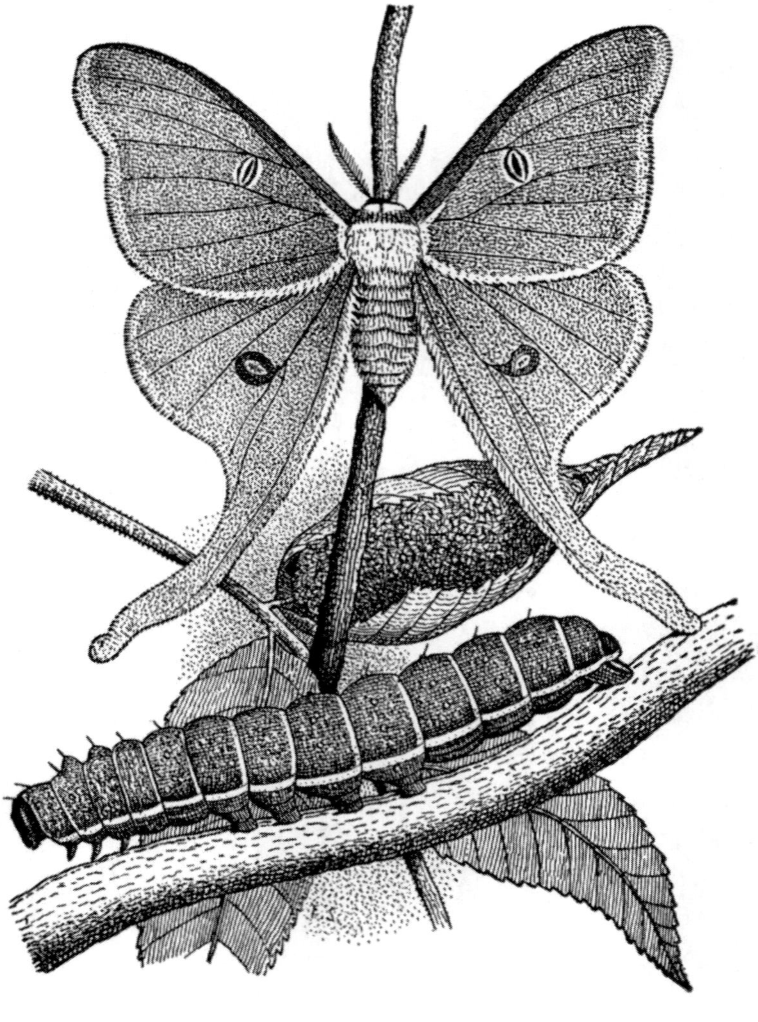

Figure 21. American Luna Moth. Larva on branch below, and cocoon on twig just above.

In June the Lunas awake from their death-like slumber, burst asunder their silken cerements, having at first made loose the compact threads by a fluid-ejection, and come out into the world in all the freshness and glory of a new and untried existence. Their wings, which expand from four and three-fourths to five and one-half inches, are of a delicate light-green color, the hinder ones being prolonged into a tail of an inch and a half or more in length. Along the anterior margin of the fore-wings is a broad purple-brown stripe, extending also across the back, and sending downwards a little branch to a glittering eye-like spot near the middle of the wing. These eyes, of which there is one on each wing, are transparent in the centre, and encircled by white, yellow, blue and black rings. The hinder borders are more or less edged with purple-brown. All the nervures are very distinct, and pale-brown in color. Near the body the wings are thickly invested with long white hairs. The under sides, excepting that an indistinct line runs along the margin of both wings, are like to the upper. As for the body, the thorax is white, occasionally yellowish or greenish, and coursed by the purple-brown stripe that traverses the entire length of the upper edge of the wings; and the abdomen, similarly colored, and clothed with white, wool-like hairs. The head is small and white, and furnished with broad, flat and strongly pectinated antennæ, which are very much wider in the male. The legs are purple-brown, and poorly adapted for walking, but this defect is largely compensated for in the wide stretch of wings, that fit their possessor for powerful and long-sustained flight.

Such is Luna in her various transformations. Notwithstanding her great size and almost matchless loveliness, her habits are not proportionally noteworthy. The gift of superior beauty, in the insect as in the mammalian world, does not often carry with it a high order of intelligence. It is true the young Luna knows pretty well the secret of dissembling. How quickly she perceives the approach of an enemy! And she knows how to deal with him, but her little trick of simulating death, or an immobile twig, does not always succeed with the wily spider, or artful ichneumon. That she is a tolerably good connoisseur of the character of foods, there can be no question. You cannot deceive her. Take from her the foods her ancestors have used for centuries untold, and substitute others she knows nothing about, and she is at once cognizant of the change. However hungry she may be, and in her early growing years she is ever a voracious feeder, she will starve rather than eat what the unwritten law of her race has strictly interdicted. I have known cases where death has ensued, or the caterpillar has pupated earlier than usual, when alien food has been given it to eat. But in the beginning of life, just after the first skin-moulting has been effected, ere the little creature has attained its seventh day of age, no trouble is experienced in changing the food, almost anything edible in the plant-line being eaten, though some things with a more decided relish than others. In the matter of cocoon-weaving, where the necessary leaves for a basis cannot be obtained, as occurs in captivity, the inconvenience is overcome, but not without difficulty. Leaves, you must know, are in Luna's way of thinking, as essential to cocoon-building as wooden or iron beams and girders to man's own constructing. Without a framework of some sort, what a sorry attempt would we make at home-building, but Luna does succeed, after a good deal of wise planning and no little worry, in producing a house which is well worthy her effort.

While the gaudy moth or butterfly, when contrasted in wisdom and sense with the dingy-colored bee, may suffer in comparison, yet she is by no means the dull, stupid creature she is pictured to be. She lives, it is a fact, as has often been said, for the increase of her race, but the interest she shows for the young she may never see, in laying her eggs upon the plant that is to serve them as food and home, puts her upon a rather high plane of intelligent existence. Luna's life, in the perfect state, is usually quite brief. It is one of the happiest of honeymoons. Love conquers and destroys all other passions of her being, while her gormandizing offspring are never troubled by the ardent flame which consumes even the thought of sipping the nectar of the flowers that rival in beauty the wings of the mother, who is the perfect representation and embodiment of elegance and grace. While the early insect lives and eats, the adult form, upon whom Dame Nature has expended so much wealth of color and such symmetry of shape, which make her a "thing of beauty and a joy forever," lives and dies, for in her seeming haste and forgetfulness the great mother of us all has made her without the essential means of tasting food, a delight and an enjoyment which the lords of creation are so wont to esteem the purpose and aim of all human existence.

Chapter 17

Basket-Carriers

You who have been to the country, in the summer, and who have kept your eyes alive to the surroundings, have doubtless seen the Basket-worm feeding upon the leaves of the quince, apple, peach, linden, and other deciduous trees, as well as upon such evergreen as the arbor-vitæ, Norway spruce, and red cedar. In Germany these worms are popularly designated *Sack-träger*, or Sack-bearer, while the mature insect is spoken of as the House-builder Moth. Scientifically speaking, the latter is called *Thyridopteryx ephemeræformis*, a name which is nearly twice the length of the caterpillar it represents.

During the winter the curious weather-beaten bags of these worms may be observed hanging from the tree-branches, apparently without a trace of the odd-looking creatures that hung them there the autumn before. If a number of these bags are gathered and cut open at this time, many of them will be discovered to be empty, but the greater portion will be found partly full of yellow eggs. Those which do not contain eggs are male bags, and the empty chrysalis of the male will be found protruding from the lower extremity. Upon close examination these eggs will be observed to be obovate in form, soft and opaque, about one-twentieth of an inch in length, and surrounded by more or less fawn-colored silky down. If left to themselves, they hatch sometime in May, or early in June.

The young which come from these eggs are of a brown color, very active in their movements, and begin at once to make for themselves coverings of silk, to which they fasten bits of the leaves of the tree on which they are feeding, forming small cones that are closely adherent to the leaf-surfaces. As the larvæ grow, they augment the size of their enclosures or bags from the bottom, until they become so large and heavy that they hang instead of remaining upright, as they did at first.

By the end of July the caterpillars become fully grown. They are now exceedingly restless, and may be seen wandering from branch to branch by means of their true legs which are projected from the mouths of their baskets, to which they keep firm hold, or suspended from a branch of a tree by a long silken thread of their own manufacture. When very abundant, as they were in certain localities during the season just ended, they become a great nuisance, as one can hardly walk beneath the trees without being inconvenienced by a dozen or more dangling into his face.

Removed from the case at this stage of existence and closely examined, that portion of the body which has been covered by the bag will be seen to be soft, and of a dull brownish color, inclining to red at the sides, while the three anterior segments, which are exposed when the insect is feeding or travelling, will be found to be horny and mottled with black and white. The pro-legs on the middle and hinder segments, which are soft and fleshy, will show themselves fringed with numerous hooks, by which the larva is enabled to cling to the silken lining of its bag and drag it along wherever it goes. The external surface of the bag is rough and irregular, often presenting a beautiful ruffle-like appearance, which is due to the projecting portions of the stems and leaves which are woven into it. During their growing-period these caterpillars are slow travellers, seldom leaving the tree on which they were hatched. When about to change into chrysalids, they fasten their bags securely to the twigs on which they happen to be, and then undergo their change, the male chrysalis being very much smaller than the female, hardly one-third its size.

When we examine the cases of the Basket-worm, hardly any two will be seen to be alike in their ornamentation. So completely is the outside covered, when made upon the arbor-vitæ, which seems to be a favorite food-plant of the species, that the silken envelope is concealed from view. The bits of twigs and leaves are probably protective, and yet one would think that the extremely tough case which covers the caterpillar would be quite sufficient to protect it against all assaults of foes and stress of weather. Nevertheless, this leafy coat of mail, which sometimes wholly covers the sac, must certainly add very much to the protective value of the covering. The caterpillar has a soft, hairless body, and is thus more exposed than many of its neighbors, and nature, it would seem, has favored it far above all of its fellows.

How the worm manages to trim its coat in this manner must seem, to the uninitiated in such matters, wholly inexplicable. To enable the reader to understand the manner of operation, it will be necessary first to explain its mode of feeding. The larva has perfect control of its own movements, notwithstanding the fact that it carries its house upon its back. It can thus thrust its body out of the sac-mouth until nearly the whole of it is exposed, and twist and bend itself in every direction. Specimens have been met with that had dropped from the trees hanging by a thread and squirming, bending and snapping their bodies in the most grotesque ways, while the case spun around like an old-fashioned distaff. Now, when the caterpillar wants to feed it stretches its head and neck out of the case and moves them about until a satisfactory place has been secured, which it clasps with its true legs, three pairs of hard, conical organs armed with sharp claws, and pulls up its body and commences to spin. The spinning-organs are near the mouth, and after several movements of the head, as though smearing the liquid viscid silk upon the leaf, the head is drawn back, drawing out with it a short thread. A similar movement is then made against one side of the mouth of the sac, the process being repeated several times until a stout stay-line is spun by which the larva hangs securely. Now the creature is ready to feed. The behavior, however, varies a great deal. In feeding upon the white pine it secures itself to one leaf by its stay-line, while it reaches to an adjoining leaf which it bites off, and sitting erect, as it were, in its house, comfortably chews off the end which is continually shored upward by the first and second pairs of true legs that stand out free and untrammelled above the sac.

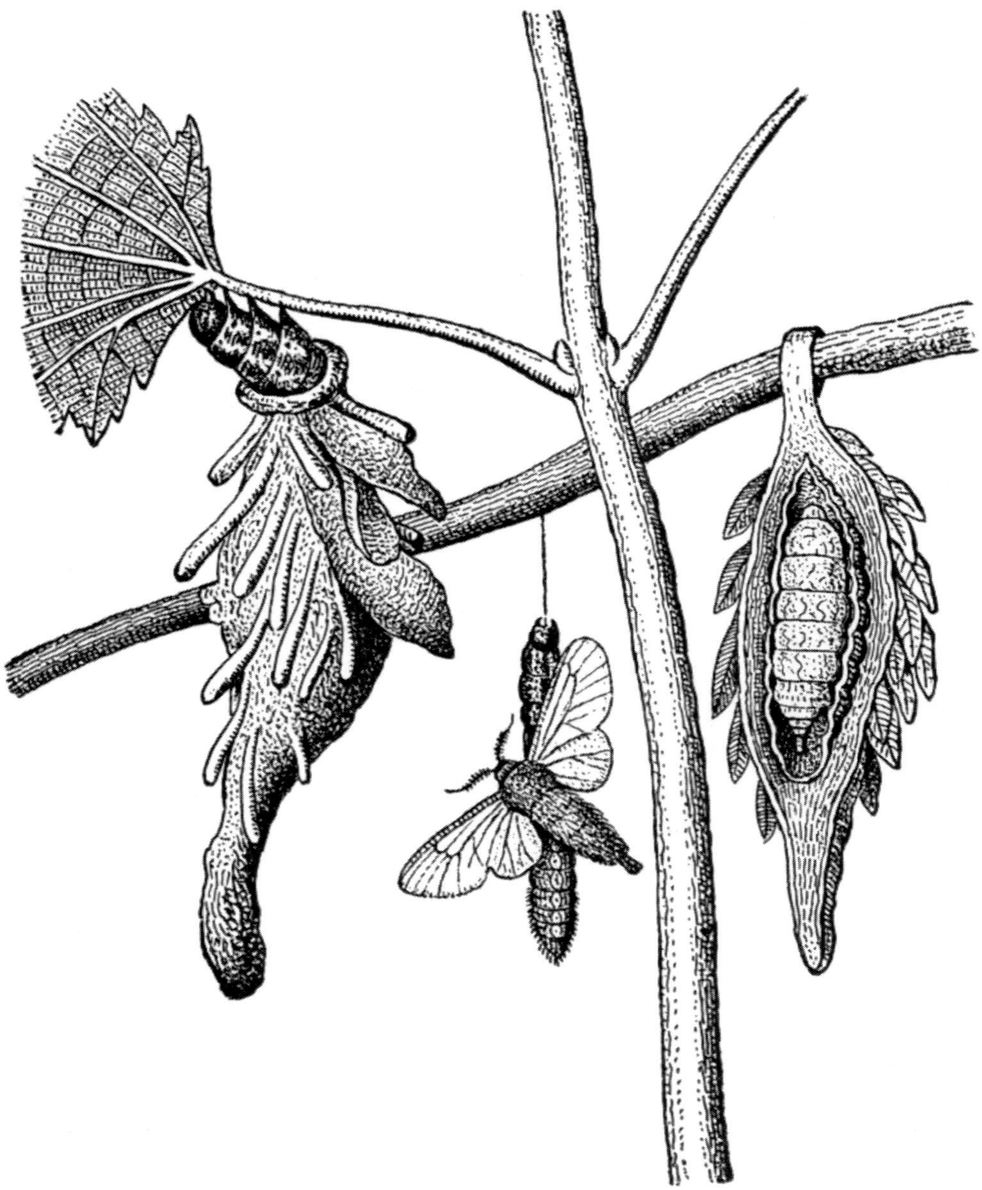

Figure 22. House-Builder Moth. Young in house, winged male, young suspended and bag-like female in longitudinally-split cocoon.

But more frequently the worm feeds without separating the leaf from the point of suspension. By making itself fast to the under part of the leaf it is thus enabled to reach the edge, which it gnaws round and round until it has completed its destruction.

So securely does the caterpillar hold on to its house, that one would suppose that its body was lashed to the inside. But no, its body is unhampered, for it can turn itself easily around in its case, and go out at either end, although the head is generally directed upward. It clings to the inside with the hooks upon its hinder feet, and so tenaciously, too, that the writer has never been able to pull one out, being checked by the fear of tearing the creature in two. And now to the mode of attaching the leaf-cuttings to the case. This is always done at or near the mouth of the sac. The Ephemeraform larva is a growing creature, unlike the moth itself, which emerges a perfect insect of full growth. It commences life as a small worm, eats small quantities, and, as may be observed, down towards the foot of the case sews on very small tags. But after it has fastened on these pieces to the mouth, it grows itself, and so also does the case, which it continually stretches and enlarges. Hence the mouth of the case is continually changing, moving upward as the worm feeds, so that the pieces sewed upon the cap of the case thus appear, in an adult caterpillar, precisely as they are seen scattered along the outside from top to bottom. And now, as to how the pieces are put into the case, I shall endeavor to explain. That the worm cuts purposely through the twig which it needs for the case, I feel certain. Of course the outer or detached part drops down. But, while eating, the worm frequently, quite constantly, indeed, spreads its viscid silk along the leaf and so keeps it attached on both sides to the upper rim of the sac, or to its own mouth-parts, and thus the tip of the twig or leaf, instead of falling to the ground when it is severed from the stem, simply drops alongside of the case, to which it is held by the slight filament that attaches it to the sac, or, as happens in many instances, remains attached to the caterpillar's spinneret. In either case the leaf, twig or stem remains, and, after being drawn up, adjusted and tightened by the worm, adheres tightly. As the creature is forever moving its spinning-tubes around the top of the sac, these fastenings are being continually strengthened, and thus one piece after another is added, and so the basket grows.

While the case of the Basket-worm, and even that part of its body which it chooses to expose to view, are known to the casual observer, yet but few persons have ever seen the mature insect. The female moth is wingless, and never leaves the bag, but makes her way to its lower orifice, and there awaits the attendance of the male. She is not only without wings, but is devoid of legs also, being, in short, nothing more than a yellowish bag of eggs with a ring of soft, pale-brown, silky hair near the tail. The male, on the other hand, has transparent wings and a black body, and is very active on the wing during the warmer portions of the day. After pairing the female deposits her eggs, intermingled with fawn-colored down, within the empty pupa-case, and when this task is completed works her way out of the case, drops exhausted to the ground and dies.

Though a Southern rather than a Northern insect, yet it is found as far north as New Jersey and New York, and occasionally in Massachusetts. It is extremely local in character, abounding in one particular neighborhood and totally unknown a few miles away. Where they occur in abundance they often almost entirely defoliate the trees they attack, but this can be easily prevented by gathering the cases containing the eggs for the next brood during the winter and destroying them. Hand-picking the cases with the worms in them, where their ravages are confined to small trees and shrubbery, will also help to hold them in check. Nature has provided two species of ichneumon for their destruction. One of them, Cryptus inquisitor, is about two-fifths of an inch in length, and the other, Hemiteles thyridopteryx, is nearly one-third of an inch. Five or six of this latter species will sometimes occupy the body of a single caterpillar, and after destroying their victim spin for themselves tough, white, silken cocoons within the bag.

Chapter 18

HONEY-PRODUCING CATERPILLARS

Late in June, growing abundantly in the edges of woods throughout this region, may be seen the Cimicifuga racemosa of botanists, popularly called Rattleweed, or Black Snakeroot. It sends up a stalk, sometimes branching, four or five feet, terminating in a spike or spikes, six to ten inches long, of round, greenish-white buds, which stand upon short stems, and are arranged in rows about the stalk, diminishing in size till they reach the pointed top. The lower buds, when they are about the size of an ordinary pea, open first, and the flowering proceeds by degrees up the spike, so that buds are to be met with throughout a period of from four to six weeks. The flowers emit an intensely sweet odor, which renders them attractive to butterflies and bees.

But should you examine these buds with care, you will find a number of small caterpillars, the larva of the beautiful Azure Butterfly, called Lycæna pseudargiolus, feeding thereon. During its younger stages it is white, and so near the color of these buds that it is well protected, and very difficult to find. Later on, it may be white or greenish, and often diversified with a few black or brown patches, irregularly diffused over the surface.

When mature the larva is one-half of an inch in length, and, like all Lycænid larvæ, is onisciform, or shaped like the little pill-bug, so common under stones and logs. The head is very small, and is placed on the end of a long, green neck, which at the junction is of the thickness of the head, but gradually enlarges, and seems to be fixed at the hinder part of the second segment, the latter being hollowed out so as to form for it a sheath. In the final larval stages this segment is elevated, transversely compressed, and inclines forward, thereby shielding the head as the larva moves about. When quiescent the neck and head are wholly retracted, and as the former, when fully extended, is very much longer than the depth of the second segment, it must possess considerable elasticity.

The larva feeds on the heart of the bud, and to reach this cuts away the surface on one side till an opening is made sufficiently large to admit its head; and as it feeds the second segment is firmly pressed against the bud so as to permit the utmost elongation of the neck. Thus it is enabled to eat out the contents of the bud, and only desists when there remains but the empty shell. When so engaged the anterior segments are curved up and the others rest upon the stalk of the plant, but very small larvæ repose wholly in the bud. Not a single instance has been observed where an open flower has been attacked, but the destruction of buds is very extensive.

But now comes the most remarkable part of the larval history of Pseudargiolus. The whole upper part of the larva is covered with small, glassy, star-shaped processes, scarcely raised above the surrounding surface, from the centre of which spring short, filamentous bodies, bristling with feathery-looking tentacles, which the caterpillar has the power of protruding at will. It throws them out like the tentacles of Papilio or the horns of snails. More singular still is an opening upon the eleventh segment, placed transversely and surrounded by a raised cushion, about which the granulations that cover the body of the caterpillar are particularly dense. From the middle of this opening, which is shaped like a button-hole, issues, at the caterpillar's will, a sort of transparent, hemispherical vesicle, from which is emitted a good-sized drop of fluid, which the animal is capable of reproducing when absorbed.

Figure 23. Pseudargiolus Butterfly. Larva feeding on bud of black snakeroot, and guarded by ants.

Four species of ants may be seen attending, not the small larvæ, but those that have attained the nearly mature condition. They are invariably found on or near the larva. Their actions, as they run over the body, caressing with antennæ, evidently persuading the larva to emit a drop of the fluid, are alike curious and interesting. Most of this caressing is done about the anterior segments, and while the ants are thus occupied, or rather, while they are absent from the last segments, the tubes of the twelfth seem expanded to their full extent, and so remain, without retracting or throbbing, until the ants come hurrying along with great excitement and set foot or antenna directly on or close by the tubes, when they are instantly withdrawn. The ants pay no heed to the tubes. They seek for nothing from them, and expect nothing. But they turn at once to the eleventh, caress the back of that segment, and, putting their mouths to its opening, exhibit an eager desire and expectancy. Suddenly a dull green, fleshy, mammilloid organ protrudes, and from the summit of which comes a tiny drop of clear green fluid, which the ants, some two or three perhaps standing about it, lap greedily up. As the drop disappears, this organ sinks in at the apex, and is so withdrawn. The ants then run about, some in quest of other larvæ upon the same stem, some with no definite object, but presently return and pursue the caressings as before. The intervals between the appearance of the globule vary with the condition of the larva. Where exhaustion by long-continued solicitings occurs, some minutes elapse before renewal is effected, the tubes in the meantime remaining concealed. Fresh larvæ, however, require little or no urging, and globule follows globule, as many as six emissions in seventy-five seconds, without even a retracting of the organ. Often the presence of the ant, when the larva is aware of it, evokes, all unsought, the sugary fluid.

Ordinarily the tubes expand when the ants are absent from the last segments, and are certainly withdrawn when they come near. These tubes, from all appearances, serve as signals to the ants. When the latter discover them expanded, they know that a refection is ready, and rush to the opening in the eleventh segment where it is to be found. The tubes certainly serve no other purpose. No visible duct appears in the dome of the tube when largely magnified, and the ants seek nothing from it or the twelfth segment. They cannot be used to intimidate, or to frighten away enemies, for in the younger stages, when the larvæ have the most to dread, neither the tube nor the organ in the eleventh segment is available. The outward openings, and the orifice in the eleventh segment, exist in the youngest larval stages, but are functionless until the larva has nearly attained maturity. Ants seldom attempt to caress or solicit young larvæ, but pass them by with indifference, seemingly knowing that they cannot emit the secretion. When an ant approaches one of these immature larvæ, the larva manifests considerable annoyance, throwing up the hinder segments, as though the ant was an enemy which it was desirous to get rid of. If the tubes could now be thrust out, the ant would be attracted, rather than repelled.

But when the period arrives that the tubes are free, and the secretion is ready to be ejected, which is perhaps just after the third skin-moulting, and it cannot be earlier, the larva grows now quiet and submissive, inviting the attentions of the ants, and rewarding their antennal caresses.

Four species of parasites affect these larvæ. Two are dipterous. These, which are of the size of the common house-fly, deposit their eggs, during the second larval stage, on the back, and near the junction of the second and third segments. In process of time the grubs hatch and eat their way into the larva, to emerge when the latter has become fully grown, thus destroying its life. Another of these enemies is a minute hymenopterous insect, whose egg is placed in the very young larva, probably in the first stage of its life. The grub, in this case, eats its way out of the half-grown larva, spins a silken cocoon, from which in a few days issues the newly-matured parasite. The destruction of larvæ by these, and very likely by other similar parasites, is doubtless immense. But no parasite attacks, it does seem, the mature larva, for, if it did, the grub of the former would live within and destroy the chrysalis, and instead of a butterfly emerging therefrom would come forth the parasite. Multitudes of chrysalids of other species of butterflies are thus destroyed, but Pseudargiolus, at this stage, appears to enjoy a singular immunity from enemies.

Why this species, and doubtless many others of its family, are thus favored, will soon be apparent. Ants may be seen wherever these larvæ may be found, ever ready to receive the honeyed secretion when it pleases the little creatures to eject it, but all the while exercising the closest vigilance lest some wary ichneumon may come along and deal a thrust of its ovipositor, which means misery and ultimate death to their helpless friends. So intent is the larva, with its head buried in the flower, upon its feeding, and so quietly and stealthily does the ichneumon approach its intended victim, that hardly a single individual would be left to tell the story of its existence were it not for the ants. The larvæ know their protectors, it would seem from their actions, and are able and willing to reward their services. The advantage is mutual, and the association friendly. No compelling by rough means on the one part is noticeable, and no reluctant yielding on the other. All demonstrations made by the ants are of the most gentle character. They caress, entreat, and as they drink in the sweet fluid, lifting their heads to prolong the swallowing, they manifest to the utmost their satisfaction and delight. It is amusing to see them lick away the last trace, caressing the back of the segment with their antennæ as they do so, as though they were coaxing for a little more.

In Pseudargiolus the tubes are white, cylindrical, nearly equal in size, rounded at summit, and studded with little tuberculations from which arise the tentacles. These last are tapering, armed with small spurs set in whorls, and stand out straight, making a white hemispherical dome over the cylinder, but none of them fall below the plane of the base of the dome, nor do they ever hang limp or lie across the dome, as is the case in a European species. When the tube comes up the rays rise in a close pencil, and take position as the dome expands; but, on the contrary, when the tube is withdrawn, the top of the dome sinks first, the rays coming together in pencil again.

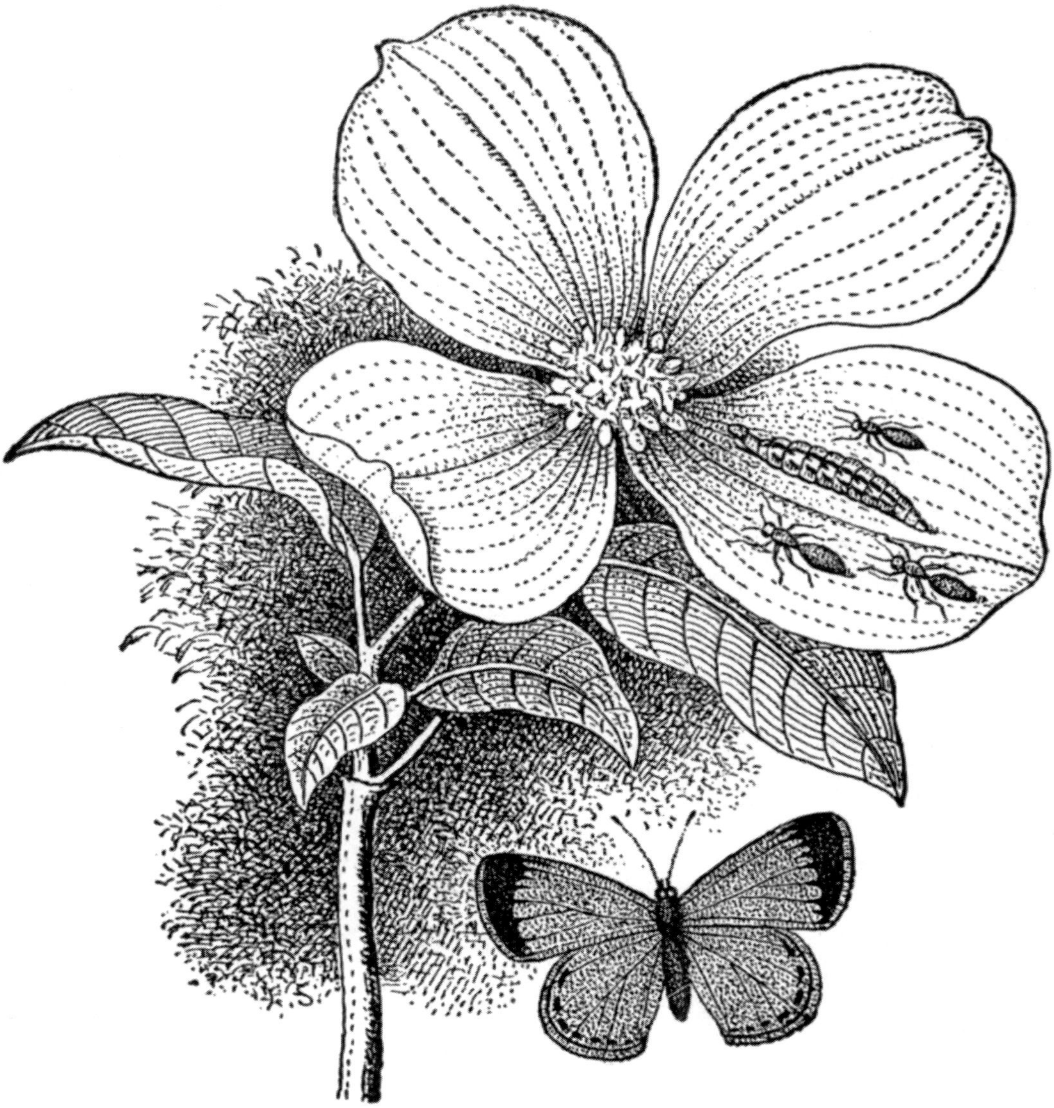

Figure 24. Violacea Butterfly. Larva, protected by ants, feeding on flower-buds of Dogwood.

Lycæna pseudargiolus is subject to great variation, and occurs under many forms, most of which having been regarded as distinct species. In the early spring Violacea appears, and is characterized by dimorphism in the female, some of that sex being blue, others black. This form, which may be called the winter form, deposits its eggs in the clusters of flower-buds of the Dogwood, the young larvæ obtaining their first food by boring into the buds, but later on eating their way into the ovaries. The flies that come from these larvæ late in May are Pseudargiolus, which, as stated before, lays its eggs on *Cimicifuga racemosa*, most of the resulting butterflies over-wintering to produce Violacea. A small percentage of the May chrysalids give butterflies as late as September, which are smaller than the parent-form, and also differ therefrom in the more decided character of the marginal crescent discal spots on the under side of the wings. There does not seem to be any regular second summer brood, that is, there are but two regular annual broods, the Violacea of March and the Pseudargiolus of May, the individuals happening to emerge in July, August and September being irregular visitants, for which the name of Neglecta has been given. The females of the last form lay their eggs upon *Actinomeris squarrosa*, and the chrysalids, thence resulting, give Violacea the next spring.

Larvæ feeding on Dogwood vary much in color from those that feed on the Black Snakeroot, few being white in the last stages, but nearly all dull-crimson or green, or a mingling of the two. Nevertheless, a small percentage of the larvæ on *Cimicifuga racemosa* are also green or crimson, though the most of them white. Ants do not seem to visit the larvæ on the Dogwood, and on being introduced to them in confinement treat them with indifference. On rare occasions tubes have been discovered in the eleventh segment, fully expanded, and accompanied by a pulsating movement, but no teasing or irritating availed to make them appear. Even severe pressure applied to the sides of the segment failed to force out any fluid. As with the fall food-plant, *Actinomeris squarrosa*, the Dogwood is neither sweet nor juicy, and it is possible that the larvæ feeding on these plants do not secrete the fluid.

Eggs of this polymorphic species are round, flat at base, the top flattened and depressed, and have a diameter of one-fiftieth of an inch. Their ground-color is a delicate green, the entire surface being covered with a white lace-work, the meshes of which being mostly lozenge-shaped, with a short rounded process at each angle. In from four to eight days the egg hatches into a larva, which is scarcely one-twenty-fifth of an inch long, and whose upper side is rounded, the under being flat. On each side of the dorsal line is a row of white clubbed hairs, with similar ones at the base and in front of the second joint, making a fringe around the body. The head is very small, obovoid, retractile and black; the legs retractile, and the color a greenish-white or brownish-yellow.

The first moult occurs in from three to five days, the larva having increased to twice its former length, while very little difference is manifest in the coloration. In from three to five days the caterpillar has again changed its skin, doubled its length, assumed more pronounced colors, which are diversified in some with mottlings upon back and sides, and developed along the back, from the third to the tenth joint, a low, broad, continuous, tuberculous ridge, cleft to the body at the junction of the segments, the anterior edge of each joint being depressed, the sides incurved. The third moult takes place in three or four days more, but there is very little change from the former period. Three or four days subsequent to this change occurs the fourth or final moult, and in five or six days from this the larva is ready to pass into the chrysalis state.

In its mature form the larva is about one-half of an inch in length. The body is onisciform, flattened at base, furnished with retractile legs, and has the back elevated into a rounded ridge, which slopes backwards from the sixth segment. The sides are rather deeply hollowed, and in the middle of each segment, from the third to the eleventh, is a vertical, narrow depression. The last segments are flattened, the last of all terminating roundly, its sides being narrowed and slightly incurved, while the second segment is flattened, arched and bent nearly flat over the head. Standing on the body is a ridge, tubercular in nature, which in each segment from the third to the eleventh is distinct and cleft to the body. In color, specimens vary. Some examples are white, others decidedly greenish, but many have the posterior slope of the second segment black or dark brown, while a few have most of the back a dark brown, irregularly mottling a light ground, or with small brown patches diffused over the back, but mostly on the anterior segments. The entire surface is velvety. This appearance is caused by minute stellate glossy processes, scarcely raised above the surface, mostly six-rayed, and sending from the centre a concolored filamentous spine a little longer than the rays. These stars are arranged in nearly regular rows, and are light, except in the brown patches, where both star and spine are brown. This velvet-like condition of the skin only reveals its true composition under a magnifying glass.

On the eleventh segment, near the posterior edge of the back, is a transverse slit, in a sub-oval spot, from which proceeds a membranous process; and on the twelfth, on each side, is a mark like a stigma, but a little larger, from which proceeds a membranous tube, ending in a crown of feathery tentacles, these three special organs being exposed or concealed at the will of the larva. The head is small, obovoid, dark brown, and is placed at the end of a long, pale green, conical neck, which is rectractile, both neck and head being covered by the second segment.

Figure 25. Neglecta Butterfly. Larva feeding on central florets of actinomeris, and guarded by ants.

Before changing to a chrysalis, the summer larvæ sometimes turn pink, and from pink to brown, or become brown without the pink stage, although others remain white or change to rusty brown. The body contracts to about three-tenths of an inch and takes on a rounded form.

The chrysalis is dark-brown or yellow-brown, but varying in color, the wing-cases being dark or green-tinted. Two sub-dorsal rows of blackish dots are found on the abdomen, and sometimes a dark dorsal line. In the few instances in which the butterfly emerges the same season the duration of this stage is from thirty to sixty days, but most chrysalids pass the winter and mature in the spring.

Now for a description of the butterfly. In general terms, the upper side of the wings of the male is a deep azure-blue, with a delicate terminal black border. On the apical part of the fore-wings the fringes are black, but white and barred with black on the rest of these wings and on the hind-wings. In the female the fore-wings have a broad, blackish outer border, in some examples extending along the costa, while the hind-wings have a blackish costa and a row of dark spots along the outer margin. Usually the ground-color is a lighter blue in the females than in the males. A pale silvery gray, with a silky lustre, is the color of the under side of the wings, which is relieved by a row of spots along the outer margin, each preceded by a crescent, a curved row of elongate spots across the disk of the fore-wings, and several spots on the basal part of the hind-wings, all the markings being of a pale brown color. Violacea, the so-called winter form, has the dark parts and crescents on the under side of the wings quite prominent, but they do not, either in the outer border or in the basal portion, coalesce. Pseudargiolus, the largest of the series, there being but three forms in Pennsylvania, expands one and four-tenths inches. The upper surface of the male usually has a terminal border to the hind-wings of the same shade of blue as is visible on the fore-wings, the middle area of the hind-wings being a little paler than this border on the fore-wings. On the under side of the wings the spots are much smaller than on the preceding form. Neglecta, which resembles Pseudargiolus, and has the spots on the under surface small, is a smaller form, never expanding more than one and one-tenth inches. It is a summer form when there is more than one generation in a season, ranging from Canada, through New England to West Virginia and Georgia, and occurring also in Montana and Nevada. Violacea has a more extended limit, being found in Alaska, British America, Ontario, Quebec, New England to West Virginia, and Colorado, while Pseudargiolus ranges from Wisconsin south to Tennessee, and on the east from Pennsylvania to Georgia.

Chapter 19

HIBERNATING BUTTERFLIES

Early in March, and often while the snow yet lingers upon the landscape, may be seen flying in and out among the forest-trees, or lazily meandering along some deserted road through a thicket, the beautiful Antiopa. Her rich crimson dress, so dark that it almost seems black, with its buff-colored, sky-dotted border, serves to distinguish her from her no less interesting, but smaller, sisters of the Vanessa family of butterflies. But the Antiopas you then see are generally ragged and shabby, which is not to be wondered at, when it is considered that it is their last year's dresses they wear, for late in the preceding August they had their being, and all through the autumn had been exposed to a hundred misfortunes or more while seeking their living.

But with the coming of frost and of cold comes the blighting of flowers. A feeling of torpor in consequence steals over their once bouyant spirits, and into some crevice in a barn or a wood-pile or stone-heap they creep, and there sleep the winter away, till the warmth of the sun from his southward-bound journey returning sets the brown buds a-swelling, when out of their hibernating retreats they leisurely crawl for a flying stroll through the awakening trees. Slow and deliberate their movements are, as though some grave and momentous event were dependent thereon.

Never have I watched such actions, so human-like have they seemed, than the conviction has gone home to my mind that they plainly evinced a thought and a purpose, which had their origin, if not in a brain, at least in one of the several ganglions which largely make up their wonderful and somewhat complicated nervous machinery.

No matter how low in intelligence she may rank, Antiopa has nevertheless, or all experience is at fault, some general ideas of the time and fitness of things. From her gloomy abode in the wood-pile she has emerged, while all the gay butterfly world, barring a few familiar exceptions, is asleep, for a tour of investigation. Her venture is seldom ill-timed, for the violets have preceded her, and from their delicately curved flagons proffer her food and refreshment.

Cool and unhealthful as the mornings are at first, it is not till the sun is nearly overhead that she leaves her retreat, for what of plant-life exists is then, under the full force of his beams, at its very best. Three or four hours a day, with few intervals of rest, she is actively on wing, regaling herself with exercise and food, thus storing little by little her body with some of the strength and vivacity which were hers when the famine of winter overtook her and forced her to retirement, so as the better to prepare for that work, the propagation of her kind, which is the principal, but not the only, aim of her existence. After four in the afternoon her presence is scarce, as she has sought her old, or some other, place of shelter and security.

But when the days have grown longer and warmer, and the trees are arrayed in their livery of green, she is in the fields bright and early, and often ere the dew has disappeared from the grass and the flowers. The most restless of beings she now is. Anon alighting upon a bush for a momentary rest, then off for a dozen or more rods, when the presence of some favorite blossom meets her quick sight and invites her to pause, which she does, but only for a second to quench her thirst. Where willows, or elms, or poplars abound, she is more frequently seen later on in May, but flying more slowly and sedately than ever before. The flowers pass unheeded. She seems in a dream, in a reverie. But all of a sudden she quickens her speed. You look for the cause. There, in the distance, another is seen, just like her in mien, some would-be suitor for her hand and affections. He enters his suit, he pleads his great love, and awaits her sweet pleasure. The answer is brief, and soon by their actions, as high up in the air they circle and circle, caressing each other with strokes of the antennæ, the story is told that his love has been requited. A brief honey-moon of two or three days and the love-scene is over, and the two settle down to the prosy realities of everyday life. The male goes back to his old-time pursuit of rifling the flowers of their honeyed treasures, whilst the female, upon whom devolves the duty of providing for the offspring whom she is never likely to see, looks scrutinizingly about for her favorite trees, the poplar, the elm, or the willow. In her selection of a tree a wonderfully keen discernment is shown, for she seldom, if ever, mistakes her plant-species.

When a choice has been made, no time is expended in fruitless endeavor. She proceeds at once to deposit her eggs. They are laid in a cluster round the twig, and near the petiole of a young leaf, upon which the newly-hatched larvæ are to feed. The eggs hatch inside of a week into small black spiny caterpillars which, in their early stages, are very social in their habits. Just before the final skin-moulting they separate, each caterpillar living alone, the necessity for food, which their very vigorous appetites now demand, being the impelling motive. In a state of maturity the larvæ are two inches in length. They are black, and minutely dotted with white, which gives them a greyish look. A row of brick-red spots are found down the back, and their body is armed with many black, rather long and slightly branching spines. The head is black, and roughened with small black tubercles.

Figure 26. Mourning-Cloak Butterfly. Larva feeding on willow leaf, and chrysalis suspended from twig.

Having completed their period of feeding, which they do in about four weeks, the caterpillars attach themselves by means of their tails to a fence-rail, a window-ledge, or some such place, and pass into the chrysalis state, which is accomplished in about four days. In this condition they present an odd-looking appearance. The head will be found to be deeply notched, or furnished with two ear-like prominences. The sides are very angular. In the middle of the thorax there is a thin projection, somewhat like a Roman nose in profile, while on the back are two rows of very sharp tubercles of a tawny color, which contrast very markedly in coloration with the dark-brown of the rest of the chrysalis. Fifteen days, when the weather is favorable, are sufficient for the development of the imago, or butterfly. As maturity approaches, the chrysalis-shell becomes quite soft, and the efforts of the imago to free itself from this covering are facilitated by the ejection of a blood-red fluid, which rots the case, while it acts, at the same time, as a lubricant to the emerging butterfly.

When these caterpillars are very abundant, as was the case in the vicinity of Germantown some twenty-five years ago, every fence-rail was hung with chrysalids, as many as a dozen being found upon a single rail. The caterpillars even climbed up the sides of the houses and suspended themselves from the window-ledges and the edges of the overhanging shingles. When the butterflies emerged, great blotches of the fluid bespattered the fences and houses as though the clouds had rained great drops of blood. The willows and poplars were alive with the caterpillars, and even the maples were overrun when there came a scarcity of the leaves of the natural food-plants. Green caterpillar-hunters were everywhere plentiful, and the writer could have taken hundreds of specimens, but these highly-useful beetles made a very sorry attempt in holding the enemy in check.

Two broods of the caterpillars are raised, one in June and the other in August, but the agencies by nature employed for their destruction so effectually accomplish their mission that hardly a season brings to my notice a dozen full-grown larvæ. *Vanessa antiopa*, as this species is called by the scientific student, or Mourning-Cloak by people and amateurs, is generally found through the whole of North America. In England, where it is popularly called the Camberwell Beauty, because specimens were first taken near Camberwell, it is the rarest of butterflies; while on the Continent, as in this country, it is a very plentiful insect.

Chapter 20

Leaf-Cutter Bee

Few hymenoptera of the family of bees are so little known as the Megachilidæ, or Leaf-cutters. They are stout, thick-bodied insects, with large, square heads, and armed with sharp, scissors-like jaws, which admirably fit them for the work they have to do in preparing materials for the building of their homes.

Our commonest species, *Megachile centuncularis*, is about the size of the hive-bee. In gardens and nurseries where shrubbery abounds, it is very prevalent, especially the female, which is readily distinguished by a thick mass of stout, dense hair on the under side of the tail, which serves as a carrier of pollen. The honey- and bumble-bees differ materially from them, for they have the hind tibiæ and basal joints of the tarsi very much broadened for that purpose.

Megachile is by no means a remarkable-looking insect. Judging from its very humble exterior, one can hardly believe it possessed of the wonderful intelligence, as shown in its wise provisions for its young, which it is found to display.

Ordinarily the female, who is entrusted with the discharge of this very essential business, places her nest in the solid earth underneath some species of shrub. A vertical hole, three inches in depth, is dug, and this is enlarged into a horizontal gallery, some five or six inches in length.

You should see the little creature in her never-tiring work of preparing material for her nest. In and out among the roses she goes, examining each leaf with the most critical care, and only desisting from her labor when a suitable one has been chosen. She scans it over and over, and at last from a position on its upper or nether surface proceeds to cut a piece just fitted for her work, which, heavy as it seems, is seized between the legs and jaws and carried on swiftly-agitated wings to her burrow.

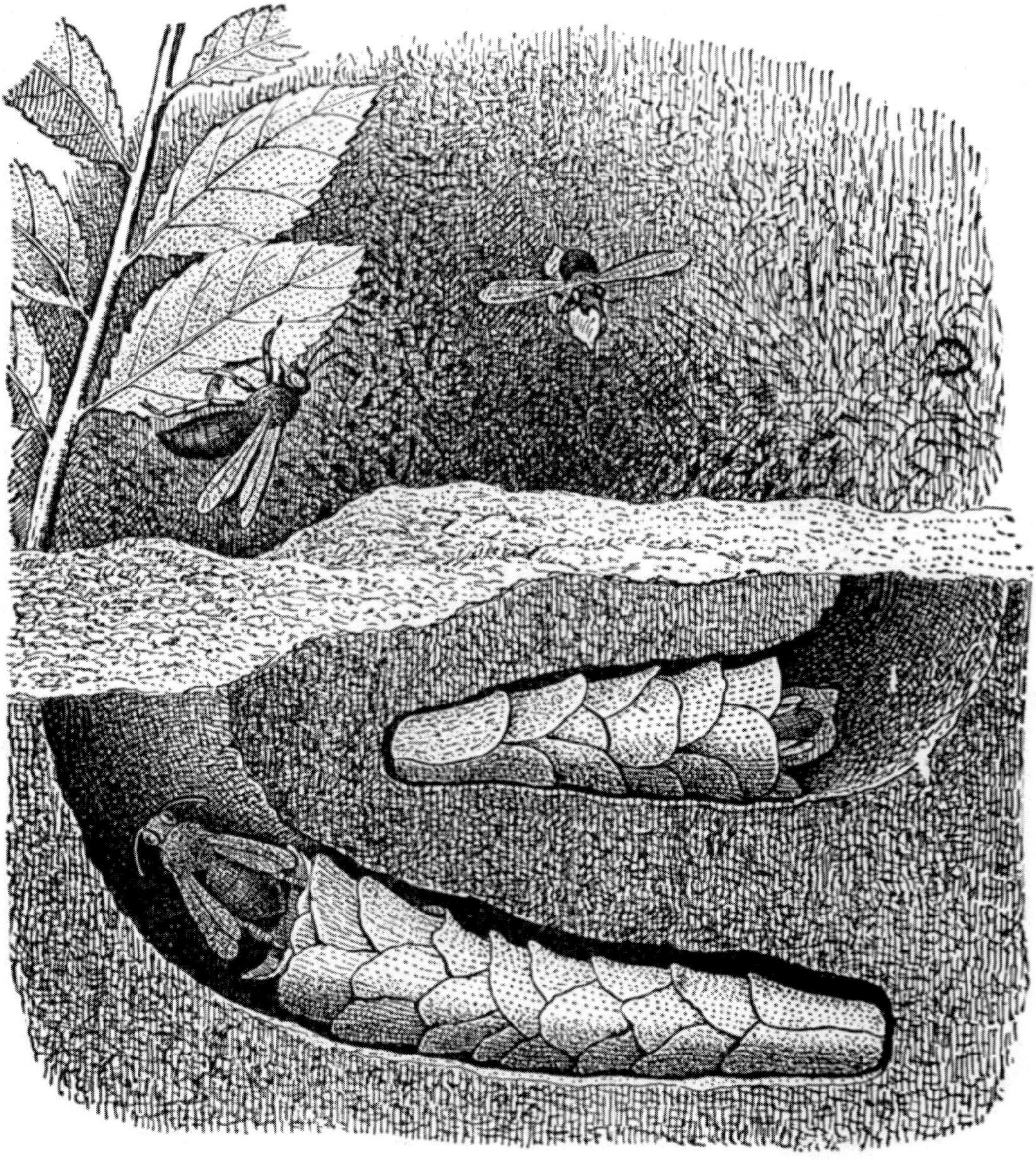

Figure 27. Leaf-Cutter Bee at Work. Two tunnels being filled with leaf-cells.

Ten pieces or more, each differing in shape, are cut and borne away, which the ingenious insect tailor twists and folds, the one within the other, until is formed a funnel-like cone, whose end is narrower than its mouth. So perfectly joined are the parts, that even when dry they have been found to retain their form and integrity. A cake of honey and pollen, for the use of some yet unborn Leaf-cutter, is deposited within, and on this, in due time, is laid a single small egg. Nought now remains but to wall up the cell. A circle of leaf, of the size of the opening, is cut, and this is closely adjusted within the wall of rolled-up leaves. Sometimes as many as four pieces are thus utilized. A second cell, similarly built, is fitted to the first, and this is succeeded by eight or ten others. When all is completed, the eggs being laid and the cells all victualled, the hole of the shaft is closed with the earth that was thrown out, and so carefully, too, that not a trace of her doings remains to tell us the story.

Like other insects, Megachile is occasionally prone to change. Some laborers while digging, one early spring-day, some thirteen years ago, about a cluster of plants of *Spiræa corymbosa*, a species allied to the roses and cinquefoils, came unexpectedly upon a dozen or more cells of this insect, arranged horizontally in layers, some three or four inches below the ground's surface. These cells were three-fourths of an inch in length, one-fourth in width, and formed of the leaves of Spiræa. Six circles, of three pieces each, constituted the cell, and these were so arranged that each succeeding circle was made to project but slightly beyond its predecessor. Six circular pieces, larger than seemed needful, closed up the opening of each cell. That there was a purpose here manifested was very apparent. This purpose, as it appeared to the writer, was the better accommodation by the hollow surface of the cell that was to follow, and the giving of greater firmness and security to the entire structure.

More curious, however, were some cells that were found the ensuing year, which, in looks, resembled very closely those of Pelopæus, a species of wasp, familiarly designated the Mud-dauber. These cells, in numbers of three, were adherent to the rafters of a hardly-used garret. In form, and in the peculiar combination of their pellets of clay, they were the exact counterpart of the Mud-dauber's. But the curious funnel-like arrangement of leaves on the inside, so strikingly characteristic of the Megachilidæ, was evidence of the most positive kind that Pelopæus had nothing whatever to do with their putting together. It bespoke a piece of work that was entirely beyond the highest capability of her being to execute.

Each of the included leafy cells was one and one-eighth inches in length, and just barely exceeding one-fourth in width. Elliptical pieces of Spiræa, less in size than those previously described, but arranged in a similar manner, composed the several structures. Within each, a dead but perfectly-formed Megachile, encased in a cylindrical bag of silk, was found, so that there could be no possible doubt of the builder. That this inner fabric was the labor of some mother Megachile admits not of a scruple, for no other bee is known to construct a nest of like character. But what of the outer enveloping fabric of mud? It was clearly impossible for the skill of a Megachile, who, while certainly fitted for tunnelling the ground and for snipping circular and elliptical pieces of suited dimensions from leaves with all a tailor's precision, would find herself woefully unadapted for the making of mortar and the building of nests, in imitations of tunnels, out of pellets of mud that had to be moulded into consistency and shape by the jaws of the builder. Pelopæus alone, of all hymenopters, possesses the ability and means of making such structures. Megachile, who is known to occasionally build under the boards of the roof of a piazza, might sometimes in her quest of a place appropriate the discarded cells of some pre-existent Pelopæus for nesting purposes, but she runs a very great risk in so doing, for the Mud-dauber does not always build a fresh home for her treasures, save when there is a lack of the last year's structures. Old nests, when found, are put in speedy repair and made to do as invaluable a service.

Chapter 21

Battle Between Ants

Whilst reclining one beautiful May afternoon in the shade of an oak that stood on the outskirts of a thicket, my attention was arrested by the activity and bustle presented by a colony of yellow ants, which proved to be the Formica flava, so common everywhere.

Scattered indiscriminately about were numberless larvæ in various stages of growth, and not a few immobile pupæ, that had been brought up from subterranean domiciles by thoughtful nurses, while here and there were a dozen or more ants, but recently escaped from their mummy-cases, basking in the sun's warmth, preparatory to entering upon the duties of the formicarium.

The very picture of restlessness and anxiety were these full-grown neuters. That something was transpiring, or was about to transpire, seemed not unlikely, for ovæ, larvæ and pupæ were being quickly carried to places of concealment in the earth, or hustled away among the entangling and interlacing grasses.

Looking about for the cause of all this excitement, the truth at once became painfully apparent. Three large, burly ants, representatives of *Formica subterranea*, a black species that is everywhere abundant in wooded regions, had intruded their obnoxious presence into the happy colony, bent, as it was evident, on pillage or slaughter.

Were plunder the inspiring motive, these giant invaders were not slow to learn that their weaker kin, though lacking their strength, could more than match them in cunning and stratagem.

Not daring to attack the foe, and being unwilling that any of their number should be led into slavery, or suffer aught at the hands of others, they immediately set to work to destroy all whom it was impossible to protect.

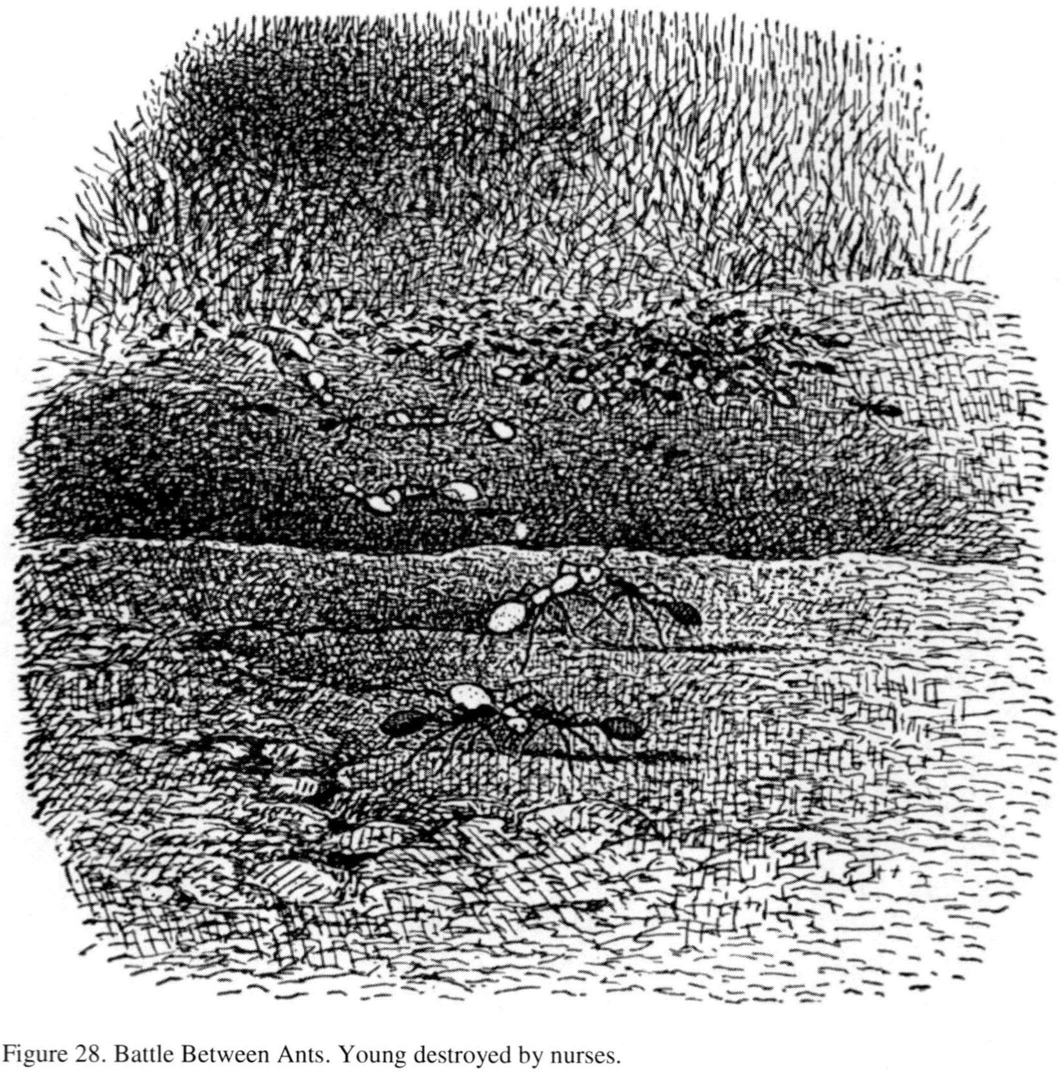

Figure 28. Battle Between Ants. Young destroyed by nurses.

Detailed as most of the neuters seemed to be in looking after the wants of the immature, there were a few observed running hither and thither and seizing in their jaws the newly-developed, not to bear them out of the reach of danger, as was at first supposed, but to kill them so as to prevent them from falling a living prey into the hands of the enemy.

Knowing the sympathy and affection which the nurses are ever wont to cherish towards the objects of their care, this act of cruelty struck me as something very astonishing and peculiar.

Prompted by curiosity to know the nature of the wounds thus inflicted, I placed upon the palm of my hand one of the wounded ants, and made, by means of a microscope, a careful examination of its injuries. Above and below the abdomen, between the second and third segments, two deep wounds, which met each other in the interior, were plainly to be seen.

Several cases of the kind were afterwards noticed. These were not accidental occurrences, made through efforts to carry the young to places of shelter. Possibly, through inexperience, accidents might happen once in a long time, but to suppose that insects, accustomed to handling their young as the neuters assuredly are, would be likely to make such blunders, is too unreasonable to be entertained. Admitting for argument's sake that such things might occasionally occur, would successive repetitions be expected? I apprehend not. But on the supposition that a purpose was thereby subserved, the object had in view warrants, it would seem, the means employed for its accomplishment.

What the purpose was it will now be my aim to show. That many animals, tame as well as wild, are wont to destroy disabled and wounded companions, is well established by history. In many instances the destruction is justified to preserve the herd or pack from the close pursuit of enemies. "Instinct or reason," as Darwin says, "may suggest the expelling an injured companion, lest beasts of prey, including man, should be tempted to follow the troop."

Audubon, in writing of the wild turkey, so abundant in his day, observes substantially that the old males in their marches often destroy the young by picking the head, but do not venture to disturb the full-grown and vigorous. The feeble and immature being an encumbrance, it is obvious that the watchfulness and attention which they would require, were sympathy and affection the emotions by which the males are actuated, would necessarily retard progress, and lead to the destruction of the entire flock. Instinct or reason here operates for individual and family good.

Granting that instinct or reason does sometimes act for individual and family preservation in the manner described, I am not willing to admit that in every case that may arise in which the weak and disabled are sacrificed, that it is done for the material benefit of the physically able and robust. How the destruction of the weak and nearly-developed ant can result in good to the colony, in view of the fact that not the slightest effort to escape the danger by flight is undertaken, the sole object being the hiding of the young, it is most difficult to conceive.

There seems to be one of two theories, in the writer's judgment, that will, in anything like a satisfactory manner, account for this strange, abnormal habit upon the part of an insect that has been proverbially distinguished for its kind and affectionate disposition towards the tender beings committed to its trust; either to attribute it to an unwillingness and dislike to see its offspring made the servants of a hostile race or the subjects of ill-treatment and abuse, or to the survival of a habit of the past when its ancestors were a migratory, or nomadic, species.

That a feeling of repugnance does sometimes take possession of animal nature when the objects of parental care and solicitude are, or are about to be, reduced to slavery or confinement, and impels to actions of cruelty, will be patent from what follows:—

A friend, several summers ago, having procured a pair of young robins, placed them in a cage, which he hung from a tree-branch close to his dwelling, where the parent-birds could have an opportunity to feed them. All went well for a few days, when the parents, who had busied themselves in the intervals of feeding in attempts to secure their release, finding their efforts unavailing, flew away, but only to return with something green in their bills, most probably poisonous caterpillars, which they fed to their offspring. A few minutes later and they lay in the bottom of the cage dead, but the parents, as if conscious of what would result, flew away, and never came back.

May it not be that the parents, finding all efforts to restore their young to freedom ineffectual, sought this method of saving them from a life to which death must assuredly be preferable? Instances of like character might be adduced by the hundred, but enough has been written to show that, in the case of Formica flava, an unwillingness to allow the humblest of the colony to be taken into bondage was the motive which prompted the sacrifice.

Chapter 22

NEST-BUILDING FISHES

Not alone in color do fishes resemble birds. In the home-life and love of offspring a close resemblance obtains. Many are nest-builders, erecting structures quite as complicated as those of some birds, and hardly less elaborate in design and finish.

Floating along some woodland stream, or strolling along its grass-fringed margin, we have watched the domestic life of the Sun-fish, the *Eupomotis vulgaris* of writers, that mottled, bespangled beauty that seems always on hand to be caught by the angler in default of more noble game.

Where delicate grasses grow, and floating lily-pads cast their shadows, there among the winding stems the Sun-fish builds its home. Moving in pairs in and out among the lilies near the shore, as if jointly selecting a site for a nursery, they may be seen. The spot is generally a gravelly one, and, once determined upon, no time is lost in pushing the work to a speedy conclusion. For several inches around the space is cleared of stems or roots, and these are carefully carried away. The smaller roots are swept aside by well-directed blows of their tails, or by mimic whirlpools which the fishes, standing over the nest, create by their fins. The stones are next taken up, the smaller ones in their mouths, the larger being pushed out bodily, or fanned away by the sweeping process, until an oval depression, with a sandy bottom, finally appears. About the sides the stems of aquatic verdure, which seem to have been purposely left, may be seen standing, and these now naturally fall over, oftentimes constituting the nest a perfect bower, with walls bedecked with buds, while the roof is a mat of white lilies floating upon the surface. Here the eggs are deposited, the male and female alternately watching them.

While the Sun-fish is always recognized as the most peaceful of the finny tribe, and only chasing in wanton playfulness its neighbors, it is otherwise when the passions are wrought to a high pitch of excitement through the play of amatory influences in the spring-time. Let a stranger, a bewhiskered cat-fish, approach the bower, and war is at once declared. The little creatures snap at the intruder with anger and defiance. Their sharp dorsal fins stand erect, the pectorals vibrate with repressed emotion, while the violent movements of their powerful tails evince a readiness and determination to stand by their home at all hazards. Indeed, so vigorous is their charge, that even large fishes are forced to retreat, and, as the Sun-fishes build in companies, the intruder often finds himself attacked by a whole colony of them.

Figure 29. Nest of Common Sun-Fish. Male and female defending it from attack of Cat-fish.

Nearly all the Sun-fishes are nest-builders, some forming arbors, as we have seen, others scooping out nests on sandy shoals, while one, the Spotted Sun-fish, is more democratic, affecting muddy streams, where, on the approach of cold weather, it makes a nest in the muddy bottom, and there it lies dormant till the coming spring.

Who has not made friends with the Dace—*Rhinichthys atronasus*? He is a veritable finny jester. We have watched him in his watery retreat, and, perhaps unseen, have played the spy upon his domestic proceedings.

Life is a gala time to these little fishes. They have seemingly never a care or a bother. In jest they join in the chase of some curious minnow that intrudes upon their presence, suddenly changing their course to dash at some resplendent dragon-fly that hovers over the leafy canopy of their home, and as quickly darting off again to attack some bit of floating leaf or imaginary insect.

All is not play, however, even among the Dace. The warm days of June usher in the sterner duties, the nesting-time. Male and female join in the preparation, and a locality, perhaps in shallow water in some running brook, is selected. Roots, snags and leaves are carried away, both fishes sometimes found tugging away at a single piece, taking it down-stream, and working faithfully and vigorously until, in a few hours, a clearing over two feet in diameter is the result.

There the first eggs are laid. The male, who has retired, soon appears from up-stream, bearing in his mouth a pebble, which is placed in the centre of the clearing. Now they both swim away, but soon returning, each bearing a pebble, that is also dropped upon the eggs. Slowly the work proceeds, until a layer of clean pebbles apparently covers the eggs. A second layer of eggs is now deposited by the female, and these are covered by pebbles as the others had been, the industrious little workers scouring the neighborhood for them, seemingly piling up eggs and stones alternately until the heap attains a height of eight inches or more. These heaps vary in shape, some being pyramidal, and others dome-shaped.

Such patience as these finny housekeepers manifest is not appreciated by man. The gleaners of the golden fields, in whose waters our little friends are found, have not discovered their secret, and think the curious piles the washes of the brook itself. But their purpose is the protection of their eggs. In swift-running streams, which these fish are so wont to affect, the eggs would be washed away, and, driven against rocks and snags, would be destroyed, or, even escaping destruction, would, by the undulating movement to which they would become subjected, be rendered impossible of incubation. Besides, were they not thus protected, even though there was no danger of being washed away, they would become easy prey to the attacks of carnivorous fishes.

Unlike as the Lamprey-eels are in structure to the Dace, yet in their habits of erecting a nest they are very similar. Upon our Eastern sea-board they are a common species, inhabiting both salt and fresh water. In the early spring they follow the shad up the rivers, occasionally preceding them, and search about for suitable localities in which to deposit their spawn. They clean away the stones as the Dace were seen to do, bending their long bodies in coils, which they use in pushing aside the accumulation on the bottom. To the unlearned the appearance of two Eels, each three feet in length, twisting and seemingly coiling about each other, would be indicative of war. But having cleaned for themselves a smooth spot, the Lampreys proceed to place stones. Irregularly-shaped stones of small size are easily and quickly transported in their mouths, but when stones that weigh several pounds are to be brought, the tactics they adopt are worthy of an engineer. As the spots chosen for the rearing of their submarine castles are ordinarily subjected to a swift current, the largest stones, which it would be thought impossible for them to move, are looked for up stream. A suitable one found, and a favorable position presented, the sucking mouth is fastened to it, and by a convulsive effort, the tail of the fish being raised aloft, the heavy stone is lifted from its place, the current pushing against the fish and stone, bearing them along several feet before they sink. Another effort of the fish, and the rock is again raised and carried down stream, until finally, by repeated liftings and struggles, the ingenious, persevering nest-builder is swept down to the nest, where the load is deposited. This laborious work is carried on until the pile has attained a height of two or three feet, and a diameter of four. No special form seems to be necessary. The nest is generally oval, compact and well devised to contain the eggs, which are carefully deposited within, thus affording protection in its numerous interstices for the young when they hatch. When about six inches long, the young Petromyzon marinus, which is a strange little fellow, is devoid of teeth, and blind, and possesses so many characteristics distinct from the parent, that for a long time he was considered a separate species, and even assigned a place in a different genus. Enormous nests are sometimes built. John M. Batchelder, Esq., describes one, which he saw in the Saco River, Maine, that was about fifteen feet long, and from one to three feet in height, its position and triangular shape in vertical section being well adapted for securing a change of water, and a hiding-place for the young. The operation of building was very methodical, a hundred and more Eels being at work upon the structure. Water-worn stones, chips of granites and fragments of bricks, sometimes weighing as much as two pounds and transported by a single individual, were utilized in the building.

More remarkable, however, than any previously described, are the nests of the Fresh-water Chub, *Semotilus bullaris*, which is known in some localities as the Stone Toter. This fish attains a length of about fifteen inches. The finest nests are on the shores of Westminster Island, but they are common on nearly every island that has a sandy, gravelly shore among the many that make up the Thousand Islands. The nest is a pile of stones, sometimes measuring ten feet across at the base, four feet in height, and containing a good-sized cart-load of stones, weighing in all perhaps a ton. Stones from small pebbles to some four inches in length were used, and as some of the nests are placed at considerable distances from the gravel-beds, and each stone represented a journey, the amount of labor performed, when it is considered that tens of thousands of stones must have been used in the building, certainly was incredible. Each stone is brought in the mouth of the Chub and dropped over the piles, one or more fishes working at the same heap. Some plan is evidently followed in the work, the first deposit of stones being small, and dropped so as to form a circle or semi-circle. The largest heaps are undoubtedly the work of successive years, the nests being annually added to during the last of May or June, when the Chubs are seen lying in the heaps, at which time the eggs are probably deposited. All the labor of piling up is to protect them from predatory fishes, a necessary and wise provision, as cat-fish, rock-bass, perch and others prey upon the eggs.

Figure 30. Black-Nosed Dace. Constructing their nest of pebbles.

In gravelly beds the Trout excavates a simple nest, a mere depression in the sand, that is not at all incomparable to the nest of some species of gulls. A furrow in the gravelly bottom of a river, often ten feet in length, the depression being made as fast as it is required, is the nest of the Salmon. In Canadian rivers these nests can be easily distinguished by the lighter marking in the bottom.

Few persons of the many who delight to drift along our sea-shores are unfamiliar with the Toad-fish. So closely does he in shape and color resemble a moss-covered stone that his enemies are deceived. Intrenched among the weeds and gravel, which the mother-fish carelessly throws aside, after the fashion of some of the gulls, the young are reared, their yolk-sacs enabling them to cling to the rocks of the nest soon after birth. There, under the watchful eye of the parent, they remain until old enough to swim away.

But the most vigilant of all nest-builders is the Four-spined Stickleback—*Apeltes quadracus*. In some neighboring stream, that sooner or later finds its way to the ocean, he may be found. There are different species of these fish, but their architectural ideas are pretty much the same. They vary mainly in the locations they select for nesting. Some place the nests upon the bottom, concealed among the sea-weed found there, while others hang theirs from some projecting ledge, or swing it in the tide from the sunken bough of some overhanging tree. As is unusual, the work of nidification is solely performed by the male Stickleback, the female taking no part in the labor. The spawning season having arrived, he, assuming a bright nuptial lustre, shows remarkable activity in selecting a site for an edifice, and transporting the building material thither. Fragments of all kinds of plants, gathered often at a distance, are brought home in his mouth. These are arranged as a sort of a carpet, but as there is danger of the light materials being carried away by the current, they are weighted down by sand to keep them in their places. Having entwined them with his mouth to his complete satisfaction, he then glides gently over them on his belly, and glues them with the mucus that exudes from his pores. More solid materials, sometimes bits of wood, sometimes bits of straw, which he seizes with his mouth, are adjusted to the sides of the floor to constitute the walls. He is now very particular. If the piece cannot be properly adjusted to his building, and he does not lose patience in his efforts to fit it in, he carries it to some distance from the nest and leaves it. After the side walls are erected, a roof of the same materials with the floor is laid over the chamber. Firmness is given to the whole structure by passing over it with his body, the light and useless particles being fanned away by the action of his fins and the vibratory movements of his tail. In carrying on his building operations care is taken to preserve a circular opening into the chamber, his head and a great part of his body being thrust therein, thus widening and consolidating it, and rendering it a fit receptacle for the female. When choosing material, the fish has been seen testing its specific gravity by letting it sink once or twice in the water, and if the descent was not rapid enough finally abandoning it.

Of the exact method used by the fish in binding the nest together we are indebted to Prof. Ryder. The male fish spins from a pore or pores a compound thread, using his body to insinuate himself through the interstices through which he carries the thread. The thread is spun fitfully, not continuously. He will go round and round the nest perhaps a dozen times, when he will rest awhile and begin anew. Its shape is somewhat conical before completion. The thread is wound round and round the nest in a horizontal direction, and when freshly spun is found to consist of six or eight very thin transparent fibres, which have alternated tapering ends where they are broken off. Very soon after the thread is spun, particles of dirt adhere to it, and render it difficult to interpret its character. The nest measures one-half of an inch in height, and three-eighths in diameter.

The time occupied in collecting materials and constructing the nest is about four hours, and when all is ready the male starts out to seek a female, and, having found her, conducts her with many polite attentions to the prepared home. The eggs being deposited, the male establishes himself as a guardian of the precious treasures, not even suffering the female to approach it again. Every fish that comes near, no matter how large, is furiously assailed. He gives battle valiantly, striking at their eyes and seizing their fins in his mouth. His sharp dorsal and ventral spines are very effective weapons in his defence. Constant watchfulness upon the part of the male is needed, for, if he go away for only a few moments, the sticklebacks and other fish lurking in the vicinity rush in and devour the eggs in an instant. A whole month he is occupied in providing for the safety of his offspring. About the tenth day he employs himself in tearing down the nest and carrying the material to some little distance. The fry may now be observed in motion. And these the male continually nurses, suffering no encroachment, and if the young brood show a tendency to stray beyond bounds, they are driven back within their precincts, until they are strong enough to provide for their own living, when both old and young disappear together.

But nothing in the lives of all these little nest-builders is more interesting than the intelligence they display and the facility with which they adapt themselves to circumstances. They seem to be able to grasp almost instantly the conditions of the environment, and to employ a wise discrimination in suiting them to their wants. Hardly two nests are alike. Marked differences in details of structure, configuration and surroundings are apparent, which prove that these creatures are controlled by reason, rather than instinct, in the elaboration of their homes. That they have some means of communicating their desires to each other cannot be doubted. When the male has laid hold of a stem, a pebble or a stick that completely baffles all effort at removal, his mate seems summoned to his assistance, and the united strength of the pair accomplishes the object to be gained. There is ever noticeable in whatever the sexes undertake some concert of action which would put to shame the boasted intelligence of man himself. The Sun-fishes, as has been said, nest in companies. When the combined effort of two individuals is unable to expel an invader, the entire community, as by a single mighty impulse, rises up against the foe. There is evidence of some form of society, even though simple in its organization, where individual members league themselves together for mutual protection and defence. Other examples might be cited to give the reader a common-sense estimate of the comparatively high order of intelligence that characterizes the actions of many of our fishes.

Chapter 23

SLIPPERY AS AN EEL

Eels are found in almost all warm and temperate countries, and grow to a very great size in tropical regions. They are impatient of cold, and hence do not exist in the extreme northern and southern parts of the world. In many islands of the Pacific Ocean they are held in considerable estimation, being preserved in ponds and fed by hand, but in many civilized communities a strong prejudice prevails against them, probably from their similarity to snakes, which prevents even a hungry man from caring to eat such wholesome and nutritious food.

Not one of our river fishes is so mysterious as the Eel, and although much is now known that was involved in obscurity, yet there is still much to learn of its habits, especially the manner of its reproduction. Difference of locality, it is likely, may influence the Eel and cause a difference of habit, an opinion which seems warranted from the various and perplexing accounts that have been given of its customs by numerous practical observers.

During the hot, still and sunny days of June they are chiefly seen on top of the water, wherever masses of aquatic weeds may be found, either in the calm enjoyment of a sun-bath, or for the purpose of feeding upon the myriads of gnats, moths and flies that seek the plants for rest or food, and which by unavoidably damping their wings become easy prey to their ambushed enemies. At night, similar retreats are affected for like purposes. Floating masses of detached weeds that the eddying stream has wound and kept in one place are sought in warm, stilly weather, but in blowing, cooler or rainy weather they forsake such places for the still, deep ditches. If a flush of water comes, and a little, shallow stream, running from or into the main river, becomes fuller than usual, there they resort in vast numbers, evidently pleased with the delicious change, only to remain as long as its freshness continues.

Like many other fishes, Eels are very tenacious of life, and can live a long time when removed from the water, owing to a simple and beautiful modification of structure, which permits them to retain a sufficient amount of moisture to keep the gills damp and in a condition to perform their natural functions. They have been seen crawling over considerable distances, somewhat snake-like in their movements, evidently either in pursuit of water, their own dwelling-place being nearly dried, or in search of some running stream in whose waters they may reach the sea after the customary manner of their race. Multitudes of Eels, both old and young, some of the latter scarcely six inches in length, have been seen crawling up the banks of a creek, apparently without any purpose, and over the smooth surface of a projecting rock, with all the ease of a fly moving over a ceiling. So active were the little ones as to defy, unless the hand was moved with extreme rapidity, their capture. Vast numbers of these little Eels are in the habit of proceeding up the rivers in the spring-time. In some places in England they are called Elvers. They are caught in immense quantities, and scalded and pressed into masses termed Eel- or Elver-cake. When dressed these little Eels afford a luxurious repast. Towards the latter part of summer these fishes migrate towards the sea, being capable of living in fresh or salt water with equal ease, the mouths of rivers constituting favorite localities. Even in our seaport towns and marine watering-places the common river Eel is caught by those who are angling in the sea for fish.

Various modes of capturing Eels are adopted by man. Bobbing, or clodding as it is sometimes called, is a very common and successful method, consisting in bunching a number of earthworms upon a worsted string, and lowering it near the place where the fishes are supposed to be feeding. So eagerly do the voracious fish seize the bait, and so fiercely do they bite, that they are pulled out of the water before they have time to collect their thoughts and disengage their teeth from the string. Night-lines, which are laid in the evening and taken up in the morning, are another plan. But the most successful method is by spearing. The spear used for the purpose is not unlike the conventional trident of Neptune, except that the prongs are four in number, flattened, slightly barbed on each edge, and spread rather widely from their junction with the shaft. This is pushed at random into the muddy banks where the Eels love to lie, and when one is caught, its long snake-like body is wedged in between the jagged prongs and lifted into the boat before it is able to extricate itself. Almost any kind of food that it can master, whether aquatic or terrestrial, is eaten to satisfy the creature's most voracious appetite. Even mice and rats fall victims to its hunger, and an Eel is recorded to have been found floating dead on the water, having been choked to death by a rat which it had essayed to swallow, but which proved too large a morsel for its throat.

So remarkable is the tenacity of life which this fish possesses, that after the creature has been cut up into lengths, each separate piece will move as if alive, and at the touch of a pin's point will curve itself as though it felt the injury. When all irritability has ceased, the portions will flounce vigorously about if placed in boiling water, and even after its influence has ceased will, upon the addition of salt, jump about as vigorously as before. There can be no real sensation, let it be understood, as the spinal cord has been severed and all connection with the brain, which is the seat of sensation, has been cut off.

How the Eel reproduces its kind has long been a subject of discussion. Some held that the young is produced in a living condition, and others that it is hatched from the egg. The matter has, however, been set at rest by the microscope, which shows that the oily-looking substance, generally called fat, which is found in the abdomen of the Eel, is really an aggregation of eggs, and that these objects, minute as they are, and which are not so large as the point of a pin, are quite as perfect in their structure as the eggs of a moth or a bird are seen to be to the naked, unaided vision.

Anguilla rostrata, as the Common American Eel is technically known, is abundant in the United States, living in fresh-water streams, but depositing its eggs, often eight millions to a single fish, in the ocean, the young ascending the rivers. Eels are devoid of ventral fins. Their scales, which are very minute, are covered with a thick, slime-like material. Under the microscope each scale is beautifully ornamented, and the exquisite pattern formed by the scales on the skin may be readily and effectively seen if a bit of it, when fresh, be placed on the window-glass and allowed to dry. The sexes are difficult to distinguish; the females have the highest dorsal fin, smaller eyes, and a lighter color than the males, while the snout is generally broader at the tip.

When contiguous to the sea, as in a pond near Wells, on the coast of Maine, the Eels invariably go down into salt water at night. As the connecting stream is narrow, the sight is remarkable, thousands filling the channel, many of whom, when alarmed, leaving the water and passing over the dry rocks to the ocean. Eels are not the silent creatures which many persons suppose them to be. They frequently utter a sound, expressed by a single note, which is more distinctly musical than the sounds made by other fishes, and which has a clear metallic resonance. They are of slow growth, scarcely reaching the length of twelve inches during the first year, but subsequently attaining to large dimensions, the preserved skins of two Eels, which Mr. Yarrall saw at Cambridge, England, weighing together fifty pounds, the heavier being twenty-seven pounds in weight.

Fish, as a rule, do not live more than a few minutes out of the water. An Eel, however, will remain alive for many hours, and even days, in atmospheric air, provided it is laid in a damp place. Now, if one be carefully watched when placed upon dry land, it will be observed to pout out the cheeks on both sides of its face. Underneath this puffed-out skin will be found the gills, and the skin which covers them will be seen to be so arranged as to form a closed sac, which the Eel fills with water, and so keeps the gill-fibres moist. This wonderful contrivance enables the Eel to come out of the water, and to travel, so to speak, by land. Thus Eels are often found in outlying ponds of human construction, where they were never placed by the hand of man. Finding old quarters uncomfortable, they take in a good supply of water, and exchange them for the better, not by repeated leaps towards the water, as some fish are known to do, but by a smooth, uniform snake-like progression.

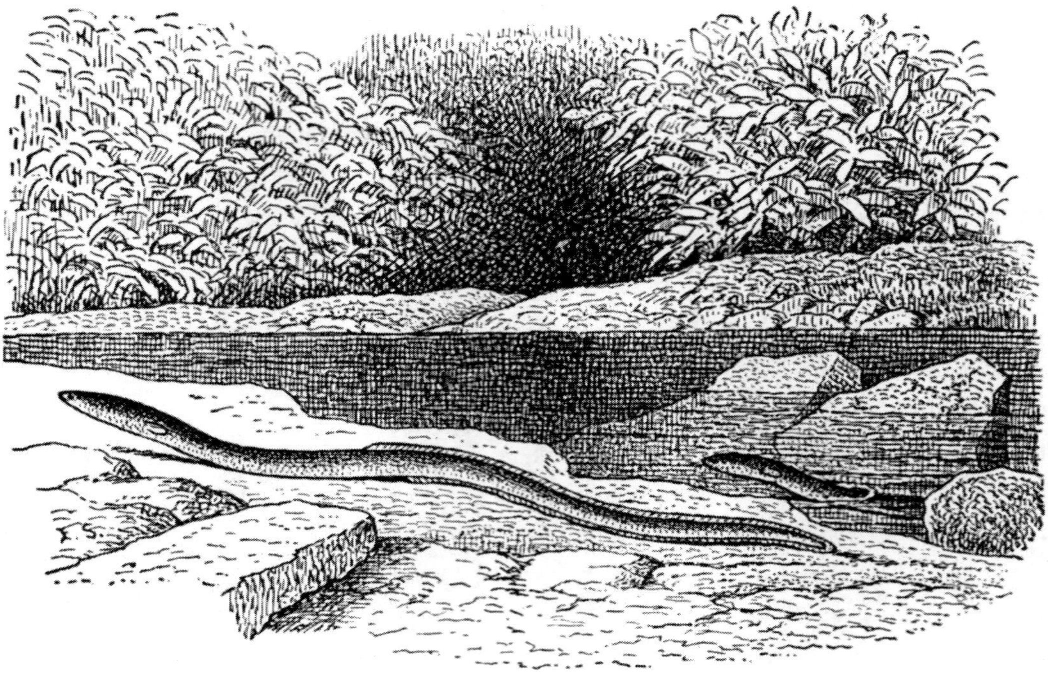

Figure 31. Common American Eel. How it seeks new feeding-grounds.

That some fishes should leave the water and travel overland is, perhaps, not more remarkable than that some birds, the ouzel for example, should leave their natural element and fly into and under the water. Whoever knows the hidden paths of the marsh has doubtless watched the brown-hued Eels wriggling their way through the grass from one pool to another, especially at night, leaving their home and wandering about, seemingly unconscious whither their pilgrimage will end.

"Slippery as an Eel" is proverbial. Many a person has, by his slick, cunning ways, succeeded in eluding the law and escaping justice, affording an apt illustration of the character of the animal about which we have been talking, but the slipperiness of the Eel is not given to it that it may take some unlawful advantage of its neighbors, but that it may the more readily slip from the grasp of a more powerful enemy, or the more easily make its way into the muddy depths of the pond or stream which it so very much affects. So it will be seen that while this slippery character in the one is protective, in the economy of nature, for a wise and laudable purpose, yet in the other it but secures to the possessor the getting of an ignoble gain and the ruin of a once proud name.

While these agile denizens of aquatic life are selfish and voracious almost beyond precedent, and apparently more concerned in feeding than in anything else, there are certainly some traits in their character which are redeeming features. Low as they are in the scale of piscine existences, occupying the very lowest family of the *Anguillidine Apodes*, they are none the less susceptible to the human influence of kindness. They grow accustomed to man when good is at the basis of his actions, and have been known to accept food from his hand. They remember the face of a friend, and when it is presented at the door of glass, so to speak, that opens the way to their home, they come without fear or suspicion showing itself in their movements. Even the sound of the voice of a benefactor awakens a sympathetic response in their bosoms.

Chapter 24

RANA AND BUFO

Belonging to the lower vertebrates is a family of animals called scientifically Ranidæ, but which are, popularly speaking, best known as frogs. They are queer-looking creatures, scarcely met with in Australia and South America, but reaching their highest state in the East Indies. They are capable of enduring great changes of heat and cold, and can live on land as well as in water, provided they have the amount of moisture necessary to preserve the suppleness of their skins. Salt water is fatal to the frog in any stage of its existence.

Rana clamata, the lusty croaker of the summer pond, is our most familiar species. He may be recognized by the colors of his dress, in which green, bronze, gold and silver play important parts, and by the ear-splitting character of his vocal intonations. The glandular ridges down the skin of his back, together with his strange coloration, singularly fit him for his home. Imitations of the stems of plants are seen in the darker ridges, and their leafage in the green color of his coat. The silver of his vest has the glimmer of the water in which he bathes, and the moist earth seems to have left its stain upon his brownish feet and markings, while the yellow of the several badges that adorn his person in being like the stamens and pistils of the surrounding flowers, and of the hue of many buds and blossoms, adds largely to his protective display. Thus is the frog in his natural haunts protected by his garments, and, unless he stirs or is betrayed by his full, bright eyes or the palpitation of his breast, he is not likely to be observed.

Four fingers or toes are found upon the anterior extremities, while those of the posterior are five in number and webbed. The front legs are much shorter, smaller and weaker than the hind ones, which are largely developed, and thus serviceable in swimming and leaping.

Though the frog is possessed of a back-bone, yet he has no ribs. Being ribless, he cannot expand and contract his chest in breathing, but must swallow what air he requires. In swallowing the air he must close the mouth and take the air in only by the nostrils; therefore, oddly enough, if his mouth is forcibly kept open, he will smother. The frog's breathing, a fact not generally known, is partly through his skin, which gives off carbonic acid gas; and moisture, therefore, is just as essential to his skin as it is to the gills of a fish. Damp, rainy weather is his extreme delight. When the rain falls, out come the frogs. Their skin absorbs moisture, which is stored up in internal reservoirs, and some of this water, when these timid creatures are alarmed by being suddenly seized, is ejected, but I do not think that it is purposely so done, as the water is not, as some people have fancied, of a poisonous nature. Frogs have no poison-sacs, and in truth no weapons of any kind.

Open a frog's mouth, and you will find but a few tiny teeth in the upper jaw and palate, which are useful for the partial grinding up of horny insects. His tongue you will discover to be a very odd affair, which is fastened at the front end of the mouth, the hinder part being free and hanging down the creature's throat. This organ is covered with a glue-like secretion. When an insect is to be captured, it is snapped forward from the mouth, and, striking the insect, which it seldom fails to do, causes it to adhere as to bird-lime.

A few thoughts now about the life-history of the frog. From egg to egg is the story. In roundish masses, upon sticks lying in water, or upon the leaves and stems of submerged water-plants, are the eggs deposited. The creature that comes from the egg is no more like a frog than a caterpillar is like a butterfly. It has a large head, small tail, branched gills, and is devoid of limbs, resembling, in this stage, more a fish than a frog. This is its early childhood, or tadpole state. It can only live in water now, and swims and feeds from the very moment it leaves the egg. Change in form almost immediately begins, the branched gills being drawn within the neck and hidden, a pair of fore-legs beginning to bud, and subsequently a pair of hind-legs, which push out much faster than the fore-legs. As the legs grow, the tail is gradually absorbed and disappears. The interior of the body meanwhile changes, the lungs and heart becoming reptilian. When the gills and tail are gone, and the legs are fully formed, the once-swimming tadpole hops out of the water a perfectly-formed frog.

When first the tadpole emerged from the egg, it ate the jelly-like cover. Then soft animal and vegetable matters, with the strengthening of its pair of horny jaws, began to be devoured. Insects later on, and even its own kith and kin, became its food. The fare of the adult frog is almost exclusively insect in character, although necessity sometimes drives him to make a meal out of some of his weaker brethren.

Seated in cool, leafy shadows, not far from his favorite stream or pool, the frog watches with his great, black, gold-ringed eyes for such insects as good fortune shall bring to his retreat. As one hovers near, out flies his limber, sticky, ribbon-like tongue, true to its mark, and the hapless insect, adhering to the viscid projected ribbon, is gently and cleverly deposited in the open throat, the frog maintaining all the while an air of calm, superior self-satisfaction, as if he had not so much satisfied an appetite as fulfilled the mission of ridding nature of a superfluous insect.

A most harmless, timid and interesting animal is the frog, and often most unfortunate. He is the legitimate mark for all the missiles that can be thrown at him by urchins wandering about his native pool. Snakes make him their prey, and he is always in mental fear lest some insidious serpent shall take him unawares, or his musings shall be suddenly cut short by the stately progress of some swan or goose, sailing over the limpid water, or searching the green herbage wherein he sits concealed.

Figure 32. *Rana Clamata*, Or Green Frog. Lusty Croaker of the Summer Pond.

That he is susceptible of being trained, there can be no question. Man is not always viewed by him as an inveterate enemy, nor does he always dive headlong into the pool when his presence is near. He has been known to cultivate man's acquaintance, and to live on friendly terms with him. Some three years ago a tiny frog was taken from a swamp by a friend and placed in a small stream of spring water that passed close to the house where the writer was summering. A dozen times a day the little frog was dipped up by the hand from the bottom of the stream, and forced to endure down the head and back the tenderest caresses. A few insects were then offered as food in conciliation for the liberty taken, which the little frog was only induced to accept after a great deal of persuasion, when he was carefully put back into his watery bath. In the space of a week, the frog had become so attached to his friend, that he would leap into his outstretched hand and take his food without the least distrust or fear. Even the voice of the master was recognized by the frog, and, when heard in the distance, was the signal for the strangest behavior. Froggie would leap out of the water upon a bare stretch of earth, peer off in the direction whence the sound came, and there await his master's arrival with restless anxiety. The strongest bond of friendship seemed to unite the two. Not only was the frog able to recognize the voice of his friend, but he knew him in person as well. Repeated efforts were made by the writer to gain the attention and good-will of the frog, but all his advances were received with the utmost indifference.

While the species which I have just described represents the aquatic Ranidæ, the Wood-frog, its near kin, represents a branch of the family which prefers dry situations, except in breeding times, when the eggs must be deposited in water. The Wood-frog is somewhat smaller than the Bull-frog, and is clad in olive-green and brown colors, which are in perfect keeping with the coloration of dead leaves and dry twigs. There is a large black patch on the side of the head around the big ear-drum, which seems still further to distinguish him from his cousin. He is a very shy and suspicious creature, and makes a prodigious jump at the first intimation of danger, his leaps being so enormous that it is very difficult to capture him. When upon the ground, he can hardly be discerned from the dry vegetation around. By hiding in damp moss or in decayed logs, and in little hollows in the ground, he is enabled to maintain the moisture of his skin. He avoids the sunshine, and keeps close to the earth.

Another curious Rana is the Tree-frog. He is smaller than any of his cousins, and may be known by his bright green dress, which is spotted with black, and by a membrane stretched between his toes, which gives him a broad, flat surface, while it helps to sustain him as he leaps, somewhat after the fashion of a flying squirrel, from branch to branch. In tropical regions, where many of the trees are bedecked with gorgeous blossoms, Tree-frogs appear in the gayest of colors, the splendor of their garb being protective in such surroundings.

Dressed in black and light brown, and living in marshes in the Eastern United States, is another species—the Swamp-frog. His voice is a prolonged croak, which, to the practised ear, can be readily distinguished from the bawl of Clamata, or the roar of the Bull-frog.

Cats, geese, hawks, vultures, owls and other animals eat frogs, and the luckless creatures can scarcely appear without finding an enemy. But nature, who is a very wise and considerate mother, provides a means for balancing this great destruction of their forces in endowing them with wonderful reproductive organs. So prolific are frogs, that when the little black tadpoles appear, so thickly are they huddled together that the pond seems literally alive with their swimming forms.

In the same class of animals to which the frogs belong, as well as to the same order, but to a different sub-order, are placed the toads, somewhat remote cousins of the frogs. As the frog is well-known about our ponds, so the toad is a constant denizen of our groves and gardens. The frog, you have been told, is a species of Rana, and now I shall introduce to you the toad as a species of Bufo. In general anatomy they are alike. Their eggs and young are closely similar, and the stages of growth from egg to adult form are nearly identical. When the adult stage is attained the frogs and toads are very tiny creatures, but, small as they are, they are readily distinguishable from each other by the conformation of the snout, and by the larger development of the hind-legs of the frog. Their chief differences will now be enumerated. The toad has no teeth, but the frog, as has been stated before, has teeth in both the upper jaw and the palate. Similarly attached is the tongue, but the free end of the frog's tongue is forked, and the toad's entire. The skin of the toad is usually warty, while the frog's is smooth. A rounder body, shorter hind-legs, less fully webbed feet and more rounded snout still further distinguish the toad from the frog. Their soft moist skin shows them to be Amphibians. The absence of tails places them among the Anuran, or Tailless Amphibians. Thus far they agree well together, but differences loom up upon careful examination, and we are compelled to say of the frog that he belongs to the Ranidæ, and of the toad that he belongs to the Bufonidæ. Of the two animals, the toad is by far the more interesting and useful.

Figure 33. Common American Toad. How he manages a difficulty.

The toad is almost unrestricted in his territorial range. He hops through the tropics and the temperate zones, and well up into the polar regions. Everywhere he is the same inoffensive, gentle, humble, useful and generally silent creature. But like his human brother he has his faults. He has a great fondness for bees. Happy is he when, brigand-like, he can stand by the highway of the bees and capture them as they return to their waxen city. Their wealth of honey he does not demand as a ransom, but swallows the little creatures themselves, alive and whole, and digests them at leisure. Bee-eating seems his only fault. Not only the hive-bee, but other insects as well, share his attention. Millions of noxious beetles and bugs are devoured, and the world is the richer by thousands of bushels of fruit and vegetables. The good he accomplishes largely outweighs the mischief he commits. So ceaselessly and swiftly he swallows his game, that a grasshopper's legs or a sphinx's antennæ may often be seen sticking out of his mouth, while the carcass itself is well down in his throat. French gardeners so appreciate his utility that he is brought to market and sold for a pittance to such as may need his services.

Toads can be tamed and taught to eat from the hand. They are easily beguiled with sugar and with bread that has been soaked in milk, but, like a captious child, they eat only the middle out of the slice, and leave the crust. We once saw a toad, a noble fellow he was, who, at a certain hour of the closing day, would come from his gloomy retreat to receive at the hands of man his supper of flies, which he had been trained to catch on the throw. So unerringly would his tongue dart out at the opportune moment, that he seldom, if ever, shot wide of his mark. It is amusing to observe him when, in his greed and haste, he has attempted to swallow a huge grasshopper whose legs will not accommodate themselves to his peculiar gape of mouth. How he swallows and twists and contracts the walls of his throat, but the legs seem unmanageable. He does not give up, or endeavor to eject the half-swallowed body, but ponders the matter over and over. A look of delight beams out of his eyes, that shows he has managed the problem. Up goes to the mouth the right fore-leg, and, in less time than it takes to chronicle the event, the obstreperous insect is pushed into the stomach.

Some curious myths are told of the toad. One says he can live for hundreds of years encased in clay or in stone. No more true of Bufo is this than of Rana, his cousin. Another asserts that his skin, when handled, is productive of warts, and that the fluid he emits, which serves but to moisten his body, for without moisture he could certainly not live, and to protect him from enemies, is poisonous in character. His power to produce warts, we cannot admit. But that the fluid he exudes, if not poisonous to touch, is offensive to animals, there can be no doubt. We are led to this conclusion from the following considerations: Dogs, young animals especially, are prone to attack the toad, but they never want to repeat their experience. The toad's exudation so affects the salivary glands of the dog as to cause him to froth and foam like an animal with rabies. A case is recalled where a dog, that had taken a toad in his mouth, became almost frantic. This dog never afterwards was well. His whole system apparently had become diseased, and, in less than a year, he had wasted to a complete skeleton, when death relieved him of his sufferings. Another allegation, that the toad has a jewel in his head, has been believed from very ancient times. The story doubtlessly originated from the beauty of the toad's eyes, the irides being a rich flame-color, which, in the dusk of the even, shine like burnished gems.

When hatched the young of the toad are of a jet black color, and are very active. Their changes are made very early and in the same manner as those of the frog. They are quite small when arrived at the perfect toad state. Their legs produced and their tails absorbed, they quit the water and set out on long journeys. Unlike the frog, which is a home-stayer, the toad is a born vagrant. They travel chiefly by night, hiding under stones and herbage during the day. If clouds cover the sky, they take heart and joyously hop forth upon their pilgrimage. During a long drouth they mysteriously disappear, but if a rain comes on they suddenly come out by hundreds, and this has given rise to the tale of a "shower of toads."

Worms, as well as flies, etc., constitute a toad's bill of fare. After a rain toads and worms, it would seem, are mutually inspired to take their walks abroad, and many an unfortunate worm makes its way into the toad's maw. Dead insects are at a discount with him, and he views with suspicion anything that shows not the active wriggling principle of life. When winter comes on the toad, like the frog, goes into winter-quarters. Since the young toad reaches its adult size in the autumn, it is forced to pass the first period of its grown-up life in a sleep, or coma, in some hole or burrow which it has found or fashioned in the earth. Sometimes toads creep into rock-crevices, or into hollows in logs and trees, and being found in these places in the early spring are hastily supposed to have been prisoners for many years.

In the process of growth the skin of the toad, as well as that of the frog, becomes too small, and hence is cast off. As the shedding-time approaches, the white, green and brown colors become dull, and a peculiar dryness appears. A new skin is now forming under the outgrown one, and presently the latter splits half down the middle of the back and along the under part of the body. By a series of violent twitchings of the toad the old skin is made to wrinkle and fold along the sides. A hind-leg is now tucked under a fore-arm, and by a good pull the animal is soon out of that leg of his trousers. The other leg is removed in similar fashion. Putting one hand in his mouth and giving a jerk, off comes the covering of that hand and arm, like a discarded glove. He has now but to take off the other, and he is free. Relieved of his dress he neither sells nor gives it away, but rolls it up into a neat solid ball and swallows it. The frog strips off and disposes of his outgrown skin in a similar way.

Strange to say, toads and frogs can change to some extent the color of their skin to suit their homes. Kept in the dark with dark surroundings, toads become darker in color, while those that are kept in light with white accessories become lighter. The color of the toad changes more slowly than that of the frog. It is not the arrangement of the color that alters, but merely a change from light to dark.

What has been said applies to our Common American Toad, the Bufo Americana of the books. Let us now look at some curious specimens of the Bufonidæ. The Pipa, or Surinam Toad, does not lay her eggs in water, but places them on her back. A fold of skin rises up and encloses each egg in a separate cell, until the young have not only been

rarely exceeds twenty inches, and he is stout in proportion. The Moccasins show the same curious care for their young already mentioned. A low, blowing noise apprises them of danger, and into the slightly-opened mouth of the mother, which is held close to the ground, they hurriedly disappear.

One of the commonest of the non-poisonous snakes is the Striped Garter Snake, ten species of which being known in the United States. Upon the earliest appearance of spring they are almost the first to roll out of their holes, where they have lain dormant in balls or clusters during the cold winter months. Though easily excited, and striking quickly, yet their bite is little more than a scratch. Their appetites are now quite vigorous, and they have been seen to chase a toad for more than fifty feet over a gravelly road, effecting its capture. They are remarkably prolific, and their numbers about pools are sometimes astonishing. It would seem that they are viviparous as well as oviparous, from the fact that some young ones have been free and others in sacs in the abdomen of the mother. With a brood of forty or fifty young, which a single female has been known to produce, it would seem that the Striped Snake would have a difficult time in protecting her offspring by taking them into her mouth. They have this habit, however, as abundance of evidence could be adduced to show. One witness observed a Striped Snake upon a hillside, and noticed something moving about her head, which proved to be young snakes. He counted twenty little ones from one and a half to two inches long. Led by curiosity, he made a move towards the spot, when the old one opened her mouth, and they went in out of sight. He then stepped back and waited, and in a few minutes they began to come out. Another witness came across a female with some young ones near her, who, perceiving him, uttered a loud hiss, and the young ones jumped down her throat, when she instantly glided away to a place of concealment beneath a huge heap of stones.

The Black Snake, *Bascanion constrictor*, the mortal enemy of the Rattlesnake, is a familiar species, and one that is widely distributed. As winter approaches, these snakes come from far and near to some apparently appointed place of rendezvous, where, rolling themselves up into a matted ball, they sleep the days and nights of winter away, and come out in the spring-time, when the common mother of us all has conditioned things to their habits and ways of life. In appearance, from a decorative point of view, they are very attractive, being of a uniform steel-blue color, with a rich tessellated arrangement of scales. They are of wild and untamable natures, powerful and active as foes, often engaging in encounters with other snakes, especially the Rattlesnakes, whom they kill or force to disgorge their prey. In their movements they are so rapid that they are often called the Racer. It is in the breeding season that they manifest their greatest boldness, and have often been known to go out of their way to attack a passer-by. They will chase an intruder for a long distance, and will even descend a tree to attack the one who is teasing them.

It is the Black Snake that appears the most frequently in the guise of a charmer. But, as has been remarked before, this power, so often imputed, is merely imaginary. The reptile preys upon birds in their nests, penetrating the thickets in quest of them, and often the cat-

hatched, but have also passed through all their metamorphoses, and come out fully formed. Another toad, abundant in Europe and Asia, is largely colored with bright crimson, and the father-toad carries the little ones in separate cells fastened to his hind-legs like chains. The young change to their perfect shape in these cells, and with the withering away of the cells the young toads hop out, able to take care of themselves.

Somewhere I have said that toads are generally silent. A little toad about three inches long, called a Natter-jack, is common in England, and is a noted singer. His "gluck-gluck, gluck-k-k," can be heard any night. The Green Toad, well known on the Continent, is not so noisy as the Natter-jack, but has a low, moaning cry.

All the Tree-toads, or Hylidæ, have clear, shrill voices, and are fond of singing serenades. In the spring the Common Toad takes to the water and there sings very loudly. The loud continuous trill that we hear in swamps in spring-time is made by toads, and not by frogs, as is generally believed. Another toad with a voice is the Spade-foot. This Toad is rare, though widely distributed. It is remarkable for its feet, formed for digging, its subterranean habits, and its queer way of appearing and disappearing very suddenly. After a rainy season the Spade-foot will emerge from its hiding-place, attract attention by its loud cries, swarm by hundreds about the ponds, lay innumerable eggs, and vanish. But while thousands of eggs are laid, scarcely any hatch, for most of them perish from being laid so near to the water's edge as to become dried up on the subsidence of the water.

Thus we find that toads have three different methods of life. Some live on trees, but seldom appear upon the ground. Others are underground dwellers, and hardly ever come to the surface. But the Common Toad, and his numerous kin, are dwellers in the ground, hiding among grass and other herbage when asleep, or when the sun is too intense for their comfort. But all toads, excepting the two varieties mentioned above, which carry their young on their bodies, repair to the water to drop their eggs, and the young live in the water until they have attained the adult state.

Chapter 25

Our Natural Enemies

No animal, perhaps, is so little known and understood as the snake. This is not because its study has been neglected or overlooked, as our scientific institutions are replete with fine collections of most of the reptiles, and exhaustive works upon their habits and customs. Yet, notwithstanding this, the snake continues to be the subject of ever-recurring stories, fabulous in the extreme, that seem handed down from generation to generation. Strange to say, many of these stories are current among those who, from the nature of their lives, would be expected to be well and accurately informed on the habits of the animals. Farmers and horticulturists are plentiful who religiously believe that the Milk Snake, the beautiful *Ophibolus clericus*, deprives milk-giving animals of their supply of milk. A statement often seen, and that has many believers, is that the Whip-snake of the South seizes its tail—which is supposed to have a sting—in its mouth, and rolls away in the form of a wheel, stopping suddenly and striking the enemy with the sting. Such fables are current by the score, and denial only strengthens belief.

More than a hundred species of snakes, nearly all having a wide geographical range, are found in America, north of Mexico. They constitute the first order, Ophidia, of reptiles, and have long, cylindrical bodies, are footless, without a shoulder-girdle, and invested with a coat of scales, which is shed in the summer months. Snakes have no eyelids in the strict sense of the term. Their eyes are permanently covered by a delicate membrane that takes the place of the lid, and this explains the stony stare, so disagreeable to many, that all snakes have.

The skeleton of snakes is so arranged as to allow the greatest freedom and flexibility. Numerous pieces of bone, hollow in front and convex behind, make up the long tapering backbone, which literally works on a ball-and-socket plan. Articular facets, that lock into each other, are found upon the processes of the vertebræ, and these strengthen and give to the backbone a greater degree of flexibility. A more remarkable arrangement, however, is found in the head, which enables the snake to prey upon animals that are larger than itself.

The jaws seem a combination of elastic springs, having no gauge to their tension, the quadrate bones connecting the lower jaw with the skull being movable, thus allowing that enormous gape with which all are familiar who have seen a snake swallow its prey. Besides this, the bones of the jaw itself and palate are more or less movable, also tending to the larger distention of the throat.

As snakes do not tear or mutilate their prey, their teeth are not set in sockets, but serve merely to poison and stupefy the prey, or to prevent its escape, acting as hooks by which the body is hauled over the victim. The bones of the lower jaw, as we have seen, are not fastened closely to each other; so in swallowing prey the teeth on one side are advanced, and then those on the other side, and so on until the victim is hauled, hand over hand, as it were, into the snake's throat.

Poisonous snakes, such as the rattlers, have two long, sharp fangs, each compressed and bent up, and forming a hollow tube, open at both ends. The upper portion of the hollow fang is fastened to a bone in the cheek, which moves with ease, so that, when not in use, the fangs can be packed away until needed.

All animals, man included, have doubtless in their saliva a deadly poison, though in the latter it is extremely diluted, and essential only to the digestion of food. In poisonous snakes, however, it is stored up in sacs, modifications of the salivary glands, and placed in each side of the upper jaw. From the poison-gland under the eye forward to the edge of the jaw, a delicate canal, which opens into the fang above the tube of the tooth, extends. Alongside of the latter may be seen rudimentary fangs, all ready to grow out should the large one be lost. To use the poison, the snake has merely to strike its prey, when the muscles of the jaw, which are admirably fitted for the part they have to play in the tragedy, press upon the glands, squeeze the poison through the little canal down through the hollow fang, and the work is accomplished.

In their actions, snakes are most graceful. The gliding motion, so characteristic, is effected by the movements of the large central scales, that are successively pushed forward, the hinder edges resting on the ground and constituting a support. These scales, or pushers, are fastened to the ribs by muscles, and by holding a snake by the hand the swelling movement can readily be felt.

Snakes vary much in color. They are generally adapted to their surroundings. Green Snakes are found in green grass and vegetation, while grey snakes affect rocky districts, where they are alike protected. Their skin is shed in one piece at various seasons of the year, being forced off by the snake forming a ring with its tail and squeezing the rest of the body through it, or by wriggling through entangled bushes. Poisonous snakes may be always recognized by their broad, flattened heads, generally short and thick bodies, and the almost invariable possession of a vertical keel along the centre of each scale. Long bodies, small heads devoid of distinct necks, and scales not keeled, characterize non-poisonous species.

Probably the best-known of our common kinds of poisonous snakes are the rattlesnakes. They belong to the dangerous family Crotalidæ, to which the copperheads and moccasins also belong, and are distinguished by the large, ugly head, absence of teeth in the upper jaw excepting the fangs, and the pit in the head.

Figure 34. Northern Rattlesnake. Prepared to attack a Song Sparrow.

Crotalus horridus, our Northern Rattlesnake, has doubtless the widest geographical distribution, being found in nearly every State in the Union, from the Gulf of Mexico to Northern New England, and thence west to the Rocky Mountains. It has a most forbidding appearance, and when once seen with its enormous head, triangular in shape, and large brilliant eyes, with fiery irides, it can never be mistaken. Between the eye and the nostril is a deep pit, a character that is peculiar to the family.

All rattlers, as the name indicates, have a horny appendage to the tail, formed of separate button-like objects, that rattle together when the tail is vibrated. This rattle not only serves to warn human beings of danger, but also to arouse in animals a curiosity that often proves fatal. The popular belief that a rattle is added every year, and that it is possible to determine the age of the animals by this means, is not borne out by facts. Sometimes two rattles are known to appear within a year, and other instances are recorded where four have been attained in that period, and others still when several have been lost, new ones taking their places. The number of rattles is also uncertain. The greatest number, as observed by Dr. Holbrook, is twenty-one, but a specimen is mentioned in the books that had forty-four.

Mild and peaceful in disposition, the Rattlesnake has never been known, unless provoked, to attack a human being, nor to follow him with hostile intention. He preys upon small animals, as rats, squirrels, rabbits and birds, and can always be approached when he is stretched out, only striking when he is coiled. He is not a climber, seldom, if ever, being found in trees. His alleged powers of fascination are purely mythical. The horror his presence inspires often paralyzes with fear his victim, who, incapable of flight, stupidly awaits his fate. Men, women and children have been known, when attacked by these animals, to become rooted to the spot, as it were, by fear and surprise. All the so-called cases of fascination can be explained by the fear which the snake's unlooked-for presence inspires.

Wonderful curative powers are imputed to the oil of the Rattlesnake. Many snakes are killed during the summer months for this oil, but the grand gathering of the crop is in the fall, when they have repaired to their dens and wintering places. Sunny days in October and November are chosen by snake-hunters for raiding them. The snakes, dull and sluggish at that time of the year, crawl out of their dens upon the rocks, huddling together by the score for the purpose of basking in the sun. Armed with old-fashioned flails the hunters, when they come upon a group of snakes, proceed at once to thresh them, but few making good their escape. The Rattlesnakes, assorted from other species that are frequently massed together with them, are carried home, when the oil is simply tried out, bottled up and is then ready for the market and the credulous patient.

No subject connected with snakes, it would seem, has attracted so much attention as the vexed one as to the care which they take of their young. Snakes would hardly be expected to show any great amount of maternal affection, but that they do, and in a most remarkable manner, by taking their young into their mouths, if alarmed, is a well-established fact. The mother, when danger is imminent, sounds her rattle as a signal, opens her very large mouth, and receives in it her little family.

The bite of nearly all rattlesnakes is extremely dangerous, though not necessarily fatal in the smaller kinds. Almost all animals succumb to their bite, and even man himself, if the proper remedy is not at hand. There is a general belief that the hog is exempt, and acting upon this belief farmers have been known, where these reptiles are very abundant, to turn in a few hogs upon them for their destruction. This animal, though it has a fondness for the reptile, and exercises a great deal of caution in its attack, has not infrequently been killed by the reptile's poisonous fangs. Large doses of whiskey have been successful in neutralizing the effects of the poison, but it has been practically and experimentally proved that permanganate of potash is the best antidote.

But of all the poisonous snakes of this country, the Copperhead, *Ancistrodon contortrix*, is the most dreaded. In the South, he is known as the Cotton-mouth, Moccasin and Red-eye, and is just as common in the Gulf States as in the Atlantic and Middle States. He attains a length of two feet, is of a hazel hue, the head having a bright coppery lustre, and loves to conceal himself in shady spots in meadows of high grass, where he feeds upon small animals, rarely, if ever, attacking large ones unless trodden on. The mother Copperhead has also been observed to shelter her young in her mouth when threatened by danger.

Figure 35. Mother Black Snake. Her affection for newly-hatched young.

Ancistrodon piscivorus, the Water Moccasin, that commands so much respect from the negroes of the South, is, from the pugnacity of his nature, equally to be feared. While the Rattlesnake will slink away from danger, the Moccasin will attack man or brute with savage ferocity. He is essentially a water-snake, chasing fishes and small reptiles in the streams of his native haunts, and may be recognized by the dark-brown colors on the upper portion of the head and the yellowish line that passes from the snout to or over the nostril. His length

bird and the red-winged blackbird, which are so prone to attack, are seen acting strangely, crying and fluttering before the reptile in fear and rage, while thus *charmed*, and frequently falling a victim in their endeavors to protect their young. At such times the cries of distress of the old birds bring birds of different genera together, who join their forces against the common enemy, finally compelling him to retreat. Like other snakes mentioned, the Black Snake has the same remarkable habit of taking her young into her mouth for protection.

Figure 36. Summer Green Snake. Manner of mimicking a vine.

Among the most attractive forms are the Green Snakes. *Leptophis æstivus*, so common in the South, and occasionally to be met with in Southern New Jersey, is of a brilliant green color, and so perfectly mimicking a vine that it would rarely be taken for a living creature when lying around the branches of a tree. They have a habit of coiling in the nests of birds, often surprising the egg-hunter by bounding swiftly away. Allied species, further to the South, have been observed, when approached, to leap twenty feet in the air, falling to the ground and making their escape. They are perfectly harmless creatures, and, like the Green Snake of the North, can be handled with not the slightest risk of danger. We once knew a gentleman who had one in confinement, whom he had trained to eat from a dish and to come to his hand at the sound of his voice. The beautiful creature, which was a female, showed the most marked affection, and would often twine her little form about his neck or glide her smooth head, lazily as it seemed, along his face and forehead.

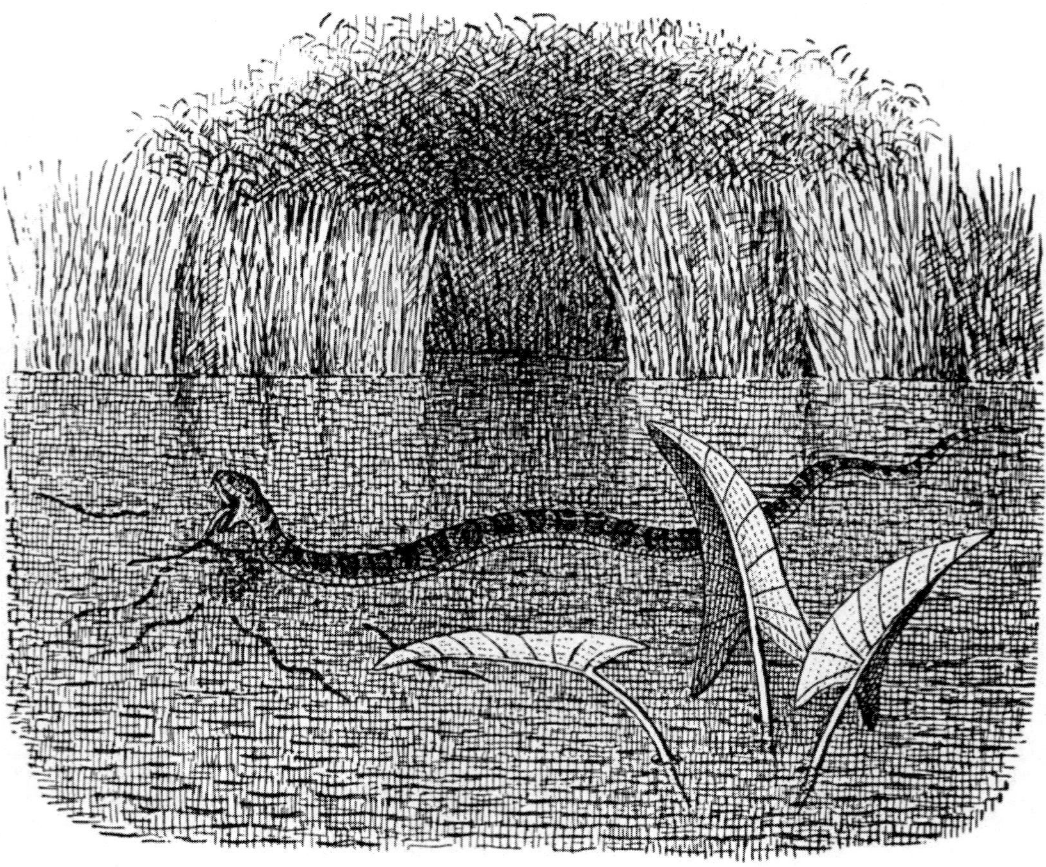

Figure 37. Water Snake. Swallowing her young.

An extremely common snake in the Eastern United States is the Water Snake. *Nerodia sipedon* is the name by which it is known to the naturalist. There is in Michigan an allied form, known as the Red-bellied Water Snake, which is quite as common, while several other species abound in other localities. They are all inoffensive creatures and prey upon small animals. The female shows the same regard for her young as other kinds, suffering them, even when three or four inches long, to take shelter in her throat, when she will clumsily turn in search of some place of concealment.

Water-snakes generally affect water-courses, often hanging from the branches of trees over streams, into which they drop when disturbed. Dr. Bell, an English naturalist of distinction, once tamed a European species of this genus. This pet could distinguish him among a crowd, and would crawl to him, passing into his sleeve, where it would curl up for a nap. Every morning found it at the doctor's table for its share of milk. For strangers it had an aversion, flying and hissing at them when any familiarities were attempted.

Were these grovelling creatures better known, there would be found much in them to admire and commend. They are not the hideous beings they are represented to be. The feeling of hatred against them, an instinctive and unappeasable enmity, is perfectly natural, and has grown out of religious superstitions. Fear, disgust and aversion are man's experiences at the sight of a snake, and there is at once a disposition to seize a stick or stone, or to make use of his heel, if well protected, to deal a fatal stroke. War to the death seems to be the cry between the highest of the mammals and the serpent tribe. It is not at all surprising, therefore, that the snake, seeing a human enemy, should either glide hastily off into the bushes, or, being thwarted, should coil itself up and hiss or throw itself forward in attack. Man would do well to protect the snakes about his domains, and treat them as friends, for they do him invaluable service in the destruction of vermin that make havoc with his crops.

Ants, bees, spiders, and many fishes, animals that are lower down in the scale than the snake, it is claimed, show far more forecast, ingenuity and architectural ability than it, but asserters of such an opinion forget that the snake is never studied under favorable conditions. Long ages of persecution have made him fearful of man, from whose presence he flees as from a pestilence or scourge, and there is consequently no chance to learn his better nature. Even man, until recently, has shown no inclination to make his acquaintance, being controlled by a dread which it appears well nigh impossible to overcome. Where the animal has been made to partake of the milk of human kindness, and has learned to regard man as a friend and not an enemy, he has shown remarkable susceptibility to culture and enlightenment. Let it be hoped that a modicum of the wisdom which has been attributed to him from the earliest of times, when he was made the object of homage and the insignia of the physician, shall at least be found to remain to the credit of science and truth.

Chapter 26

HOUSE-BEARING REPTILES

Turtles are four-legged reptiles, with short, stout, oval-shaped bodies encased in bony boxes, from which they are able to protrude their heads, legs and tails, and into which they can withdraw them, at pleasure. Considerable diversity exists in the size and shape of the box-like covering in the different species. The Box Tortoise can retire into his shell or house, closing the under part or plastron into a groove of the upper edge of the carapace, as the upper part is called, thus constituting for his security an impregnable retreat. There are species only partly enclosed by the shell, which cannot bring their heads and feet under cover.

With his house upon his back the turtle wanders about as the snail does, and against his enemies can close its doors and be emphatically not at home. He has acute sight and hearing, but is devoid of teeth, the jaws being, like those of birds, simply cased in horn. Turtles are not altogether silent creatures, for many of them are capable of producing very loud sounds.

Their eggs, which have a parchment-like covering, are buried in earth or sand, and left to themselves to hatch. The sea-turtle, our largest variety, is sometimes found to lay as many as two hundred eggs in a heap, and in tropical regions has been known to attain a weight of a thousand pounds. Even on the Atlantic Coast of the United States individuals, weighing upwards of eight hundred pounds, have not infrequently been captured.

In the four species of sea-turtles, the feet are flat and paddle-shaped, and the shell of one rather leathery than horny. Some of these marine forms are carnivorous, living on fish, mollusks and crustaceans, while others are strictly vegetarians, subsisting only on roots and the various sea-weeds. The flesh of some is rich and delicious, and a favorite and costly article of food, but of others it is coarse and ill-flavored, and necessarily not edible. The eggs, however, are always sweet, good and wholesome food. Valuable articles of commerce, such as boxes, cases, knife-handles, jewelry and other delicate ornaments, are made from the shell, for it is susceptible of a very high polish, which brings out with surprising clearness its rich brown and golden shades and markings.

Next to the sea-living turtles, come the fresh-water species, which eat both animal and vegetable foods. They enjoy much better than aught else a bed of soft mud, their heads lifted above the surface of the stagnant water, their long necks moving snake-like as they gulp in mouthful after mouthful of air. They are generally gregarious in habits, large numbers often being found huddled together in the sun on logs or banks, close to the water, into which they quickly slide upon the first intimation of danger. Timid as they are, yet they will snap and bite most furiously when taken in the hand.

Salt- and fresh-water terrapins are varieties of turtle, although some scientists restrict the latter term to marine animals that do not hibernate, and that cannot draw their head and feet inside the shell. The tortoise never goes to sea they say, can draw himself within his shell, although the Box Tortoise only can close the shell fast when thus withdrawn, and finally, that the tortoise hibernates. Some of the best and latest writers on the subject call all these animals turtles, applying the name tortoise only to the familiar Box Tortoise of the wood.

Awkward as turtles appear in their box-like covering, yet they can walk rapidly on land, are climbers of some note, and all are able to swim. The head, neck, and legs of a turtle are of a bronze, blackish green, or deep-brown color, and the shells are beautifully marked, glossy, ridged, or carved, and made up of closely-united, many-sided plates, arranged upon a thickened, lighter-colored and apparently uniform bony plate, which is capable of being separated into many independent pieces. The shell, or epidermic covering, is not brittle and lime-like, as the shells of all mollusks are, but is of the nature of horn. In general the plastron is of a lighter color than the carapace, being light-brown, yellow or cream, with yellowish lines dividing the plates, and with bordering bands of red, yellow and purple. The upper plate is usually of a very dark color, marked and lined with darker and lighter tints, and often displaying a bevelled yellow edge. *Chrysemys picta*, the Painted Turtle, receives his name from the beauty of his many-colored shell, while the Spotted Turtle, *Nanemys guttatus*, which is often called the Wood Turtle, is distinguished by the round yellow spots that are regularly distributed over his dark-colored carapace.

But of all our turtles none is so well known or so interesting in his ways as the Common Box Tortoise—*Cistudo clausa*. He affects dry woods, and dislikes the water, and is a long-lived creature, some individuals having been known to live more than a hundred years. Box Tortoises in confinement have been found to eat meat, insects and bread and milk from the hand, but if berries were put into their mouths they wiped them out in a very funny manner with their front feet, which they used after the fashion of a hand.

Figure 38. Box-Tortoise Feeding On Fungus. Copyright 1900 by A. R. Dugmore.

When foraging in the woods, especially during the rainy season, at which time manifold varieties of fungi prevail, they make their meals largely upon these plants. We have seen a huge toadstool that had been gnawed off so evenly, the central pillar only being left intact, that appeared as though it had been cut away by a knife. This had been the work of the Box Tortoise, for on looking around we soon descried, moving leisurely over the leaf-strewn earth, the creature himself making a fresh attack upon another species in a little opening in the woods.

Very amusing it was to watch him, as with praiseworthy deliberation he ate round after round of the cap of the fungus. He would bite off a mouthful of the toadstool, chew it carefully until he had extracted the whole of the juice, then open his mouth and drop out the masticated fibre, and take a fresh mouthful, not biting inward toward the stem, but breaking off the morsel next beside that which he had just eaten. He paced round and round the fungus as he took his bites, and as the fungus decreased in regular circles, the chewed fragments increased. In less than an hour he had eaten all the disk of the fungus to the stipe, and then walked slowly away to seek for another. The discarded parts of the fungus appeared quite dry when examined, nothing nutritious being left in them. There must have been some very good reason for rejecting the central part and the stem, which were left in every instance, but what that reason was we could not imagine. If a decayed or wormy portion of a toadstool was encountered in the feeding process, he did not bite round it, but abandoned the plant altogether, and went off in quest of a fresh specimen.

Figure 39. Common Box Tortoise. Breakfasting on a Toadstool.

Coming, in his travels, to a steep gully or ravine which he desires to cross, he does not attempt the undertaking without counting his chances of success. He seemingly revolves the matter over and over for some time in his mind, and, when at last he has reached a conclusion, draws his head and feet under cover, and by some quick, sudden jerk flings himself down to the bottom, trusting to good fortune and his own wits to making his way over the further incline. Observation teaches that his deliberations are generally attended with the accomplishment of the result to be attained.

There is a very common turtle, quite abundant in the small lakes and streams of our Western States, where he is trapped in great numbers for the market, which country people dub the Snapping Turtle, or which, from the resemblance which the head and neck, when stretched out, bear to the same parts of the alligator, takes the name of Alligator Turtle, or *Chelydra serpentina*, with the more learned naturalist. He has a shell too small to close over him and hide him completely, but nature, to make up for this deficiency of covering, has given him a bold and hasty temper, which leads him to snap vigorously when disturbed. Snapping Turtles live rather harmoniously together, even when confined in the same pen, and only manifest their ugly dispositions towards each other when excited by causes from without, with whose origin they have nothing to do. Contests of a very vicious character are often thus precipitated, which sooner or later end in the death of one or more of the belligerent parties.

Down in the pine countries of our Southern States lives a large, stout animal, with a shell fifteen inches in length, which is denominated the Gopher, or Testudo Carolina. These animals dwell in troops, several families digging their dens or burrows near together, the entrance thereto being about four feet long and expanding into a spacious apartment. In each burrow resides a single pair of Gophers. By day the Gophers keep close house, but by night they wander out in search of food, devouring yams, melons, corn and other garden produce. They dislike wet weather, and always go in-doors when it rains. Gophers' eggs, which are as large as pigeons' eggs, and also their flesh, are highly esteemed as articles of diet by the negroes.

In Europe, a near cousin of the Gopher is kept about the house for a pet. If allowed, in the autumn, to find his way into a garden, he digs a hole and hibernates, coming out in the spring. An English lady had one of these animals which lived in the kitchen. He was fond of creeping into the fireplace and getting under the grate, where he would contentedly lie until the hot coal and ashes dropped upon his back and burnt his shell. When winter came this little creature wanted to take his long sleep, and dug so persistently into baskets, drawers, boxes and closets, that finally a box of earth was given to him, into which he worked his way until out of sight, and there he remained until April sun and showers called him from his winter retreat. His fare was potatoes, carrots, turnips and bread and milk, which he especially liked.

Chapter 27

Summer Duck

Perhaps no species of North American water-bird is more highly esteemed by lovers of the beautiful in nature than *Aix sponsa*—the Summer Duck, or Wood Duck—and, when obtainable, is one of the first to find room in the collection of amateur naturalists. With the epicure, however, he is of rather inferior standing, lacking as he does the delicacy of flesh which makes the green-winged teal and others of his tribe of such immense gastronomic value.

Though truly an American species, yet this bird is more generally found throughout the United States than any other, nesting wherever suitable localities present themselves. North of the Potomac River, and in the various States situated above the parallel of latitude which cuts its head-waters, at least so far as the country east of the Rocky Mountains is concerned, it is chiefly a migrant, arriving towards the close of March, or in the beginning of April. South of this line, from Maryland to Florida, and thence south-westerly through the Gulf States into Mexico, the birds are found in more or less abundance during the entire year.

Pairing commences in the South Atlantic and Gulf States, we are told, about the first of March, but in New England and the Middle States in favorable seasons from the first to the fifteenth of April, whereas in the country bordering on the Great Lakes and in the Western States generally, it does not take place till the last of May or the beginning of June.

Upon their arrival in our Northern States these birds, strange to say, unlike many of their numerous family connections, seldom frequent the sea-shore or the adjoining salt marshes, but manifest a predilection for the ponds, mill-dams and deep muddy streams of the interior; and the same is true in more southern latitudes, for they prefer to place their nests along the creeks and bayous of the land where the orange and palmetto charm the eye with perennial verdure.

Between the time of their appearance in March and the plighting of their vows at the accustomed trysting-places, the sexes consort together in flocks of four or more, but never in very large numbers, and fatten on acorns, the seeds of the wild oats, and such insects as they can procure from the tree-branches or the muddy borders of the streams and ponds which they so delight to visit.

On each recurrence of the mating season there is reason to believe that the same couple come together and pledge anew their fidelity and affection, unless debarred by death, or some other of the many vicissitudes to which life is prone. The troth-plight sealed, and a union effected, the happy couple soon start off in quest of a spot for a home. In the case of old birds the same locality, where no interference has been experienced from beast or man, has been known to be visited for four successive years. For obvious reasons, Wood Ducks delight to live in close proximity to bodies of water, such places affording conveniences to the young, when they are sufficiently matured to betake themselves thither, for food and exercise. Situations remote from water entail unnecessary labor upon the female, who is then required, at considerable risk and peril, to carry them one by one to the pond or stream in her bill. When the distance is not too great, and the ground underneath the nesting-tree is amply covered with dry leaves and grasses, the young scramble to the mouth of the nest, drop themselves down, and under the maternal leadership wend their way to the much-loved fluid. Often the tree or stub which contains their home is found to overshadow the water. All that is necessary then is for the tender little creatures, after reaching the entrance, to spread their ill-feathered pinions and oar-like feet and fling themselves down, a feat which can be performed without jeopardy to life or limb.

Almost any tree, or tree-branch, containing the essential hollow, and suitably located, is utilized. Broken branches of high sycamores, seldom more than forty or fifty feet from water, are, according to Audubon, favorite places, while Wilson claims to have met the home of a pair of these birds in a fork composed of branches, and built out of a few rude sticks. In the South, the forsaken retreat of the gray squirrel and the hole of the ivory-billed woodpecker are common nesting-places. Often the entrance to the nest is apparently so small when compared with the bulk of the occupant that it is a matter of surprise to many that she can manage to make her way into it without suffering bodily injuries. But she does, nevertheless, which is an evidence that she either knows how to conform to circumstances, or is a better judge of dimensions than many of the would-be-wise lords of creation. All nests of our finding have been wide enough at their mouths to admit of easy passage, and have been from four to six feet in vertical direction. Soft decayed wood, and a few feathers, doubtless plucked from the breast of the builder, were their only contents. Dry plants, down, and feathers of the wild turkey, wild goose and the common barnyard fowl, have been observed, in addition to the foregoing articles, by other writers. The height of the entrance above the ground varies from fifteen to thirty feet, but probably a less, or even a greater elevation, may sometimes be attained.

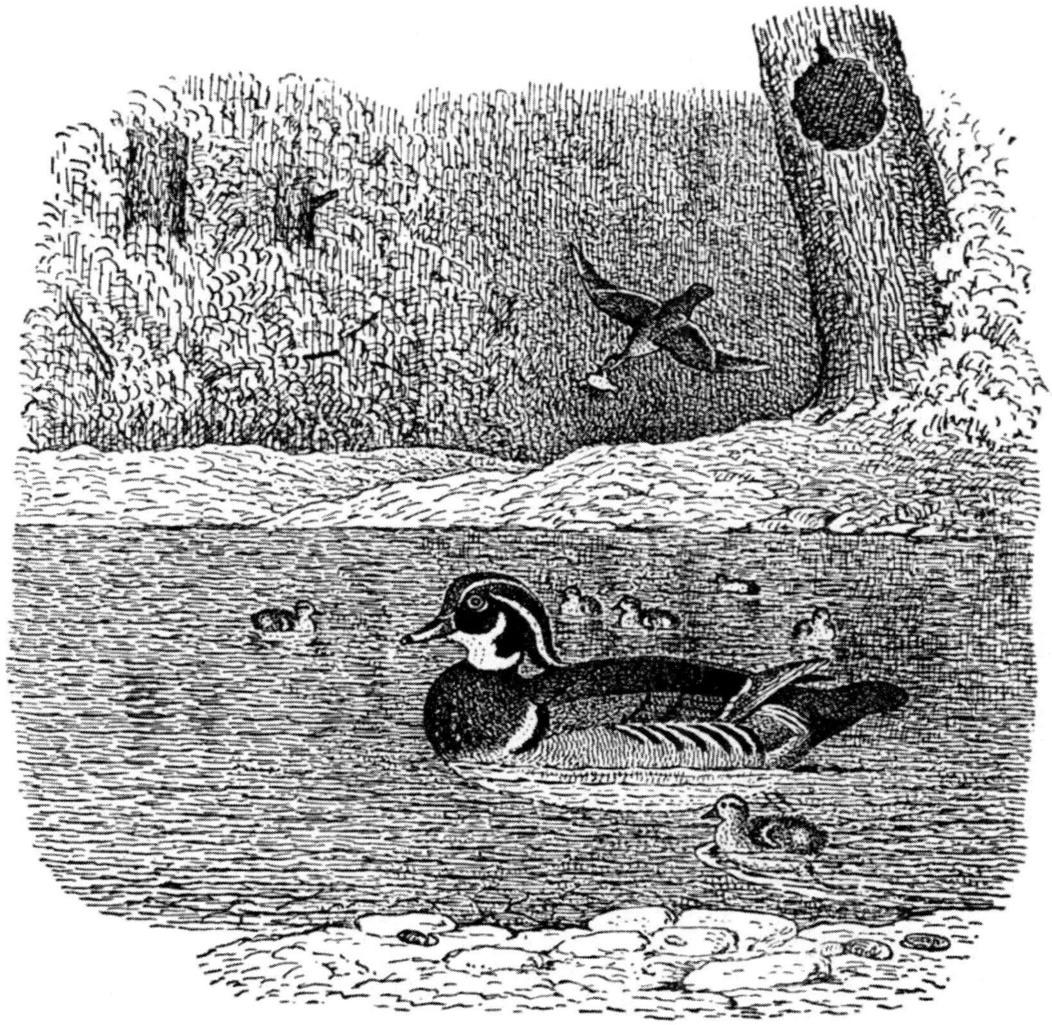

Figure 40. Summer Ducks And Young. Female carrying young in her bill from nest in hollow tree.

Wilson speaks of a nest which he observed in an old grotesque white oak, which stood on a slope of one of the banks of the Tuckahoe River, in New Jersey, just twenty yards from the water's edge, that had been occupied for four consecutive years. At the time of his visit the nest contained thirteen young birds, which the maternal head was engaged in carrying down to the water to give them, perhaps, their first experience in the art of swimming. So carefully, and yet so adroitly and quickly, did she perform this seemingly difficult task, that she was less than ten minutes in its accomplishment. Although the male usually stands sentry while the processes of laying and sitting are going on, and signals the approach of enemies by a peculiar cry which has been likened to the crowing of a young cock—œ-ēēk! œ-ēēk!—yet from the silence of one writer upon the subject we infer that the duty of rearing the rather numerous family is left to the mother, while he—her friend and consequential partner, as though disdaining such ignoble and degrading work, because of its slavish character—is off with his gay companions, disporting themselves in mid-air, or trimming, while perched upon some sheltering bough, their rich and varied plumage. So intent, however, was the mother-bird upon the faithful discharge of her home-duties, that she heeded not the stately sloop, then nearly completed, as it lay upon the stocks close-by, with its hull looming up within twelve feet of her home, darkened with the presence, and reverberating with the noise of workmen, but continued to pass in and out as though utterly unconscious of the so near approach of danger. Audubon claims that the male deserts the female when the period of sitting commences, and joins his sterner brethren, who unite into flocks of considerable numbers, and keep apart from their partners until the young are fully matured, when young and old of both sexes come together, and thus remain until the return of another breeding-season.

The female, it is evident from what has just been said, assumes the entire charge of incubation. For more than twenty-one days she is thus busied, with nothing, it would seem, to relieve the monotony of her task. How often she despairs and bewails the hardship of her lot, none can know. It is the inexorable decree of fate that she should perform the duties alone and unassisted, and most willingly she submits. But the *ennui* of the labor is, in a measure, forgotten in the vision that hope holds out to her patience, for her persistent assiduity is ultimately rewarded by a whole nest-full of happy ducklings. While the hatching process is going on the patient housewife only leaves the nest when pressed by the pangs of hunger, and but for a short time. Before leaving, however, she takes the precaution to see that her creamy-white, elliptical treasures, to the number of ten or thirteen eggs, are carefully covered with down.

Like the young of our domesticated species, the little Wood Ducks follow the mother almost as soon as they are hatched, and gather whatever of vegetable and insect food they happen to encounter. They are passionately fond of the water, and best show their real character when gracefully floating upon its glassy bosom, or diving into its azure depths. At an early age they respond to the parent's call with a soft and mellow *pee, pee, pee-e*, which is uttered quite rapidly, and at repeated intervals. The call of the mother, when addressing the young at such times, is rather low and soft, and resembles that of the young, being only a little more prolonged.

These beautiful birds have often been domesticated. They become at such times so unsuspicious and familiar as to allow themselves to be stroked by the hand. No handsomer bird could be chosen for introduction into our yards. The male, some nineteen inches in length, and with a scope of wing of two and one-third feet, is a being of no mean proportions. But it is the richness and variety of his colors that render him an object of admiration. A conspicuous green and purple crest adorns his head, while the sides, which are iridescent purple, are relieved of their monotony by a streak of white from base of bill to occiput, and by another, back of the eye, of a pure white color, which is continuous with that of the throat. The sides and front of the lower neck and the forepart of the breast are a bright chestnut, with five white spots, while the lower parts are generally white. Beautifully iridescent metallic hues set off the upper surfaces of the wings, which show most effectively in the blaze of the noonday sun. To the female nature has not been, it would seem to the casual observer, quite so propitious. Her grayish head, with lengthened hind-feathers, white throat, brownish-yellow fore-neck, upper breast and sides, striped with grayish and generally dark-brown upper parts, glossed chiefly with purple, contrast most markedly with the rich, gorgeous attire of her other half. While less showy in dress and lacking the dignity of demeanor that characterizes her lord, she is none the less fitted to perform her part in the drama of life. Her dress, sober in color, and with just enough of ornament to relieve the oppressiveness of its sameness, is so accordant with her home-surroundings as to afford her the protection and security she requires in the trying and perilous duties of brood-raising.

Chapter 28

American Woodcock

Quite as interesting in habits is the American Woodcock, the *Philohela minor* of Gray, which belongs to the grallatorial, rather than to the natatorial, family of birds. In distribution he is somewhat restricted, differing in this respect from his numerous congeneric brethren, which have a wide dispersion. He is chiefly a denizen of the eastern parts of the United States, and of the British territory immediately adjacent. Fort Rice, in north-western Dakota, and Kansas and Nebraska in the West, appear to be the limits of his range in these directions. In the Middle and Eastern States Woodcocks are found in greater abundance than anywhere else. While the bulk pass North to breed, a few remain in the South and raise their happy little families in spite of the ardor of the climate.

Few migrants arrive earlier at their breeding-grounds. They usually appear from the fifth to the tenth of March in New England and the Middle Atlantic States, although instances are recorded where they have been observed as early as the twenty-fourth of February. These cases are rare, however, and only happen, if at all, when the weather has been remarkably auspicious for a lengthy spell. As a few birds have been known to winter in the North, when the season has been unusually mild, their emergence from sheltered localities so early might be construed by persons not cognizant of their presence, or of their occasional winter sojourn, as a case of recent arrival. In view of this fact, it would be difficult to prove that a bird seen in winter had just come from the South, unless discovered *in transitu*.

Small companies, from four to six in number, start together upon the migratory tour. Low, swampy thickets invite their presence upon reaching their destination. Here they conceal themselves during the day, but when night has gathered dark they come out of their grassy retreats and wander about in quest of food. From the setting of the sun behind the western hills to the appearance of the first streaks of dawn in the east, they pursue their nocturnal rambles. Few persons have visited these birds in their accustomed haunts while foraging. Let me take the reader to some neighboring swamp, or by the side of some lonely woodland, which these birds delight to frequent. The utmost silence must be maintained, or they will be frightened away. While it will be difficult to see the creatures that have called us hither, yet we know they are not far away by the rustle they produce among the dry leaves, and by the peculiar notes they emit. *Chipper, chip-per, chip* may be heard from the right, and almost at the next instant it is varied to *bleat* or *bleat ta bleat ta*, produced in the contrary direction, or off in the distance, showing that the authors of these sounds have changed their positions. While these birds have a habitual fondness for humid thickets, they not infrequently betake themselves to corn-fields and other cultivated tracts in close proximity, and even to elevated woods.

For more than a fortnight after their arrival the sexes, though feeding in company, do not apparently manifest a disposition to assume conjugal relationship. The desire for food seems to outweigh every other consideration. The inclemency of the weather, and the coldness of the earth, may have much to do with holding the amatory forces in check. But when the opportune period arrives, which it does in the course of events, the sexes desist in a measure from their riotous living and give the nobler instincts of their being a chance to assert their power. The males are the first to feel the changes which are being wrought in their natures. For more than a week from the incipiency of this feeling, in the early morning and evening hours, they may be seen exercising themselves by means of "curious spiral gyrations" in mid-air, and uttering, as earthwards they descend, a note which has been likened to the word *kwank*. This note may be a call to the female in the spring, but as it is often uttered in the fall after the breeding-season is past, it may also be a summons for the gathering together of the members of the same household. The production of these sounds seems a labor of very great effort. But the movements of the males at these times must be seen to be appreciated. The head and bill are bent forward until the latter comes into contact with the ground, and, just as the sound is being emitted, the body is urged violently forward. These spasmodic exertions having ceased, the actor in this drama twitches his abbreviated, half-spread tail, assumes an erect attitude of listening, and, if no response is elicited, repeats his characteristic cry with all its accompanying movements. If the call awakes an answering note, the happy lover flies to the presence of the one he seeks and lavishes upon her the most endearing caresses. Sometimes, as Audubon affirms, the male awaits the arrival of the dear one, and does not fly to meet her. The summons, according to the same eminent authority, seem sometimes to be replied to by one of the same sex, which is always the prelude to a fierce encounter between the two, for, on such occasions, when the feelings are in a high state of tension, the most intense enmity exists between the males. But these contentions are ordinarily short-lived, and cease with the assumption of matrimonial relations.

The happiness of the male is now complete. With his homely but prepossessing bride by his side, he soon journeys off in search of a home. This is a matter of great consequence, and tasks the patience to the utmost. But their labors are eventually crowned with success. The most secluded resorts are visited, and in some low, dense and swampy woods or brake, difficult of access, and one that none but the cruel collector would be likely to find, they hide away their nest. The structure is generally placed on the ground, at the foot of a bush or tussock, in the midst of small birches or alders, or on a decayed stump or prostrate log. In certain localities, it is snugly nestled in the midst of a meadow. It is by no means an elaborate affair, but merely consists of a few dried leaves or grasses which are scratched together by the female, and the work of a few brief hours at the most.

Figure 41. Woodcock on Nest (showing protective coloring). Copyright 1900 by A. R. Dugmore.

Being ready for occupancy, the female soon commences to deposit her eggs. These, to the number of three or four, are laid one at a time on consecutive days. Oviposition, in the Southern States, commences in February or March, while in the northern limits of the bird's range from the tenth to the fifteenth of April, seldom later. Both birds perform the labor of incubation, and so attentive are they to the business that it is an unusual occurrence to find both absent from the nest at the same time. When the female is sitting the male busies himself in attending to the demands of hunger; and when her turn has come the care of the nest is resigned to her noble, conscientious lord. So faithfully do they keep to the nest that nothing short of the most menacing danger will compel them to leave. The approach of a team or a pedestrian, even when within a few feet of its location, has not been known to startle them. But when the danger is quite imminent the sitting bird slips out of the nest and makes her way into the tall grasses, where, hidden from view, she becomes a silent and sorrowful witness of any disaster that may befall her home. Should no destruction be perpetrated, and the intruder has gone his way, she cautiously comes out of her place of concealment and resumes her labors. But she has learned a very impressive lesson, for on a second visit to the nest no bird is to be seen. Apprised of the coming of danger, she has slipped out in time to escape detection. Thus, patiently, persistently and unweariedly these faithful creatures apply themselves by turns to the task of sitting until success has crowned their willing labors. The time spent in hatching is, under the most favorable conditions, from seventeen to eighteen days.

Figure 42. American Woodcock. Mother flying away with young between her feet.

The young are very timid creatures and keep close to their parents. Considerable solicitude is shown by the latter for their well-being. Their helpless infancy, so to speak, is watched over with all the care that a human mother bestows upon her offspring, and when their lives are endangered recourse is had to many a *ruse* to deceive their enemies and bring them into places of security. When severely pressed by foes, the mother, by a peculiar alarm, warns them of the state of things, and while they are scattering in different directions seeks to attract attention to herself in many a well-feigned artifice. In her anxiety for their safety, she has even been observed to seize between her two feet a youngling and fly with it away—a behavior whose purpose seemed to be the diversion of the enemy from the rest of the brood, thus giving them a chance to flee from impending peril to places of security in the surrounding verdure. After all danger has disappeared, she summons them together again by a familiar call, and doubtless relates to them the story of her adventures and the dangers from which they were saved. Worms, animalcula, ants and other soft-bodied insects, which the parents assist them in procuring from the soft earth, and from beneath the grass and dead leaves that abound in the places they frequent, constitute their food. Later on they are able to obtain their subsistence, with all the address of older birds, by thrusting their bills into the soil and in such other places as would be likely to contain the objects desired. Their tongues, covered with a viscid saliva, adhere to the food, and when drawn into the mouth carry it with them without danger of being lost. All who have made these birds a study have often discerned holes made in the soft mud by their bills. The presence of these "borings," as they are called, is always an indication that game is not far distant, which a careful exploration of the locality soon verifies. The young, when matured, continue to occupy the same haunts with their parents, and, unless brought to an untimely death by the merciless gun of the hunter, repair to the warm, sunny, smiling South with the return of frost. In the Middle States—and the same is doubtless true of other sections of our great country—there is never more than a single brood raised, although the early breeding of the species would certainly afford time for a second hatching before the close of the season. Less pyriform are the eggs of the Woodcock than waders' mostly are, being, in some instances, almost ovoidal. Their ground-color varies from a light clay to one of buffy-brown, and the markings occur in the form of fine spots and blotches of chocolate-brown, interspersed with others of obscure lilac, more or less thickly scattered over the surface of the egg, their size and intensity of color bearing, in general, a direct correspondence with the depth of the background. Remarkable variations of size exist throughout the species' range, some being short and broad, while others are long and narrow. A set of three from Pennsylvania, which the writer carefully measured, showed an average measurement of 1.54 by 1.21 inches.

So familiar a bird as the Woodcock, which is sometimes termed the Bog-sucker or Wood-snipe, hardly needs description. He has a thick, heavily-set body, short and thick neck, and large head, bill and eyes, and ears beneath the visual organs. His wings are short and rounded, the first three primaries being very narrow and shorter than the fourth, and

The extremities of the tail-feathers, as well as the feathers overlying the base of the tail, are in the latter creamy or fulvous white, while in the former they are of a decided chestnut-brown color. Other characteristics exist, but these only become evident to the keen-sighted ornithologist.

The difficulty experienced in establishing a cross between our wild and tame birds, shows that they are not as closely related as was once supposed. Did a near kinship subsist, interbreeding could most readily be accomplished. With the Mexican Turkey, matters are otherwise. That a relationship does obtain between the domestic bird and the latter—its wild original—there can be no question, as specimens of the naturalized species are often met with which are nearly the counterpart of its Mexican progenitor, differing only in the greater development of the fatty appendages of the head and neck, differences which may be accounted for as the effects of the influences to which the birds have been subjected by man. No well-authenticated instances of similar reversions to our once familiar Eastern bird have been known to occur, which would necessarily have been the case had they been so closely related as was once maintained.

Meleagris Mexicana affects sparsely-overgrown savannas, and occupies in Mexico the region of the oaks and the coast—the *tierra caliente* of geographers. It is a very wary bird, and lives in families. Insects of diverse kinds, but chiefly of a coleopterous character, as well as the seeds of grasses, constitute its bill of fare. When searching for food, especially in perilous localities, a sentinel is stationed on the outskirts of the flock, whose duty it is to announce the presence of danger. Flight is seldom resorted to at such times, for these birds, being fleeter of foot than the swiftest dog, are able to escape their enemies by running.

Toward the close of March, or in the beginning of April, the hens separate from the males, and seek for themselves nesting-places in secluded localities. The nest is anything but an elaborate affair, consisting of a few dry leaves or grasses scratched into a depression by the side of a prostrate log. Here the eggs—twelve beautiful, oval, speckled treasures—are laid, and for thirty long, weary days and nights they are sat upon by their author in her efforts to warm them into life. When she leaves them, which she does a short time each day for food, she always takes the necessary precaution to cover them with leaves, as a protection against cold and intrusion. Nothing will tempt her to quit the nest when the young are about to be hatched. So absorbed does she then become that she has been known to submit to capture rather than endanger the lives of her offspring.

No human mother manifests deeper affection for her children than does this bird of the prairie for hers. She fondles and dries them after they have escaped from their prison-houses, and tenderly helps them out of the nest. It is now that her cares may be said to commence. Where their interest and well-being are concerned, hardly any responsibility is too great for her to assume. She leads them into pleasant pastures, teaches them to know good from bad foods, and acquaints them with all the devices and subterfuges practised for eluding man and other enemies. But it is not long that they are thus subservient to maternal wisdom and forethought, for in fourteen days they are old enough to scratch for a living,

the fourth and fifth the largest. The tarsi are about one and one-fourth inches long and rather stout, the tibiæ feathered to the joints, and the toes long and slender, and without marginal membranes or basal webs. More than two and a half inches in length is the bill, straight, tapering, and stout at base, with ridge at base of maxilla high, and the upper mandible a little larger than the lower, and knobbed at the end. Three long grooves, one on ridge above, and the others on each side of maxilla, complete the structural details of the bill. The sexes are alike, the female being larger than the male. Adult specimens vary from ten to twelve inches in length, and have an expanse of wings of from fifteen to eighteen inches, and a weight ranging from four to nine ounces. The eyes are brown, legs and bill of the dried skin pale-brownish, upper parts black, gray, russet and brown, chin whitish, and rest of under parts different shades of brownish-red.

So exquisitely sensible is the extremity of the bill, as in the snipe, that these birds are enabled to collect their food by the mere touch, without using their eyes, which are set at such a distance and elevation in the back part of the head as to give them an aspect of stupidity. The eyes being situated high up and far back is a wise provision of nature, as, by this peculiarity, they escape many of their enemies, their field of vision being greatly augmented by such an arrangement. Obtaining their sustenance, as they largely do, by probing with their bills, so amply endowed with nerves, they have comparatively little use for their eyes, unless to keep watch for their numerous foes.

Though well known to the sportsman, yet by the casual observer this bird is frequently confounded with the Wilson's snipe. But the error can readily be avoided, if it is borne in mind that the Woodcock has the entire lower parts, including the lining of wings, a reddish-brown color, while the snipe has the abdomen white, the throat and upper parts of the breast speckled, and the lining of the wings barred with white and black.

Chapter 29

PIPING PLOVER

Have you ever been to the sea-shore? Then, of course, you have met the Piping Plover, but, perhaps, not to know him. He is of the size of the robin, not quite so robust, but stands much taller, being mounted on rather long, stilt-like legs, which admirably fit him for the life which he is designed to fill in the world. He belongs to the family of wading birds, and seeks the principal part of his food in or by the water, which could not possibly be were his walking appendages curtailed the least bit of their fair proportions. But to be more precise in my word-picture, let me describe him to you as of a pale ashy-brown color, fading into grayish upon the under parts, and as having his head set off with some narrow black bands, that on the neck rarely, if ever, forming a perfect ring. His bill will be found to be short and stout and blunt, and there will be an appreciable lack of webbing between the middle and inner front toes.

Now that it is plain what the bird looks like, you are certainly prepared, more than ever, to take some interest in him in his brief stay by the sea. So strongly is he attached to the scenes rendered dear by past associations and memories that, from his winter home in the sunny South, and even from over the waters beyond our southern borders, he hails with delight the return of the vernal equinox, for he knows full well that it brings with it the summer's heat and all its varied, priceless wealth of insect life.

So with the first spring signs of open weather he quits his brumal retreat, winds his way up along the trend of the Atlantic seaboard, and at last reaches in the nights of early April the sandy beaches of our Jersey coast. In flocks of a dozen individuals they run about the sand in a most lively manner, and utter all the while a variety of notes more or less pleasing, blending as they do with the deep-toned bass of the ocean. When this sound, welling up from a dozen throats, is heard in the dark it is particularly striking, as wild and weird as the whistling of a wind at sea through the rigging of a ship.

But these flocks soon disperse into pairs to breed. Slight depressions in the dry sand, and always in the midst of groups of broken colored shells, but out of the reach of the maddened waves, rather than in muddy, marshy places back of the beach-line, serve them for nests. This nesting among clustered shells seemingly points to a love for the beautiful. But may it not be that the shells but mark the various nest-positions in the unbroken waste of sand? We incline to this opinion. There is so much diversity manifested in the size of the groups and in the arrangement and coloration of the individual shells that comprise them, that no very great difficulty should be experienced by the several pairs nesting in the same locality in knowing each other's nest.

While the birds are concerned with the cares of brood-raising, which is usually towards the close of May or the beginning of June, they confine their feeding to the damp, wet sand. Between it and the dry a clear line of separation is plainly noticeable. It is only when they are ready for the home duties that they are seen to resort to aerial navigation. Even when on the very boundary-line of the two stretches of sand, the wet and the dry, and with the nest almost in sight, they are known to assume wing, taking due care, however, to alight before they have fairly reached the spot. In flight an advantage, that of a more commanding view, is acquired, which walking does not give. But in leaving the nest for food, or for any other purpose, they, as before, walk some distance away before they venture to fly. There is a seeming purpose in so doing, the object to be gained being the deceiving of man and other enemies as to the real location of the nest.

Figure 43. Female Piping Plover. Nest in midst of broken shells.

All these precautions are undertaken for the sake of the eggs, although in color and markings these so closely resemble the dry sand and intermingled bits of foreign substances, that such actions seem all unnecessary. When birds have been flushed from the nest, and its exact position has been noted with the greatest care, I have failed, after several minutes of the closest searching, to detect the eggs, so true has been the color-harmony between them and the surrounding sand. This resemblance in coloration must be seen to be fully appreciated. In ground the eggs are the palest possible creamy-brown, but marked all over, quite sparingly, with small blackish-brown dots and specks, the largest hardly exceeding a pin's head. Four is the usual number, and these, from their peculiar pear-shaped form, are placed with their points together in the centre of the nest. They are objects of more than ordinary solicitude, the little Plovers making most violent demonstrations and pleading piteously when they are approached. The mother employs all the well-known artifices, such as lameness, inability to fly, to draw the intruder away from the nest. The young run as soon as they leave the egg, and are great adepts at hiding, squatting, and remaining motionless. Their downy plumage so assimilates them to the sand that unless they reveal themselves by moving, it requires a very keen eye to distinguish them from the numberless tufts that are scattered about the higher reaches of the beach.

Although so essentially a bird of the sea-shore, yet in August many scores of these birds come up the Delaware River as far as tide-water extends, feeding upon the mud-flats and gravel-bars, and occasionally wending their way up along the courses of the creeks until they find themselves well into the country. It is interesting to watch them as they run in and out among the little hills and hollows of the mud in quest of their prey. They are happy, light-hearted fellows, who do not begrudge, when some racy tidbit has rewarded their hunting, to pipe a few notes of thanks to Him who watches as tenderly over them as over the mighty lords of the earth.

Chapter 30

Bob White

Somewhat related to the grouse is the Quail, as he is called in the Northern States, or "Bob White," his universally recognized appellation. His scientific name is *Ortyx Virginianus*. Differing from the Old World partridges, he has been assigned a place in the sub-family Odontophorinæ, of which five genera are said to exist, most of them being restricted to the extreme south-west of our country. His habits and history are full of interest to everybody.

Quails are restless, uneasy birds, attached to one place while rearing their family, but immediately upon the brood becoming able to travel, commencing their wanderings. There is no accounting for these movements, which sometimes deprive a whole district of their presence for a time, to populate a neighboring region previously without them. When such journeys are undertaken, a large number of birds participate, travelling on foot, and passing steadily through districts where food is plentiful, and seemingly without any definite destination in mind, so loath are they to use their wings, that in attempting to cross wide rivers and inlets immense numbers are said to perish. A limited and partial migration, it is highly probable, takes place annually from the more northern to warmer latitudes, influenced in its extent by the comparative severity of the seasons, being more distinctly migrating west than east of the Delaware River.

About the middle of March the winter flocks break up, and the mating begins. Although not indulging in the noisy and seemingly meaningless antics of the grouse to call attention to his personal attractiveness, Bob White, it would appear, becomes suddenly conscious of his comely looks and excellent voice. In a dignified manner, with head erect, he walks proudly about, inviting the opposite sex to view him at his best. From the orchard gate he calls a saucy good morning to the farmer, knowing that the law holds its *ægis* over him at this time, but he keeps an eye to hawks, cats and other predatory animals that respect neither time, place nor season. He is polygamous, willing to assume any amount of family responsibility, and will help to rear two, or even three, broods a year, a successful pair often turning out twenty-five young in a season. It is not an uncommon occurrence to find a covey of little cheepers, scarcely able to fly, as late as November.

Although paired so early, the Quails do not proceed to the business of nidification in the central part of their range until about the middle of May. The leeward side of some dense tussock of grass, a mouldering stump in a wild, matted meadow, the woody margin of a clover field or orchard, or an old pasture overgrown with bramble thickets, are situations commonly chosen, the female, as is her undoubted right, taking the lead in fixing upon the site. An artificial bed of grasses and vegetable trash, filling a shallow depression, is the nest. Sometimes it is placed so as to be concealed by overarching grasses, through which a regular tunnel, several feet in length, conducts to the sanctum; and, at other times, is only covered with leaves and straw, which the birds themselves have rudely adjusted. The nest, which is constructed solely by the females of the family, varies in dimensions according to the number of this sex that anticipate using it, the male in the meantime going about in quest of food, or sitting upon a low twig close by, cheering his wives by his trisyllabic note, and very faithfully warning them of the imminence of danger.

The work is prosecuted with considerable zeal, three days at farthest sufficing to make the nest ready for the first egg, which is immediately laid, and which is followed by one on each consecutive day, until seven or eight have been deposited. As many as thirty eggs are sometimes found in a single nest, which is due to the polygamy of the male. Two, and often three and four females, are taken by a male, and two have been known to occupy simultaneously the same nest.

When a pair of birds has established itself in a locality from the first, and has been successful in rearing a family of young during the ensuing spring, if the females are in the majority the unprovided ones still continue, as a general thing, to linger with the parents after their more specially favored companions have mated and moved elsewhere. This is particularly noticeable in a new locality where the covey consists entirely of members of a single family. In cases where several families congregate in the fall, the chances are greatly in favor of monogamy. Small flocks are more decidedly polygamous than larger ones. We have never observed the converse—that is, more than one male to a female—but where several pairs are found in the same field, at slight distances from each other, there is sometimes a noticeable tendency to associate.

The eggs of the Quail are crystal white, sometimes slightly tinged with yellow, and pyriform in shape. Eighteen days are required for their hatching. Where the father is not fortunate enough to possess a harem, a part of the work devolves upon him, while the mother seeks food and recreation; but where there are several females, the work is divided very amicably among them, each sitting about half a day at a stretch, then calling her relief with a low note, if there be only two, the male taking no part in the labor of incubation whatever. Should the family be larger, two females will sit side by side on the eggs, there being too many in number for one breast to cover. Meanwhile the husband remains close by, chirping encouragement in a low tone, and betimes making the field vocal with his loud, clear whistle. He is exceedingly vigilant, and if a human being approaches the nest gives the alarm to his partners, who secretly withdraw from the nest, while he, thoughtful husband as he is, flings himself upon the ground in front of the intruder, feigning lameness or injury, and seeking by every device known to him to attract attention and pursuit, till having beguiled the enemy far away from his home he seeks safety for himself in flight. The experienced zoölogist pays no regard to this deceit, seeing in it only a sign of the nearness of the coveted prize, but patiently continues his search until he has discovered its whereabouts.

Figure 44. Home of Bob White. Two wives on same nest.

Two broods are invariably raised and often a third, but the last appearing late in the summer, and scarcely attaining their growth before the coming of snow. If unmolested, it is evident, therefore, that the species would increase with great rapidity, as shown by the celerity with which regions, where the birds had been well nigh exterminated, have been replenished when a period of quiet for a season or two has been allowed them. The young run about in a very lively manner as soon as they have left the shell, and in a few days are given over to the care of the father, whom they follow and obey as readily as they did the mother, possibly because they do not recognize the change of guardians, while she returns to the cares of rearing another family.

During the spring and early summer both old and young find an abundance of food for themselves in the larvæ of various insects, the succulent shoots of growing plants and such seeds as abound. Later on, strawberries, blueberries, huckleberries and other wild fruits supply their demands. In August they grow fat upon grasshoppers, and as this is the time when seeds ripen, acorns and beech-nuts fall, and the stubble-fields are full of scattered wheat, rye, barley and maize, and insects are plentiful upon the ground, they feast themselves to satiety before the winter begins, until they have reached that delectable plumpness so highly esteemed by bon vivants. Attaining their full growth by the end of September, at least in the case of the earlier broods, the season of play for the partridges and sport for the gunner has come. Quail-shooting is regarded as a test of marksmanship in the United States. So rare and wild have the birds become by reason of incessant hunting, that it certainly requires skill and fine shooting to make a bag. Bred in the open fields, and feeding early in the morning and late in the evening, a man may beat a field all day, and put up only one or two birds, when he is certain that twice as many lay concealed, huddled up in little knots in out-of-the-way places, which the best of dogs might easily pass without discovering. Their inconspicuous colors, too, which are in keeping with the objects around them, so conceal them from the vision of the hunter, that, trusting to them, they will sit immovable until he has gone some distance beyond, when they will spring up and away like so many arrows, requiring a quick eye and a steady hand to turn and drop a brace.

When ultimately flushed, they fly to some particular covert, and so long as this thicket or fern-brake remains undiscovered, will repeatedly repair to it for safety and security. A rather curious circumstance, which has created no little discussion among American sportsmen, materially aids their concealment. When alighting, after being flushed, the Quail is said to give out no scent for some little time. This is supposed to be a voluntary act of retention of odor on the part of the bird, as a conscious method of protection. Some, while admitting the fact, believe it to be a power belonging to particular bevies, at least in a far greater degree than to others, like the custom of alighting upon the branches of trees when frightened, while others restrict the faculty to particular individuals rather than bevies. Our earlier ornithologists do not mention the retention of scent. It is probable, as claimed by a few, that Quails' swift running over the dry leaves of upland woods or meadows allows little time and a poor surface for the transmission of the scent, and that when they drop suddenly and remain quiet no effluvium escapes, but which only becomes disseminated the very instant they move.

The open fields being smitten by the wild winds of November, and the reeds bruised and broken, the Quail retreats to the depths of the swamp or the shelter of a dense thicket, where he keeps life in him as best he can during the cold, stormy days, hunting the stubble and swamp for soft-shelled nuts and seeds, torpid beetles, and the hard fruits and seed-cases of grasses and weeds, some of which, the skunk cabbage for example, tainting his flesh with their flavor. Huddled together the forlorn covey allow the snow to cover them, trusting to shake it off on the return of the morning, but occasionally a crust freezes upon the surface, and the poor birds find themselves in a prison from which they cannot break out before they starve to death. The habit of huddling is peculiar to Quails the whole year round. They select at evening some spot of low ground, where the long grass affords shelter and warmth, and there they encamp, sleeping in a circle, shoulder to shoulder, with heads turned out, keeping each other warm, and ready to escape at a moment's warning without stumbling over one another. A suitable roosting-place once found, night after night they repair thither, leaving it in the morning before sunrise to seek their breakfast.

Unless the winter be unusually mild, they may be seen associating in the pasture with the cattle, and even following them home to glean the grain that falls into the barnyard, and pick up the scraps that are thrown to the chickens. This delightful confidence is not always abused, for many persons take pains to foster the bevies they find spending the winter in some brushy hillside near the house by daily scattering grain or clover-seed upon the snow where the hungry birds may come and get it. The pert air with which one of the cocks will perch himself on a fence-rider or walk sedately along a stone wall in the early sunlight of a glistening January morning is reward enough to the benefactor, if he cares not to preserve them for the selfish pleasure of shooting them the following autumn.

As a delicate article of food the Quail is highly esteemed, and during the time the law allows the markets are filled with bunches of them. Various devices in the form of snares, nets and traps are called into service to effect their capture, and in some parts of the country, New England especially, fresh importations have been necessary to preserve a sufficient number for sport. Bands of beaters in the Southern and Western States cautiously drive immense flocks into nets, but there is less danger of exterminating this than almost any other species of game-bird, it would seem, on account of its sequestered habits and prolificacy.

Taming and domestication is an easy matter with these birds. In all cases, however, where the eggs have been hatched under a hen at liberty, the Quail chicks have run away to the woods as soon as the leaves have turned sear in the fall and never come back. They sang their "Ah, Bob White!" just as clearly before they had ever heard one of their kin as any woodland-bred Quails could do. It is quite common to re-colonize portions of the Eastern States when they have become depopulated, and an effort made some years ago to introduce these birds into the Salt Lake Valley of Utah was eminently successful. Within the past few years some of the West India Islands have been colonized, but attempts to acclimatize the birds in England and Ireland have proved most signal failures.

Chapter 31

RUFFED GROUSE

Considerable misapprehension exists in relation to the popular appellation of this species. In some parts of the country it is dubbed the Partridge, while in others it goes by the name of Pheasant. It is neither. All its affinities point away from these families, in the direction of the True Grouse, of which it constitutes a useful and interesting member. Pheasants are never found in the United States, but are indigenous to Southern Asia. Their nearest representative here is the Wild Turkey. Almost as much may be said of the Partridge, a group of birds which are exclusive denizens of the Old World.

But now to our subject. Few Grouse are so well known as the Ruffed Grouse, the *Bonasa umbellus* of Stephens. Everywhere throughout the timbered regions of Eastern North America it is more or less plentiful, ranging from the Atlantic seaboard to the Rocky Mountains, and from Georgia to Nova Scotia. In all our Southern States, Louisiana excepted, these birds exist to some extent, and are also to be found over limited portions of the Missouri region, but, doubtless, more especially about the mouth of the river, and in the contiguous country. In the western parts of the region it is represented by a form which passes with ornithologists as a well-defined, genuine variety. It seems to be wanting in California, but in the wooded sections of the Cascade Range, as well as in the valley of the Willamette in Oregon, where it exists under a new varietal name, it is by no means an uncommon occupant. In the New England, Middle Atlantic and Northern Central States it is that these birds are to be seen to the best advantage, and in the greatest numbers. West of the Mississippi, if we exclude Eastern Kansas, Southern Iowa and the whole of Missouri, they occur, if at all, in comparatively small and isolated parties.

In regions which these Grouse inhabit, they are permanent residents, and are never known to move southward with the retreat of warm weather. They are capable of adapting themselves to climatic variations with ease, but not so readily to surface irregularities and their natural concomitants. Dense woods, craggy mountain-sides and the borders of streams are noted places of resort. Lowlands, especially such as are invested with thick growths of small bushes and tall, rank grasses, are not infrequently chosen. When in search of food and gravel, they are known to quit their favorite haunts and betake themselves to the open road, where groups may be seen absorbed in feeding, but not to that extent, however, when the rustle of a moving leaf or the crackling of a twig would pass unnoticed. The slightest noise causes a temporary suspension of labor and a momentary shudder of surprise. All of a sudden, and in the most perfect harmony, all heads are raised and pointing in the direction whence the noise emanated. The keen vision of these birds is not slow in discerning, through the gloomy recesses, the presence of danger; but should nothing of an alarming nature manifest itself, a short parley ensues and business is resumed, though not with the same earnestness and lack of care, however, as before. Greater caution is now observable, and every effort taken to prevent an ambuscade. But let the cause of the alarm, a dog or a man, be close at hand, and the birds immediately strike for the cover, either on foot or by means of flight, the latter method only being adopted in extreme cases, when the other course would be attended by disaster and probable ruin. In the exercise of their cursorial powers, they move with remarkable swiftness, as with head depressed and tail expanded they run for their lives. A pile of brushwood or an impenetrable jungle, when near, is rendered subservient. There they manage to conceal themselves for a time and thus recover breath. Closely pursued, and in danger of being trampled upon by the foot of the huntsman or lacerated by the fangs of his quadrupedal friend, they await the opportune moment, when, with sudden whirring wings, they cleave the elastic ether and vigorously press forward to some transitory haven of security, but only to fall once more in the way of their relentless persecutors. These flights are so well timed and so unexpected that many an experienced gunner is thrown off his guard, and when, at last, he has recovered from his surprise and collected his thoughts, feels vexed at himself for allowing his equanimity to become unsettled by so familiar a stratagem. He finds it useless to repine, but endeavors to choke down the bitter sigh of disappointment that arises as he presses forward to further adventures.

Like the common barnyard fowl, these Grouse are strictly gregarious, especially during the autumnal and winter months. The flocks they form vary in numbers, and when disturbed, while feeding, scatter in all directions, each member seeking only its own individual safety and well-being. But after the lapse of a few minutes, becoming reassured, they gather simultaneously about the same spot, travelling the entire distance on foot. The utmost circumspection and vigilance are always exercised in these backward movements. Scarcity of food occasionally causes these birds, where very numerous in mountainous districts, to migrate to other places. These journeys are usually undertaken about the middle of October, they then being in excellent order and in great demand for the table. Audubon witnessed, in the fall of 1820, an immense number *in transitu* from Ohio, Indiana and Illinois to Kentucky, many of whom became a prey to man. This disposition to lead a roving, migratory life is, as a general thing, not hereditary, and consequently is seldom undertaken, plenty of food usually being found in localities which these birds affect.

Where there is a paucity of food-materials, such as acorns, the seeds of the beech and of the various species of birch, they do not hesitate to devour the buds of the mountain laurel, which impart a poisonous character to their flesh. When severely hunger-pressed they feed upon dry bark, the insects that harbor in the creviced trunks and branches of trees, and even stray to the roads that wander through their gloomy retreats and peck at the hard, frozen horse-droppings they chance to encounter. But when spring returns and renews her bond of faith with Mother Earth, they more than make up for their scanty winter fare and feast with fastidious appetites upon the now tender and juicy buds of the black birch, which give a peculiar and toothsome flavor to their flesh that has acquired for them in some localities the name of Birch Partridge. For a brief spell every other interest is now absorbed in that of unrestrained feasting, to which the sexes submit themselves with all the *abandon* of civilized humanity. The middle of March, or the close of the month dedicated by the ancient Romans to purifications and fastings, when the weather is favorable, marks a change in their life. This era is announced by a loud drumming noise, which is everywhere heard. Standing upon a tall rock or a prostrate log in some secluded woods or other locality, the author of this noise may be found. His attitude and demeanor needs must be seen to be appreciated. Once seen, he can never be forgotten. Arrayed in a brand-new spring suit, he is a being not to be despised. But this is not all. His beautifully-contracted neck, broad, expanded, fan-like tail and elevated feather-tufts that ornament both sides of his neck, as he struts about with all the grace and dignity of some pompous lord or duke, render him of no mean importance and greatly add to his attractions.

But it is his final actions that impress the beholder with wonder and admiration. The hitherto trailing wings now assume a condition of rigidity, and commence a firm, but slow, downward and forward movement, which steadily increases in power and rapidity, until the swiftly-vibrating wings appear only as a semi-circular outline of mist above the bird, thus giving rise to a sound which may be appropriately likened to the reverberations of distant, muttering thunder. These sounds are most generally heard during the cool hours of the morning, when his spirits are buoyant after a night of refreshing slumber. But as the day advances, they are less frequent, and irregular. So nicely can they be imitated, that many a bird is drawn to his doom, when advancing, as he supposed, to meet an antagonist.

As the drumming is as often heard in the fall as in the spring, it has long been a mooted question as to its significancy as the call-note of the male during the period of breeding. But there can be no doubt of the correctness of this interpretation, for incontestable proof exists of it in the responsive actions of the female. Nuttall is probably correct in ascribing the autumnal exhibition of the power to self-gratification, and in affirming it to be, in many instances, "an instinctive expression of hilarity and vigor."

Besides the peculiar drumming sound which the males produce during the love-season, they give expression to other vocal utterances no less remarkable. These are generally enunciated when about to arise from the ground, and consist of two well-defined and characteristic notes. The first may be described as a sort of cackle, repeated several times in lively succession; and the other, which closely follows in its wake, as a peculiar lisping whistle, which has not inaptly been compared to the cry of a young bird. These notes doubtless play a considerable part in the reconciliation and bringing together of the sexes after their temporary separation.

Figure 45. Ruffed Grouse in Spring-Time. Two Males Displaying their graces before a lone female.

While the courting-season lasts, it is not an uncommon occurrence to find a single male in the midst of several females before whom he is engaged in showing off his many good qualities and graces, or two males displaying, upon the same fallen log, the excellent beauties of their person and movements. In the former dilemma, enamored of so many, he is sometimes disposed to be gay and trifling, dallying with the affections of some pure, simple-minded female. The most cruel flirtations are often indulged in. But when he does bring himself earnestly down to the business of choosing a partner, he does not go about it in an uncertain, hesitating manner, but makes his selection with promptness and dispatch. The successful female, proud of the honor conferred, at the call of her lord, forsakes the group of her unmarried sisters, and follows wheresoever he leadeth. The warmest tokens of affection and regard are lavished upon the elected bride, and woe to the rival who should appear upon the scene while these amours are being enacted. Should this event occur, the intruder is instantly assailed, and a long and bloody battle ensues, which results in the death of one or other of the combatants, but never in the complete vanquishment of the defensive party. Instances are known where males have treated their first loves with cruel indifference, and subsequently deserted them, but such things could not otherwise be, as will be seen when the question of polygamy comes to be considered, for it is a fact, not generally known, that both birds are slightly promiscuous, the tendency being more pronounced, however, upon the part of the male. In the case where a single female is courted by two males, the successful competitor for the honor of her hand, so to speak, is he whose movements are marked by the greatest elegance and grace. So intense does the desire to please become, that the slightest disposition upon the part of the lady to favor one of the rivals rather than the other, leads to the most unhappy consequences, a quarrel being precipitated, the contestants seeming determined to settle the result by the gage of battle.

The time of mating varies somewhat with climate and with the conditions of the season. In the warm, sunny South it occurs late in March or early in April. But further North, where winter still lingers with frosty coldness, the latter month is well nigh verging to its close, or gliding into the succeeding, before this essential business is thought of. When, however, it does happen, the female, with but little waste of time, withdraws from the society of her partner, and repairs to a secluded spot in the midst of a woods, where, usually beneath a clump of evergreen, or a pile of brush, or perhaps a fallen log or projecting rock, she hastily scratches a few dry leaves together for a nest. There she deposits, one by one, on as many consecutive days, her complement of six to twelve eggs, and immediately enters upon the duties of incubation. In this she is alone, the male lending no assistance, not even indirectly by attending to her demands for food. While she is thus occupied he seeks the company of others of his sex, with whom he remains until the young are nearly full-grown, when he joins the family, and dwells with it until spring. The period of incubation ranges from nineteen to twenty days.

When first hatched the young follow the mother, and soon learn to comprehend her clucking call, as well as to act responsively thereto. Few mothers are more devoted to their children, and it is rare to find one more courageous and wily in their defence. Let the family be surprised by friend or foe, a single note of alarm is all that is necessary to cause the brood to scatter, and with the most clever adroitness to hide themselves beneath a bunch of leaves or grass. So successfully is the concealment accomplished, that a careful and protracted search is often necessary to discover their whereabouts. Often, when squatting by the roadside with her brood, the parent is taken unawares. This is the trial which she of all others seems to dread. To save her little ones she perils her own life by venturing upon an assault. Her first impulse is to fly at the face of the intruder, but sober thought comes to her rescue and teaches her the folly of such a course. She yields to the thought and the very next moment we find her tumbling over and over upon the ground, apparently in the deepest distress, but soon to recover her self-possession in time to carry out the final piece upon the programme, a ruse in which lameness is imitated with wonderful ingenuity. While the mother is thus agitated, the birdlings are seen to scamper in every direction to places of shelter. Having accomplished her part, the happy mother now flies away, and by her well-known cluck soon gathers her brood together. The cry of the young is a simple peet, which is heard repeatedly during feeding, but only occasionally while nestling. Their food consists of the seeds of various plants and berries. While able to search for their own food, they derive, however, considerable assistance from the mother.

Such cunning, wee creatures, when first they leave the egg, can only be compared with the young of the domestic hen. Dressed in a simple garb, they look but little like their parents. Above they show a reddish-brown or rufous coloring, which fades into a rusty-white below. Excepting a dusky streak which starts from the posterior part of the eye and crosses the auricular regions obliquely downward, and a whitish bill, they have nothing to diversify the monotony of their plumage. But when they have attained the age of four or five months, they show their heredity so plainly that their identity cannot be disputed or mistaken.

In the adult, the tail is reddish-brown or gray above, with narrow bars of black. Terminally, it is crossed by a slender band of pale ash, which is preceded by a broader one of black, and this by another of an ashy color. The upper parts are ochraceous-brown, and finely mottled with grayish markings. The lower parts are chiefly white, with broad transverse bars of light brown, which are mostly hidden from view upon the abdomen. Upon the shoulders the shafts of the feathers have pale streaks, which also exist in those of the wing-coverts. The upper tail-coverts and the wings are marked with pale, grayish cordate spots, while the lower tail-coverts are pale ochraceous, each being provided with a terminal delta-shaped spot of white, which is bordered with dusky. The neck-tufts are brownish-black. The male measures eighteen inches in length, and has a breadth of wings of seven and two-tenths inches. The tail is about seven inches long. The female is smaller than the latter, with similar colors, but has less prominent tufts upon the sides of the neck.

The eggs of this species are usually of a uniform dark-cream color, but sometimes show a nearly pure-white ground. In most specimens there are no markings at all, but when they do occur, are either quite numerous and conspicuous, or few in number, and obscure. They are usually ovoidal, but forms are occasionally met with which are nearly spherical. Their average dimensions, as obtained from specimens from the most diverse localities, are about 1.64 by 1.18 inches. As far as known the species never produces more than a single brood annually, usually nesting, as has been previously stated, on the ground, but instances are recorded by Samuels, where the female has occupied a crow's nest, or the shelter of some tall broken trunk of a tree.

Chapter 32

AN OLD ACQUAINTANCE

Little is known of the early history of the domestic Turkey. Writers of the sixteenth and seventeenth centuries seem to have been ignorant about it, and to have regarded it as the guinea-fowl or pintado of the ancients, a mistake which was not cleared up until the middle of the last century. The name it now bears, and which it received in England, where it is reputed to have been introduced in 1541, was given to it from the supposition that it came originally from Turkey. As far back as 1573 we read of it as having been the Christmas fare of sturdy British yeomanry.

Oviedo, a Spanish writer, speaks of it as a kind of peacock that was once very abundant in New Spain, as Mexico was called in his day, and which had already, in 1526, been transported in a domestic condition to the West Indies and the Spanish Main, where it was maintained by the Christian settlers.

Among the luxuries possessed by Montezuma, the proud, dignified, semi-cultured monarch of the Aztecs, was one of the most extensive zoölogical gardens on record. Representatives of nearly all of the animals of the country over which he reigned, as well as others, brought at great expense from long distances, were to be found within its walls. Turkeys, it is said, were daily supplied in large numbers to the carnivores of his menagerie.

Respecting the time when this bird was first reclaimed in Mexico from its wild state, there can be no idea. Probably it has been domesticated from remote antiquity. No doubt exists, however, as to its being reared by the Mexicans at the period of the Spanish Conquest, and of its subsequent introduction into Europe, either from New Spain, or from the West India Islands, into which it had been previously carried.

Figure 46. Mexican Wild Turkey. Ancestor of the domestic bird.

Audubon, one of the early pioneers of American ornithology, supposed our common barnyard Turkey to have originated in the wild bird so prevalent in the eastern half of our great country. But it has always been a matter of surprise to naturalists that the latter did not assimilate, by interbreeding and reversion, more intimately in color and habits to the domestic form. No suspicion, until recently, appears to have been entertained that the two birds might belong to different species. That such is the true *status* of things, there is now no reasonable doubt.

Our common Wild Turkey, once so plentiful in Pennsylvania, is now restricted to the more eastern and southern portions of the United States, while in the parts of Texas, New Mexico, Colorado and Arizona adjacent to the southern Rocky Mountains, and thence stretching southward along the eastern slope of Mexico as far as Orizaba, there exists another form, essentially different, which, by way of distinction, has been popularly called the Mexican Turkey. It is from this species, and not from the other, as has been erroneously supposed and taught, that the domestic fowl has been derived. Even in this enlightened age, with so many ornithological teachers on every hand, we see this mistake propagated by such as know better, and whose business it is, or ought to be, to have a care that truth shall prevail.

Between the wild bird of Eastern North America and the Mexican and typical barnyard fowls there are differences which must be apparent even to the most superficial observer.

Coming like whirling leaves, half autumn yellow, half green of spring, their colors blending like the outer petals of grass-green daffodils, no more sociable and confiding little creatures are to be found in our midst than the Yellow Warblers. They are as much at home in the trees by the house as in the fields and woods. Wherever they wander, the glints of sunshine that flash from their backs should make the most miserable complainer feel the summer's charm. But in spite of their seeming preference for man, they are prone to build in lonely fields and by-ways. In such places it becomes one of the especial victims which the Cow Bird selects to foster its random eggs. But the Warbler puts its intelligence effectively to work, and builds a second story to its nest, thus flooring over the unwelcome eggs. This expedient is repeated as long as the Cow Bird continues her mischief, until sometimes a three-story nest is achieved. The outside of the nest, composed of glistening milkweed flax, is pressed into a felt-like case, the fibres serving at the same time to lash the nest to its support. Within, to the depth of an inch, is a soft sponge-like material, which the birds have made from the wool they have gathered from the stems of young ferns. A few horse-hairs, to give shape and stability to the nest, are to be found in the inside of the felt-like lining.

Figure 59. Three-Story Nest of Yellow Warbler. Showing the builder's manner of out-witting the cow bird.

and to seek shelter and security from lawlessness and cruelty. Their *menu* consists of wheat, berries, grasses, earth-worms, and all kinds of terrestrial insects.

When summer is over, the different families of the same neighborhood come together, unite in one large flock, and travel over the country for food. The males emerge from their meeting-places and join the moving army, and parents and young have nothing to do but to feed vigorously and grow fat. Late in October, or early in November, they begin to attract the attention of gunners, and thousands are killed for the market, where they are in eager demand by all lovers of good living.

Chapter 33

American Osprey

Pandion haliætus, as the Fish Hawk or Osprey is called in ornithological language, is found from the fur region surrounding Hudson's Bay to Central America, and from Labrador to Florida, excepting Boston Harbor, on the Atlantic Coast, and almost from Alaska to the southern extremity of the peninsula of Lower California on the Pacific seaboard. Birds have been known to nest on the rocky islands of California, and about Sitka, according to Bischoff, as well as along the small streams in the vicinity of Nulato. From Long Island to Chesapeake they breed in vast communities, which often number several hundred pairs, but away from the sea-coast they are only occasionally met with on the margins of rivers and lakes. Dr. Hayden found several pairs nesting on the summit of high cottonwood trees in the Wind River Mountains, and Mr. Allen observed the birds particularly abundant about the lakes of the Upper St. John's River in Florida, six nests being noticeable within a single circle of vision. Salvin claims that they nest on both coasts of Central America, but more especially about Balize, although on the islands of Trinidad, St. Croix, Jamaica and Cuba they are seen at all times except during the breeding-season.

Below Philadelphia, and in the south-eastern counties of Pennsylvania bordering on the Delaware, individuals have been occasionally observed. Their arrival is about the beginning of March, often when the streams which they frequent are fettered with icy bonds, and their departure occurs about the twenty-fifth of September, and frequently, especially when the weather is remarkably fine, as late as the fifteenth of October. Well-established communities, numbering more than fifty pairs, have been met within the swamps of Southern New Jersey, among whom the best order and most perfect harmony prevailed. Few species display less shyness and greater confidence, or are more eminently social, as is particularly shown when these birds take up their quarters in close proximity to occupied dwellings, or by the side of frequented by-paths and highways. Where undisturbed, the same localities are visited year after year. Their exclusive piscine habits secure for them free and unlimited sway in their carefully chosen abodes, for the poultry has nothing to fear, and the smaller birds are not intimidated by their presence and sent screaming to their coverts as they do even when pursued by the little sparrow hawk. Wilson cites a case where four nests of the common purple grackle were built within the interstices of an Osprey's nest, and a fifth on an adjoining branch, and the Osprey was quite tolerant of such intrusion and freedom. The writer has observed a nest of the grackle built in a similar position, while all around the great Hawk's home, and scarcely five rods distant, were nests of the robin, wood thrush, red-winged blackbird and others, and no annoyance was known to occur, the Ospreys carefully attending to their own business and scarcely noticing their more humble brethren.

Their bitterest enemy is the white-headed eagle, against whom the united attacks of many of these birds are concentrated when he has the audacity to venture within their hunting-grounds or breeding-quarters, for they are too familiar with his powerful muscularity and courageous disposition to attempt a single attack. When an Osprey is pursued by this king of the forest and hunting-ground, his loud, vociferous cries of distress, resounding far and near, evoke an army of defenders, who come with all possible speed to wreak vengeance upon the great arch-enemy of their pleasures and happiness. These attacks are made for the purpose of compelling the Osprey to drop his prey or disgorge, which the superior bird, if his efforts have been successful, pounces down upon and seizes before it has had time to reach the water or ground.

Powerful as the flight of the Fish Hawk is, yet it is never very high, nor much protracted. While skimming over the water's surface, even at a moderate elevation, his quick eye soon descries his quarry, and, in an instant, he pounces down with tremendous force below the water's level, often to a great distance, but seldom missing his prey. Arising from his watery bath, he moves off to a suitable perch to digest his meal at leisure. But should the movement attract the keen vision of the bald eagle, who is generally waiting in some secret covert, or sailing so high up in the air as to be almost invisible, the Osprey swallows his victim if small, or seeks to bear it away in his talons to a position of shelter and safety, but, rather than endure the too near approach of his more powerful relative, drops it to the infinite delight and great satisfaction of the latter. Where a suitable tree, or a commanding stump, presents itself by the side of his chosen fishing-grounds, he may be seen perched thereon for hours together, gazing into the liquid depths below for the finny tribes that sport therein, and ever and anon swooping down with amazing velocity and bearing up in his resistless talons the squirming victim. In shallow places his mode of capture is regulated in conformity with their character, gliding over their surface and clutching at his victims as they come within sight.

Figure 47. Nest of American Osprey. Manner of securing food for young.

The food of the Osprey consists mainly of fish, although the reptiles and batrachians that inhabit the swamps and marshes wherein he builds do not escape his vigilance. Almost all kinds of fish, except the very largest, which would be more than a match for his strength, are captured and devoured with avidity. We have watched with a great deal of interest and pleasure his piscatorial pursuits on the shores of Delaware Bay, and have often seen him bear from great depths fish much larger than the common shad. The latter, together with the herring, striped bass and black bass, are favorite articles of diet, while his catchings from fresh-water streams, the larger cyprinidonts, cat-fish and pumpkin-seed, are quite as great luxuries.

When the nesting-time comes around, the last of April or the beginning of May, these birds are not so engrossed with the thoughts of feeding as to be utterly oblivious of the duties which it imposes. Generally the same nest is selected year after year, but when a new one is to be constructed it is not uncommon to find many pairs engaged in its building, the friends of the destitute assembling and laboring with the most determined energy till its completion. A more sociable disposition could hardly be conceived. The spirit which would lead these birds to fly to the assistance of a distressed companion would certainly induce them to co-operate with their brethren in the difficult task of nest-building, especially when such a bulky structure as the species is known to construct would severely task both the time and patience of the pair which is to occupy it. The vast amount of labor and time expended in rearing such a fabric is a sufficient inducement for them not to want to indulge in such employment any more than is absolutely necessary. Hence these nests are constructed for durability. Unlike his European congener, whose nest is placed upon a high cliff, the Osprey almost invariably builds on trees. All nests taken by the writer have seldom been at a greater elevation than fifteen feet, although instances have been recorded where they were twice that height. It is a remarkable fact that the trees supporting these nests are always dead and generally placed in the midst of marshy ground, either completely isolated or surrounded by a dense growth of bushes. At all events, they occupy rather conspicuous positions. It is probable that the excrement of the birds or the saline character of their food has much to do with killing the nesting-trees. Trees which seem vigorous and thrifty at first manifest after a year's occupancy unmistakable signs of death. Not always are trees selected for nesting purposes, for a Mr. W. H. Edwards describes a nest built on a tall cliff on the banks of the Hudson River, not very far from West Point.

Externally the nest is composed of large sticks piled to a height varying from two to five feet, and measuring fully three feet in diameter. Corn-stalks, mullein-stocks and bark are occasionally intermingled with the sticks, but within there is a rather profuse lining of sea-weed and the long grasses which grow so luxuriantly in salt-water marshes. The cavity ranges from fourteen to fifteen inches in diameter, and is unusually shallow in proportion to the size of the nest.

Three eggs constitute the usual nest-complement, although two are sometimes laid, and rarely four, and these are deposited on consecutive or alternate days, at the rate of one egg a day. They measure about two and one-half inches in length and one and three-fourths in width, and are of a yellowish-white color, thickly covered with large blotches of different shades of brown. Incubation follows close upon the last deposit, the task being begun by the female, and devolving principally upon her, although the male occasionally relieves her for a brief spell each day. While she is on the nest, he is a jealous husband and a most faithful provider. The choicest catch of his piscatorial exploits is carried directly to the nest and ungrudgingly administered to the patient sitter. When not engaged in providing for their wants he stations himself upon an adjoining tree, if such should happen to be present, or somewhere in the immediate neighborhood, and exercises the closest surveillance over the nest and its occupant. All attempts at intrusion are most summarily punished. Dr. Brewer mentions a case where a lad essayed to reach the nest in order to rob it of its eggs, when he was assailed with so much violence that the male's talons were driven through a cloth cap that he wore and laid bare the scalp. Experience has proved the risk incurred in visiting these nests with hostile intentions. You may pass and repass underneath the nest, the authors criticising your every movement the while, without calling forth the slightest opposition. When, however, you attempt to mount the tree that contains their cherished treasures, you virtually invite the full measure of their wrath. That the male is affectionately devoted to his partner is shown by Wilson in a case which he cites of a female who was prevented from fishing by a broken leg and that was abundantly supplied with food by her mate.

When the young appear they are objects of more than common parental solicitude, the parents vying with each other in rendering them every needed attention and in providing them with a plentiful supply of suitable food. But one parent is absent from the nest at a time, the other remaining at home to guard against danger. They are ravenous feeders, and soon attain to full development, when they resemble very closely in dress their parents, having the upper parts spotted with pale reddish-brown or white. Adult birds are dark-brown or grayish-brown above, with most of the head, neck and under parts white, the chest in the female, and sometimes in the male, being spotted with brown. The tail, usually paler than the back, has six or seven dusky bars, and is tipped with white.

That these birds may be fitted for powerful flight they are provided with long and pointed wings, the second and third quills being the longest. They have a stout bill, with a very long hook and sharp end. Their feathers are oily to resist water, those of the head being lengthened and pointed, and of the thighs and a little of the front parts of the tarsi short and close together. The legs, tarsi and feet are very strong and robust, and the claws all of the same length and very large and sharp. Rough scales completely invest the tarsi, and the toes are padded below and covered with numerous hard-pointed projections to aid in holding their slippery prey. The iris in some specimens is reddish, but mostly yellow; the bill and claws blue-black, and the tarsi and toes grayish-blue. Male birds are not so large as the females, the latter measuring twenty-five inches in length, and with an extent of wings of fifty-two inches.

Chapter 34

Turkey Buzzard

Few species, if any, have a wider distribution in America than the Turkey Buzzard. It is found more or less abundantly to the Saskatchewan throughout North America, from the Atlantic to the Pacific Coast, and in all portions of South America as far south as the Strait of Magellan. Individuals have been met with in Nova Scotia and New Brunswick, though these birds are generally not common north of Central New Jersey. From Eastern Maine, in the neighborhood of Calais, to Connecticut, specimens have occasionally been captured. In a single instance, Mr. Lawrence observed a small company of nine at Rockaway, Long Island. West of the Alleghenies, from Central America nearly to the Arctic regions, it occurs more abundantly. Without exception, it is found in greater or less numbers in all the Middle, Western, Southern and North-western States. From Lower California to Washington, along the Pacific, numerous parties attest to its common occurrence. The West Indies, the islands of Cuba, Jamaica and Trinidad, the last-named in particular, include it within their faunæ. Honduras and Guatemala, as well as the Falkland Islands, off the eastern coast of Patagonia, are permanent residing-places.

In Pennsylvania and New Jersey, where the writer has had abundant opportunities for studying the species, these Vultures summer quite plentifully. From their first appearance, in March, large numbers may be seen high up in the air, moving in large circles, apparently exploring the ground below for their favorite articles of food. In rural districts they are, however, more frequently observed than in the vicinity of densely-populated towns, the greater abundance of carrion to be met with in the former places doubtless being the cause of this preference. But in California and Oregon, according to Dr. Newberry, they are quite as common near towns as about the large rivers. In our Southern States they visit cities and large villages, and play the part of scavengers in company with the black vulture. They are said to be so tame and unsuspicious in Kingston, Jamaica, that they roost upon the house-tops or prey upon offal in the streets. In country places they are no less familiar and trustful, as is evidenced while feeding. So intent upon their business are they at this time that the presence of human beings is unnoticed, and even when forced to forsake their booty they sullenly repair to a short distance, only to resume their feeding when the annoyance has passed. The common crow has been observed to gather around the same food, and the utmost good-feeling prevailed. A small flock will often settle down upon a dead horse around which several dogs have gathered. The snapping and snarling of the dogs, when they approach them too closely, do not cause the Vultures to retire, but only to step a few paces aside, when, nothing daunted, they continue their feeding, apparently oblivious of their whereabouts and surroundings.

Although the sense of sight is rather keenly developed in these birds, yet that of smell is none the less so. This is an advantage, for both the visual and olfactory organs seem requisite in the determination of the presence of decaying animal matters. As a proof that smell leads to food-detection, Dr. Brewer cites an instance, on the authority of Dr. Hill, where several of these birds were attracted to the house of a German emigrant who was prostrated by fever, being drawn by the strong odor escaping from his neglected food which had become putrid. Mr. G. C. Taylor, whilst a resident of Kingston, sufficiently tested their power of smell. He wrapped the carcass of a bird in a piece of paper, and flung the parcel into the summit of a densely-leaved tree, that stood in close proximity to his window. A moment or two only elapsed, when the keen smell of these birds scented something edible, but they were unable to find it, obviously for the reason that it was hidden from view by the enveloping paper.

Generally their food consists of all kinds of animal matter. They are often accused of sucking eggs, and also of eating the young of herons, as well as those of other birds. In Trinidad, they are said to live on friendly terms with the poultry. As no breach of faith has been reported to have occurred in this instance, it is not likely that they would molest in any way our smaller birds, at least we are not cognizant of any such cases of interference from our own observation, nor do we find them in the recorded experiences of friends. They are worse-disposed, it seems to us, to their own kith. When several are feeding together, most violent wrangles occur over the booty. Each strives to get the lion's share. It is amusing to witness their manœuvres. Some luckless fellow has just discovered a choice and racy bit, which he is endeavoring to make off with in a somewhat hurried manner, when instantly he is beset by a near companion, who has scarcely swallowed his own morsel. A conflict ensues. The latter, being the stronger, succeeds after a little in defrauding the other of his rightful property. When gorged, these birds are stupid and indisposed to exertion, the period of digestion ordinarily being passed in a motionless, listless attitude, with half-opened wings.

Recovered from their semi-stupid condition, they do not at once go to feeding again, but spend a long time in the healthful exercise of their volant appendages. Few birds are more graceful, easy and dignified while on the wing. On the ground they may seem awkward, but it is while soaring at a great height above the earth that they are seen in all their glory. When prepared for lofty flights, they spring from the ground with a single bound, and, after a few quick flappings of wings, move heavenward. Attaining a great elevation, they cleave the ether in ever-widening circles, or sail on nearly horizontal wings, the tips being slightly raised, with steady, uniform motion. These aerial diversions, for such they seem to be, are never performed singly, but in small parties of a dozen or more, being more common in early spring, and at the close of the breeding-season, than at any other time. It is to be observed further that these movements are executed in silence, the only sounds which the Buzzards are capable of producing being a kind of hiss, which has not inaptly been compared to the seething noise emitted by plunging a hot iron in a vessel of water.

When ready to breed these birds look about for a hollow tree, or some stump or log in a state of decay, either upon the ground, or slightly above it. Generally, there are no indications of a nest. In occasional instances a few rotten leaves, scratched into the hollow selected for the deposition of the eggs, constitute the nest, these treasures being laid without any previous care for their preservation and shelter being taken. In Southern New Jersey the nest has been inadvertently strayed upon in the midst of a deep and almost impenetrable morass, where it was found placed upon a hollow stump. Within the rocky caverns along the wide, shallow Susquehanna, as many as a dozen nests have been counted in a few hundred yards of space, often as early as the last week of March in favorable seasons, but generally not till the middle of April. When the winters are not extremely rigorous, a few individuals remain in the vicinity of their breeding-quarters throughout the entire year. We have found the birds breeding in Delaware County, Pa., towards the latter part of April or the beginning of May, but in Philadelphia they rarely do, if they breed at all. In Southern Ohio they are a common summer sojourner. Speaking of the birds in Jamaica, Mr. Gosse says they nest in depressions in the rocks and in the ledges thereof, in retired localities and also upon inaccessible cliffs. On Galveston Island Audubon found the birds nesting in great numbers, either under wide-spreading cactus branches or underneath low bushes in the midst of tall grasses in level saline marshes.

In the vicinity of Cheraw, S. C., Dr. C. Kollock met with the black vulture and our present species in swamps and dense forests, where they congregate in vast numbers throughout the entire year. These places are commonly designated Buzzards' roosts. Audubon once visited one of these roosts in the vicinity of Charleston, which covered more than two acres of ground, and which was completely denuded of vegetation. On the banks of many of the rivers of Southern Texas, Mr. Dresser found them nesting in large numbers, the timber along their borders constituting comfortable and secure shelter; but, contrary to what has always been entertained, he affirms that they build large and bulky nests of sticks, which they place at great heights in an oak or cypress, close by the river-banks. Captain C. C. Abbott says that in the Falkland Islands the eggs are deposited in the midst of bushes beneath high banks, or on the summits of decayed balsam logs, during the early part of November. In certain localities, where the birds are not very common, paired individuals are not infrequently found.

Two eggs generally constitute a nest-full, although instances are known where but a single egg was deposited. On the Falkland Islands they are said to lay three occasionally. In the West Indies, especially in the Bahamas, the complement is the same as in the United States, and there does not seem to be any difference in the habits of the birds in the two places. Specimens from New Jersey, Texas, Florida and South Carolina are creamy-white in ground, and are variously marked with shades of brown, intermingled with splashes of lavender and purple, which are often so faint as to be perceptible only upon close examination. Brewer mentions a variety from near Cheraw, S. C., that was nearly pure white, and which showed but a few small red and slightly purplish lines and dots about the larger extremity. Recently we have met with some from Texas answering the same description. In dimensions these eggs vary but little, and have, on the average, a length of 2.78 inches, and a width of 2.00, or rather less.

Cathartes aura, as the Turkey Buzzard is known by the scientific naturalist, is far from being demonstrative in the expression of her feelings. When her home is assailed, she makes no ado, but quietly slips out, and seemingly contemplates its desecration with indifference.

Though manifesting a passive disposition in the face of human interference, yet she is not always the gentle being she would have us believe, as shown by the spirit of dominancy she displays over her own household.

Unlike many of her neighbors, she does not entirely assume the responsibilities of brood-raising, permitting her partner the happy enjoyment of a life of luxurious ease, but, believing in the doctrine of a proper division of labor, forces upon him his share of the work.

Whilst she thus appears unduly exacting towards him, she is equally so to her offspring. Few mothers know better than she the right training of their children, so as to fit them to become useful and respectable members of society.

This is no figment of the imagination, as will presently be seen. It was while exploring a section of Delaware County of this State for minerals in the summer of 1894 that some interesting facts were learned of the relation subsisting between her and the rest of her family.

Having accidentally strayed upon a young ground hog which had but recently been killed, the writer resolved to carry it home and place it where it could be seen or scented by the Buzzards, so that he might have an opportunity of making a more intimate acquaintance with these birds than he had ever before been able to make.

Accordingly the dead animal was transported to a meadow overlooked by the house he was occupying. The resolution was well taken, for on the fourth day after the deposit had been made several Buzzards were seen circling high overhead, mere specks against the blue dome of the sky, evidently scanning the earth beneath with their telescopic vision for the presence of food, or endeavoring to scent it with their keen sense of smell.

Nearer and nearer the flock drew earthward, till finally, a full hour being spent in graceful manœuvring, the birds settled down upon the green-carpeted meadow, but a few yards from the carrion that lay festering with vermin.

Their feathers adjusted, and folded to rest their wide-spreading pinions, the young, in obedience to orders, as it seemed, leaped on to a huge pine log that lay near by, while the old folks surveyed, wistfully and long, from their standpoint of observation on the ground, the odorous carrion a few feet away, as if whetting their appetites for the feast they were soon to enjoy.

With a few quick steps, that were meant to be graceful, the female drew near, but the male lingered doubtingly behind. In a trice she was busy at work, tearing with claw and with bill the daintiest morsels. Rendered mad by the smell of the food the male, no longer seeming backward, pressed forward to her side, but only to retreat before her savage assaults. Again he essayed the attempt, and was beaten back as he had been before. Convinced that further effort would be useless, he strode sulkily to a distance, where, in moody contemplation, he nervously awaited her ladyship's sweet pleasure.

Being filled to the full the female now moved lazily away to a clean patch of grass, where she immediately set to work to arranging her toilet,—wiping her bill and her claws upon the green carpet before her, craning her neck and stretching her pinions, yawning and gaping and gaping and yawning,—and finally ending all by seeking the topmost rail of a near-by fence for rest and composure.

With nothing to fear, the male now stalked complacently forward, and was soon hard at work at what was left of the carcass. His appetite less capacious than that of his lady, his dinner was soon over, and off strode he too to a fresh spot of grass, where he went through the same process of wiping his mouth and stretching and yawning, which, being finished, he mounted the rail by the side of his mistress.

More interesting far than either the parents were the three black creatures that stood upon the pine log. Fixed to the spot as though they had grown there, with scarcely moving heads and downcast eyes they eagerly watched the food disappearing, wondering, mayhap, as children are prone to do, if it would all disappear before they had a chance of testing its virtues, but maintaining their souls the while in perfect serenity of repose. But their time had at length arrived, and down from the log they cast themselves *instanter*, three lusty fellows as large as the parents, but one of them, from his limping gait, proving to be lame. Great consideration was shown the disabled one by the others, who permitted him to feed first, while they stood aside until he had satisfied his hunger, when, without the least bit of ceremony, or the least indication of ill-nature or selfishness, they too set to work, finishing in quick order whatever edible was left of the dead animal. Their actions after feeding were exactly the counterpart of those of the parents. Having finished their toilet, the three sought the rail by the side of the father, where, like their illustrious heads, they were soon occupied with the most self-satisfying thoughts, utterly oblivious, as it seemed, of time and surroundings.

Figure 48. Female Turkey Buzzard Dining. Male and young awaiting her ladyship's pleasure.

More than an hour was thus spent in drowsy meditation, when, as by common consent, they all, one after the other, leaped to the ground, where they busied themselves preening their feathers and preparing for departure. The time being ripe, the female set the example. With a run of a half-dozen yards to gain a good start, she was soon on the wing, and in fifteen minutes or more was lost in the ether. The male followed suit, and when he had vanished from sight, the young, one after another, mounted the atmosphere, and gradually circling their way through its limitless depths, were also soon lost to the earth-chained beholder.

Concluding this brief chapter of bird-history, we have a few brief comments to make. To the uninitiated in science matters, the statements just made must seem well nigh incredible. But there were other witnesses of the facts, just and reliable observers, too, whose testimony could be appealed to, to settle all doubts of their authenticity. From all that has been said, it cannot but be evident that the female was the acknowledged head of the family, a sort of feathered autocrat, whose will was the law by which the family was governed. Even the male, who did not always respect her authority, especially where her interests conflicted with his own, was made to see that might makes right when confronted with her stronger and more powerful nature. But it was the patience and orderly behavior that characterized the nearly-grown young, and their sweetness and gentleness of disposition under the most trying circumstances as well, that impressed us as extremely wonderful, and led to the opinion that man-born offspring might here learn a lesson of filial obedience and respect that would greatly redound to the honor and glory of the race.

When captured, these birds offer no active resistance, but very effectually warn off their aggressor by vomiting up the half-putrid contents of their crop. They will often simulate death at such times. On one occasion an individual having been shot by Dr. Coues was picked up for dead. While being carried to the Doctor's tent, it was perfectly limp. On reaching his quarters, he carelessly threw it upon the ground, and went to work at something else. After a little, upon looking around, he beheld to his great surprise that the bird had changed position, and was furtively glancing around. On going up to it, its eyes instantly closed, its body became relaxed, and it lay perfectly motionless, and apparently lifeless. After compressing its chest for several minutes until he fancied life extinct, he dropped the bird and repaired to supper. Upon his return the bird was gone, it evidently having scrambled into the bushes as soon as he had turned his back upon it.

The young, when first hatched, are covered with a whitish down, and are fed upon half-digested matter which is disgorged by their parents. When taken from the nest and kept in captivity until fully grown they become exceedingly tame, and will feed on fresh meat, earthworms, crickets, grasshoppers, and other large insects, which they apparently relish, and oftentimes will also eat bits of bread, cake and particles of apples or pears which are thrown before them. The benefits which these scavengers render are too well known to need any comment. In the mature state the plumage of the Buzzard is brownish-black, and more or less glossy, the quills being paler on the under surface. The skin of the head and neck is red and wrinkled, and with scattering bristle-like feathers, the bill whitish, legs and feet pinkish, iris grayish-brown, and nostrils large and oval. Their length is about thirty inches, extent of wing seventy-two inches, wing being about twenty-five, and tail twelve.

Chapter 35

Rare and Curious Nests

From time immemorial it has been the current popular belief that birds of the same species never varied their style of architecture, but constructed the same form of nest, and out of the same materials, as their remotest progenitors did, instinct being the principle by which they were guided. This opinion, though long since exploded by scientific research, is still, I am sorry to say, entertained by persons who should know better. An examination of nests from different and widely-separated localities affords evidence of the most convincing character of its erroneousness. Most marked differences will always be found to exist in composing materials, as these are sure to vary with environment, and in a wider degree in the nests of some than in those of other species; even configuration, which is less prone to change, is often influenced by circumstances of position and latitude.

Among the Thrushes, the nest of the Robin is the most addicted to variation, and this is not wholly restricted to the constituents of its usually mud-plastered domicile, but is quite frequently observed to occur in the arrangement of materials, and in contour and position as well. Where low marshy woods abound on the outskirts of towns and villages, as is the case in Southern New Jersey, nests of this species have been taken that contrasted in a most wonderful manner with those one is accustomed to see in more northern localities. The great masses of grayish-green fibrous lichen, which depend from shrub and tree in sylvan marshes, are most freely used, and from its very nature to mat when pressed together all necessity for mud is precluded.

But the most curious nest I have ever met with was built upon a railroad cutting, where the ground had a slope of more than forty-five degrees. Such a position for a dwelling of the kind the Robin is known to build, to one not conversant with the facts, must appear incredible. But that it was accomplished, the nest itself was the monument of the builders' thoughtful skill and labor. A semicircular wall of mud, eight inches in diameter and five inches in height, constituted the groundwork, and within the cavity thus formed was reared a coarse, substantial, bulky fabric, that was entirely composed of the stems of grasses, leaves and roots, loaded down and held in place by pellets of mud.

Figure 49. Nest of the Robin. Built upon a railroad cutting.

Figure 50. Nest from the Top. Section of Two-Storied Nest. Copyright 1900 by A. R. Dugmore.

Red-Eyed Vireo's Two-Storied Nest with Cow-Bird's Egg Beneath

A more remarkable position, and one that seemed as difficult to manage, I shall now relate. Few birds care so little for position as the common House Wren. Almost any place answers its purpose. Near the little town of Thornbury, in the State of Pennsylvania, a pair of these birds, in the summer of 1882, took possession of a derrick, and, as a site for a home, selected the space over a sheave in one of the stationary blocks, where, in due time, was deposited their rude, yet comfortable, nest of sticks and feathers. A similar structure occupied the spot the previous year, and a brood of eight birds was raised. It was not the elements of composition of these nests that excited interest and surprise, for they are not materially different from the usual form, but the strange, anomalous situation which they occupied. So dexterously were the materials arranged within the space that the revolution of the wheel was not in the least interfered with. The nest was approached on the side facing the rope that moved the pulley. The opposite side could have been used for this purpose, and doubtless with less danger to life or limb, but preference seemed to be shown for the other. Why this was so was for some time a mystery. But when the birds were seen to alight upon the rope at the top of the derrick and ride down to the nest, the explanation at once became apparent.

Never did linnet enjoy the rocking twig, or bobolink the swaying cat-tail, with half the zest than did these eccentric creatures their ride down the rope. A hundred times a day, when necessity arose, they treated themselves to the pleasure, the rope all the while moving at the rate of thirty feet in a second. Six of the seven days, from early morn till night, they availed themselves of this strange conveyance, and never a danger occurred to mar their delight. In due time a family of happy, rollicking children was raised, and the nest on the derrick deserted.

More beautiful are the nests which the Red-winged Blackbirds build. These are the birds that affect our swamps and marshes, and make the air ring with their loud, clear, resonant notes. Before me is a nest that surpasses in beauty the average structure. It is a bulky affair for the species, but so symmetrical in contour, and so quaintly, deftly woven, that the eye never tires in looking at it, nor the mind in contemplating its wonderful mechanism. Broad ribbons of grasses are its composing materials, and eight of them are so woven into the nest as to securely fasten it to the tall typhas in the summit of which it was placed.

But a more clever nest of these birds, and one that is as unique in shape as it is in texture and composition, was found in the summer of 1879 in the vicinity of Philadelphia. It was built in a field of timothy, many of the stalks of which being wrought into the fabric. Its shape is that of an inverted cone, and so beautifully, symmetrically and compactly put together is it, that one could hardly credit the builders with the possession of the skill necessary to the production of so perfect a domicile. Externally the nest is formed of grasses and rushes, neatly and intricately interwoven, with here and there a head of the dry pappus of some species of hawkweed. Sedges and fine grasses make for it a cosy and comfortable lining. This nest shows quite conspicuously in the drawing, but in its natural position, in the centre of a large field, the authors had spared no pains to have its concealment as perfect as possible.

Figure 51. Red-Winged Blackbird's Nest. Located in a field of Timothy.

Typical nests of these Blackbirds are somewhat irregular in outline, and rather coarsely and rudely built of stubble and broad grasses, variously intermingled, and lined with soft meadow grass. Usually they are placed in clusters of weeds or in the tops of small bushes alongside of streams of water. High positions are seldom chosen for nesting purposes, as they offer poor facilities for food-collecting, the aquatic larvæ, may-flies, dragon-flies and mosquitos, which constitute a prominent part of the diet of these birds, being only found in marshy situations. Small bushes along the margins of streams, from the double advantage which they possess, are almost exclusively adopted in certain localities. Being convenient to appropriate food-stuffs, they are, at the same time, out of the reach of snakes, especially water-snakes, which have a decided fondness for young birds.

Of the sub-family of Orioles, to which the Red-wing belongs, no member, unless it be the namesake of Maryland's distinguished proprietor, builds a more magnificent nest than the one that inhabits the orchard. In the books it is known by the no means euphonious title of *Icterus spurius*. Its nest is shaped like a pouch, and generally pensile. Soft, flexible meadow grasses, neatly and compactly woven together, make up the outer fabric, while within is a lining of vegetal or animal wool, or one of fine grasses intermingled with horse-hair. But the handsomest ever seen was one that was found in the vicinity of Nazareth, Pa., by Richard Christ, in the summer of 1883. It is of the usual size, five inches in height, three in external diameter, but differing from the normal form only in materials of composition. The proverbial meadow grasses are absent, and in lieu thereof are the headed stems of such as grow by the roadside, notably conspicuous for their golden brightness in a state of desiccation.

Figure 52. Double Nest of Orchard Oriole. Female sitting, male standing guard.

Figure 53. Female Baltimore Oriole. Nest the exclusive work of her bill.

More noteworthy, however, than the Nazareth nest, is one that was removed from a silver maple-tree. It is a double structure, composed of long, flexible grasses, and is firmly bound by the same to several small, slender branches. The larger nest, inversely sub-conical, is joined to the smaller, somewhat similarly shaped, but less compact in structure, by ribbons of the same kind of grass that composes the nest. A circular opening, one inch in diameter, is a noticeable feature of the smaller. That the additional structure served some useful purpose there can be no doubt. I am inclined to believe that it was built for the accommodation of either parent while the other was sitting. The aperture was a convenient outlook for the non-sitting bird, who, from this position, could with little difficulty, like a sentinel from an outpost, detect the approach of an enemy.

But nothing can exceed in beauty and skill the nest of a female Baltimore Oriole in the writer's possession. It was built under peculiar circumstances, the builder being a prisoner, having been taken from home when quite a fledgling. A male companion was brought away at the same time. These birds, the property of Dr. Detwiler, of Easton, Pa., in 1883, were a source of considerable pleasure to that elderly gentleman in his leisure moments. Under his careful, kindly management, they became quite tame, the female manifesting greater familiarity than the male. That either would become so accustomed to confinement as to evince a desire to build never entered the mind of the Doctor. They had, when he was alone, the freedom of his studio. One lovely June morning, the outside world brimming over with life and joy and sunshine, the door of their cage was thrown open, and the Doctor settled himself into a soft easy-chair to read. Hardly had he scanned a dozen lines, when something pulling at his hair caused him to drop his paper and look around. He was not slow to detect the offender in the person of his female feathered friend who was seen flying towards the most distant corner of the room with something, resembling hair, in her bill. The reading was resumed, and again the culprit stole cautiously to where he was sitting, and, seizing another hair, was off in a twinkling.

Permitting for a while these liberties, and noticing that bits of strings were, when placed in positions to be seen, quite as much the objects of interest as the hairs of his head, the Doctor was not slow in divining the motive which led to this strange and unexpected behavior. Convinced by actions, as significant as words themselves could be, that his little friend was desirous to build a home, he began to cast about for a corner where she could be free to carry out her intentions without fear or interference. The attic furnished the place, and after fitting it up with a large tree-branch for a perch, and plenty of new white strings for building purposes, he bore his favorite and her partner to their new quarters. Soon the female became at home and entered into her voluntarily-imposed labor with alacrity, and at the end of a week had constructed a domicile which her untamed prototypes of field and roadside would strive in vain to excel. But the male would have nothing at all to do with the matter, but remained the same cold, indifferent being as I found him to be upon my first introduction.

Some nests are curious on account of shape. The birds often, it would seem, try their very best to see how oddly they can build their homes. The little Acadian Flycatcher, so common in Eastern Pennsylvania during the breeding-season, sometimes appears to be controlled by cranky ideas with regard to building. Dry blossoms of the hickory are the materials it ordinarily uses, and they can always be obtained whenever needed, but in a nest discovered by the writer in 1882, not a blossom was to be found, but in place of them there were long, stringy fibres of the inner bark of some species of herbaceous plant, which the birds had modelled into a compactly-built, shallow, saucer-like cavity, and from which they had caused to depend a gradually tapering train of the same for nearly nine inches.

Figure 54. Acadian Flycatchers. Nest curious on account of its train.

The King Bird, a distant relative of the Flycatcher, often displays as much eccentricity. Once upon a time a pair of King Birds took a fancy to an old apple-tree that stood within a few yards of my Germantown home. It was certainly not a place of quiet and retirement, for scores of noisy, dirty children daily resorted to its leafy shelter for coolness and pastime. But the birds were not the least disquieted. They had fixed their minds upon the spot, and build they did. The nest was posed between a forked branch, just out of the reach of the urchins. It was a crazy affair. Black, slender roots, wrinkled and knotted and tendrilled, made up the body of the fabric. As it was nearing completion, the opportune discovery of a bunch of carpet rags was hailed with delight. They were instantly appropriated, and promptly adjusted to the outside, but in such a manner that long ends, some fourteen inches in length, were made to project from the sides and bottom. Whether all this was for ornament or protection, or for both, I could not say, but I am inclined to think that safety was uppermost in the minds of the builders, for, looking from below at the nest it seemed but a mass of rags that had been thrown into a tree-crotch, which, the birds perceiving, and its close resemblance to an entangled bunch, had utilized.

Figure 55. Long Billed Marsh Wren's Nest. Copyright 1900 by A. R. Dugmore.

Certainly no more beautiful nests in shape exist than the spherical in form. The Long-billed Marsh Wren builds a nest of this type. Upon its arrival in the spring it seeks the inland swamps, or the brackish marshes of the sea-shore, where, amid the splatterdocks of the former and reeds of the latter, it finds suitable shelter and protection. There, day in and day out, during its entire summer stay, it pursues the even tenor of its life, happy and contented, never caring, like many of its remoter kin, for the charmed circle of man. Active, energetic and buoyant with hope, it skips about the tall rank herbage, in every direction, in quest of insects, making its presence known and felt by the lively chattering song, which resembles more nearly the sounds of an insect than those of a bird, which emanates from its grassy haunts. As these birds reach their breeding-grounds early in May, nest-building is soon begun, but so secret and mysterious are their movements at first, that we hardly know anything of their presence, except when they are colonized for the summer. The labor of building is entered into with considerable alacrity, and is mainly the result of the combined labor of both birds. Their nests are usually placed in low bushes, a few feet above the ground, or woven into the tops of sedges out of the reach of ordinary tides; but in very rare instances upon the ground in the midst of a clump of grasses. Ground nests are loosely-constructed affairs, which is not the case with those that are elevated to the tops of tussocks, or to the branches of shrubs and trees, which require more compactness and a better finish. The most beautiful, as well as artistic, nest which I have ever seen is the one shown in the cut. This nest was discovered in the vicinity of Philadelphia in the summer of 1878. A willow-branch, some fifteen feet above the ground, which was bifurcated, was made to do service. No ordinary skill was that which surmounted the seemingly insuperable difficulty of building a nest, not pensile in character, to such a swaying branch. That the birds accomplished the feat the nest itself was the evidence. In form this nest was nearly globular, four and a half and five inches in the two diameters. It was woven of the broad leaves of a species of scirpus, closely and evenly, and had its interstices well seamed with brownish cottony down. A thin delicate curtain of gauze, of the same material, hung around the opening, and this was continued within, forming a thick bedding of the softest, fluffiest nature, of which the most voluptuous sybarite might envy its fortunate possessor.

But the little Golden-crowned Kinglet, a mere mite of flesh and feathers, but with a great deal of spirit, builds a much handsomer nest. It is the perfection of symmetry. Man could not make with all the appliances at his command any thing more nearly globular. But its beauty! It looks like a ball of green moss, the delicate patches of moss being so artfully arranged as to completely hide the dry stems of grasses that constitute the walls. No moss ever spread itself over the ground, or over a stump or tree-trunk, more evenly. When it is known that this Kinglet builds its nest among the slender feathered branches of the hemlock spruce, there is manifestly a reason for the fern-like tracery upon the exterior, so necessary for the preservation of its home. Such a handsome and imposing structure would be far from complete were the inside not in keeping with the outside. But the birds have left nothing to be desired in this particular. The softest and purest of down lines the little bed-chamber, and even swells in its lightness till ready to overflow the neat circular door-way.

Figure 56. Long-Billed Marsh Wrens. Nest placed out of the reach of tides.

Figure 57. Golden-Crowned Kinglets. Nest the perfection of beauty and symmetry.

Perhaps the most graceful thing you may ever expect to find when on the quest, fitted to be considered the work of the fairies, is the pretty lace hammock of the Parula Warbler. You must search for it early in June, in remote but thin woods, but never far from running water. Often you will see it upon a branch overhanging a stream. The slender twig of a birch is sometimes chosen for its suspension, the terminal spray of a hemlock spruce, or a horizontal branch of a white oak. Like a watch-pocket, with the opening in the side, it is lightly suspended. It is made of a delicate lace-work, the grayish-green usnea moss, that grows on old trees. The whole fabric is the work of two little birds with slate-blue backs and yellow breasts. No other bird of our fauna builds a nest akin to its swinging, eery nest. It is true much of the material is found in position when the builders commence their labor, but the exquisite outline and finish, as well as the cozy interior, are due to the skill of the birds themselves. Even when the structure is just so far completed that occupancy by the female is possible, the male never wearies of its adornment by additional filaments of usnea brought from a distance. He is the happiest of fellows, for his little beak always finds something to do while his patient wife is busy with the duties that lead to maternity.

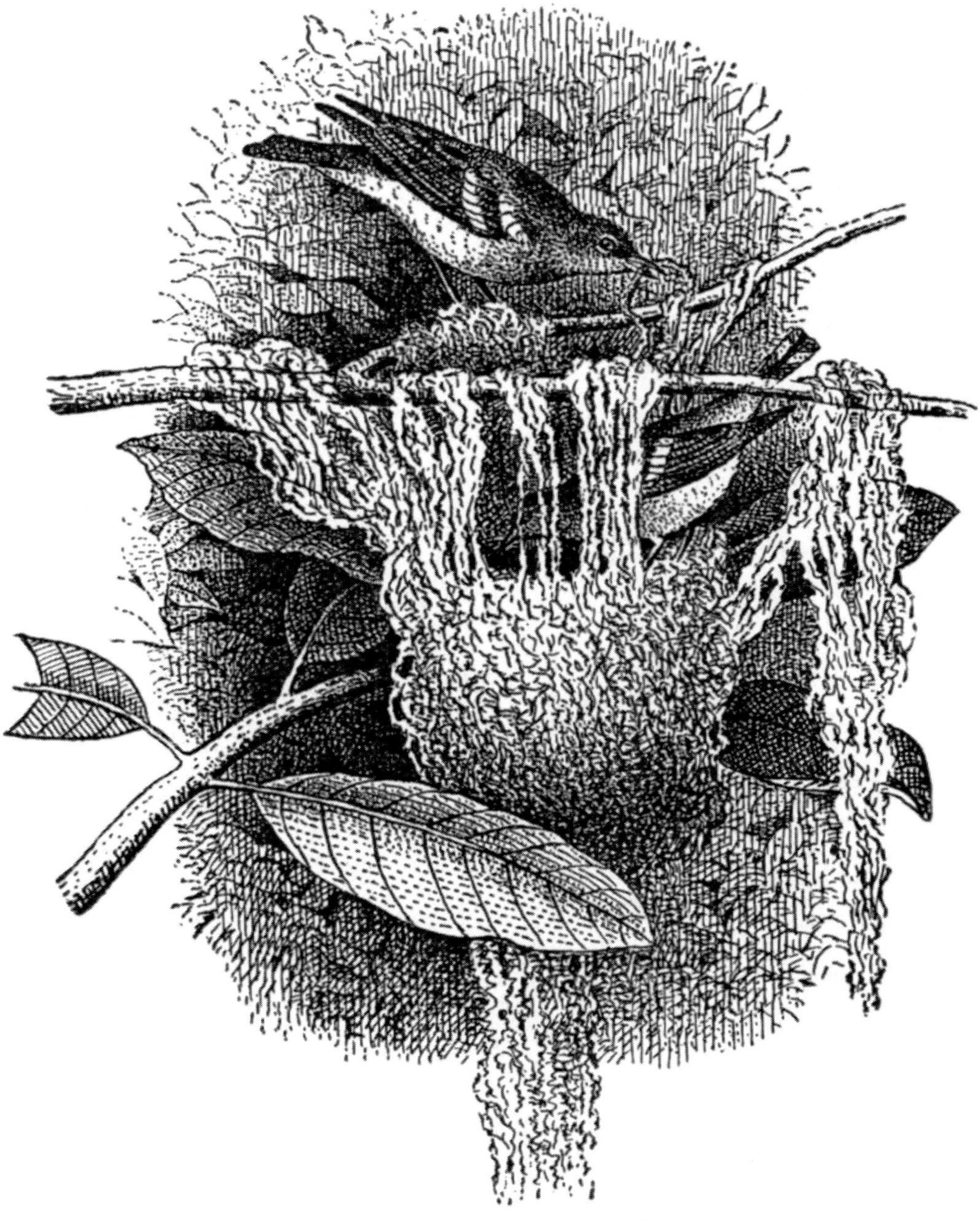

Figure 58. Lace Hammock of Parula Warbler. Female entering nest and male adding to its adornments.

Hundreds of nests, quite as novel as any that have been described, might be instanced, showing varieties from so-called normal forms, but I shall content myself with only another example. Everyone is familiar with the Ruby-throated Humming Bird, so common in the eastern half of the United States. It is the smallest of all our birds. But its nest, which is by no means scarce, is a rare sight to the average man and woman. No nest can be compared with it. It is a thing of beauty and a joy forever. A mass of cotton, with a hole in the top, and thatched all round with blue-gray lichens, and just as big as a walnut, conveys a good idea of its appearance. But all nests are not made of cotton. The yellow wool that forms the dress of the undeveloped fern-frond, or the red shoddy that is wind-swept into heaps outside some woollen factory, is often made to take the place of the down of the seed of the poplar. Not to be mentioned in the same breath with these, is the nest I am now about to describe. It was saddled upon the horizontal bough of a small white oak-tree that grew on the side of a thicket, and was peculiar from the nature of the material that composed its inner fabric. This substance resembled burnt umber in color, and was as soft as the finest wool, or the fluffiest down, and proved, upon examination, to be the mycelium of a fungus which the builders had gathered from decaying stumps or mildewing tree-branch.

Chapter 36

STRANGE FRIENDSHIP

Somewhat widely distributed throughout temperate North America, but nowhere very abundant, is the little Acadian Owl, or Saw-whet Owl, as he is popularly designated. In Eastern Pennsylvania he seems notably scarce, but this may be attributed to his pre-eminently nocturnal and secluded habits. Being a denizen of dense pine forests, and only venturing abroad in quest of food at the close of the day, his presence and numbers remain to many a mystery. Hollow trees, and the dark caverns of rocks, are his natural retreats, and as these are to be met with largely in densely-timbered regions and sequestered localities, he is seldom, if ever, seen in close proximity to human habitations. He seemingly shuns rather than courts the society of man. When routed from his burrow in the broad glare of day he becomes very much bewildered, and is scarcely able to escape the approach of danger.

The common appellation of Saw-whet Owl, which is applied to the species, owes its origin to the close resemblance which the notes of the bird bear to the noise produced by the filing of a saw. These notes are so deceptive, that persons unacquainted with their source have fancied themselves in the vicinity of a saw-mill, or in near presence to a woodman occupied in whetting a saw. Audubon, hearing these notes in a thicket for the first time, was thus deceived. The same distinguished writer gives, on the authority of Mr. McCullock, an interesting description not only of the notes of this Owl, but also of his remarkable powers of ventriloquism. On a certain occasion his informant was aroused by what appeared to be the feeble tones of a distant bell. But in nearing the spot whence the sounds emanated, they apparently shifted from point to point, being heard at one time close by, and at the next moment in the distance, now on the left, then on the right, and as often in the rear as in the front. Finally the author of these sounds was discovered at the entrance of his burrow in a birch-tree. Stationing himself at the base of the tree in full view of the bird who was calling to his partner, Mr. McCullock had a splendid opportunity of observing an exhibition of his singular and exceptional ventriloquial powers.

Smooth, gliding and noiseless is the flight of this Owl, and but slightly elevated and protracted. When seeking for food he may be seen sailing over low meadows in the neighborhood of his accustomed haunts, or, perched upon a stump or fence-rail adjoining thereto, quietly gazing in every direction for whatever of life may chance to manifest itself, which he seizes with remarkable adroitness, even sometimes venturing to assail prey larger than himself. The smaller birds, awakened from their perch by his cries, fall ready victims to his rapacity.

Hollow trees, or the deserted nests of other species, are selected for breeding-quarters. The eggs, varying from four to six in number, are pure white, sub-spherical, of crystalline clearness, and measure one and one-eighth by one and seven-eighths inches. The food of this interesting little Owl, which is not so large as a robin, though appearing bulkier, consists of small quadrupeds and birds, but chiefly of various species of insects.

When taken quite young, and held in confinement, this Owl becomes quite tame, permitting strangers as well as his keeper to handle him with the utmost freedom, without so much as resenting such familiarity. But a greater attachment is manifested towards the master whom he is able to recognize by the sound of his voice, and in whose presence he is peculiarly fascinating and agreeable.

Like *Scops asio*—the Red Owl—he leads a solitary existence, save on the approach of warm weather, when the sexes are discovered together, or are heard calling one to the other. Mating commences early in April, and about the middle of the month the birds have located their nests in the hollow of a tree, about twenty feet from the ground, where the female lays her complement of eggs. The entrance to the hole is very small, scarcely two inches in diameter. Upon the female devolves the whole work of incubation, although the male takes a hand in raising the young. The latter leave the nest about the first week of May, and when disturbed make a noise that sounds like a dog sniffling the air, which, when heard, especially at night in heavy timber, is quite certain to startle one and make him fancy a bear or some such animal up a tree near by.

Some years ago there lived in the hollow of an oak tree, not far from Germantown, a common Chickaree Squirrel—*Sciurus Hudsonius*—with this little Owl as his sole companion. This association reminded me of the connection of the burrowing owl of the West with the singular settlements of the prairie dog, the life-relations of the two creatures being really intimate in very many localities, although the owls are simply attracted to the villages of the prairie-dogs as the most suitable places for shelter and nidification, where they find eligible ready-made burrows and are saved the trouble of digging for themselves. Community of interest makes them gregarious to an extent unusual among rapacious birds, while the exigencies of life on the plains cast their lot with the rodents. That the owls live at ease in the settlements, and on familiar terms with their four-footed neighbors, is an undoubted fact, but that they have any intimate domestic relations is open to question. That the quadruped and the birds are often seen to scuttle at each other's heels into the same hole when alarmed is no proof that they live together, for in such a case the two merely seek the nearest shelter, independently of each other. In the larger settlements there are thousands upon thousands of burrows, many of them occupied by the dogs, but more, perhaps, vacant. These latter are the homes of the owls. It is possible that the respective retreats of a dog and an owl may have one common vestibule, but this does not imply that they nest together. There are fewest owls in the towns most densely populated by the dogs and the greatest number in the deserted villages, and this is strong evidence in point. But the owls are by no means confined to the dog-towns, nor even to the similar communities of other gregarious spermophiles. They sometimes occupy the underground dens of wolves, foxes and badgers. When the subject has been carefully investigated, the owls never appear to enter the same hole or burrow with a squirrel, and a squirrel is never seen to enter a burrow that was occupied by owls, however strongly he may be tempted by fear to enter the first hole he should come to. The spermophile never likes to enter any burrow but his own, and has been known to run past any number of inviting entrances in order that he may hide himself in his own domicile.

In the case of the Chickaree Squirrel and the Saw-whet Owl, they occupied the same hole together in perfect harmony and mutual good-will. It was not an accidental occurrence, the Squirrel merely seeking the cavity to escape a danger that impended, for the bird and the Squirrel had been repeatedly observed to enter the hole together, and in the most amicable manner possible, as though they had always shared the apartment. Ordinarily the Chickaree is a very pugnacious creature, attacking with the greatest fierceness the gray and black squirrel whenever they had the temerity to cross his path. He seems to be ever bent upon blood. Though strictly by nature a rodent, subsisting principally upon nuts and the bark of trees, which his powerful incisors enable him to manipulate effectively, yet he has not always remained true to his instincts, for he has been frequently detected in eating the eggs of birds, and also in the seizure of the feathered denizens of our lawns and woods, which he will capture with all the skill of the blood-thirsty weasel. His method of operation is peculiar. He will lie in wait, concealed from view by the dense foliage of the trees which he is wont to affect when in quest of game, and when some unsuspecting bird hovers near pounces upon it with unerring precision, and effecting its capture proceeds to suck, sitting up in true squirrel fashion, the life-sustaining fluid through a wound inflicted in the side of the neck. Having satiated his thirst, which may have been the prime object of the capture, the dead body of the bird is dropped, and the little monster, upon erect haunches, poses, the embodiment of perfect contentment.

But in the case of the Owl it was otherwise. Perhaps it was too large for the monster to attack, or, knowing from rumor of gossiping friends the reputation of the former for cruelty and murder, a conciliatory spirit was thought the best to adopt. No one knows the bitter character of the first interview, or whether a liking for each other sprang up from the beginning. Be this as it may, there can be no denying the fact that a friendship was cemented between the two animals, widely divergent in structural peculiarities as they are known to be, that gave hope of becoming long and enduring.

Figure 60. Saw-Whet Owl and Chickaree Squirrel. Living together in perfect harmony and mutual good-will.

D. 948, the wild Cat was rife in the British Islands and was considered as a vicious animal, which must be destroyed, and not a useful one to be protected by the law. How we came into the possession of the Cat is a matter of conjecture, the current belief being that it was imported from Egypt into Greece and Rome, and thence into England.

Chapter 37

NATURE'S LITTLE STORE-KEEPER

One of the most familiar of North American quadrupeds is the Hackee, or Chipping Squirrel, as he is sometimes termed, from the strange, quaint utterances which he emits while rollicking with his fellows or in quest of something to eat. He is a beautiful little creature, notable alike for the dainty elegance of his form and for the pleasing tints with which his dress is arrayed. His general color is brownish-gray upon the back, warming into orange-brown upon the forehead and hinder quarters. Five longitudinal black stripes and two streaks of yellowish-white adorn the dorsum and sides, which render him a most conspicuous being and one readily distinguishable from any other animal. His abdomen and throat are white. He is slightly variable in color according to locality, and has been known to be so capricious of hue as to become a pure white or a jetty black. But for the commonness of the species, which is found in great numbers in almost every place, his fur, from its extreme beauty, would long since have taken nearly as high rank as sable or ermine.

No quadruped is so brisk or so lively. His quick, rapid movements have not inaptly compared him to the wren. As he whisks about the branches of the brushwood and small timber among which he is chiefly met, or shoots through their interstices with his peculiar jerking movements, and his odd clicking cry, like the chip-chipping of newly-hatched chickens, the analogy between himself and the bird is strikingly apparent. Occurring in great plenty, and being a bold little creature, he is much persecuted by small boys, who, with long sticks, and well-directed blows, manage to fell to the earth many a luckless fellow as he endeavors to escape his pursuers by running along the rail fences.

Figure 61. Chipping Squirrels Feeding. Copyright 1900 by A. R. Dugmore.

Hackees delight in sequestered localities. There they tunnel their homes, preferring some old tree, or a spot of earth sheltered by a wall or a bank. Their burrows are rather complicated affairs, running often to great lengths, so that the task of digging the animal out of his retreat becomes one of no easy accomplishment. Sandy patches of ground, on the outskirts of a woods, are not unusually chosen for burrows. A hole, almost perpendicular, is drilled into the earth to a depth of three feet, and is thence continued with one or more windings, rising a little nearer the surface until it has advanced some nine or ten feet, when it is made to terminate in a large circular nest, made of oak leaves and dried grasses. Small lateral galleries branch off from the main burrow, in which these provident little creatures lay up their winter's provisions. Wheat, Indian corn, buckwheat, hazel-nuts, acorns and the seeds of grasses have been found in their underground receptacles, a proof, were further evidence lacking, that they do not pass the cold famine months in a sluggish and benumbed condition. Several layers of leaves, aggregating nine inches in thickness, are often found over the entrance, as a protection from frosts, which are further prevented from intrusion by the sealing up of the mouth from within.

Everything is done by the Hackee in a business-like manner. In gathering his food, lest the sharp beak of the nut may injure his cheeks when he places the fruit in his pouch, he nips off the point, and then by the aid of his fore-paws deliberately pushes the nut into one of his pouches. Another and another are similarly treated, and taking a fourth between his teeth, he dives into his burrow, and, having packed them methodically away, returns to the surface for a fresh cargo. Four nuts are his load at each journey. With his check-pouches distended to their fullest capacity, and laboring most truly under an embarrassment of riches, the little fellow presents a most ludicrous appearance.

When menaced by foes, by which so defenceless and conspicuous an animal is sure to be surrounded in great numbers, the Hackee makes at once for his burrow, and is there secure from the attacks of nearly all enemies. One foe there is, however, that cares naught for the burrow, but follows the poor Hackee through all of its windings, and never fails to attain his sanguinary object. This remorseless foe is the stoat, or ermine, whose only *penchant* is the blood, and not the flesh, of his victim.

Early in November the Hackee moves into his winter-quarters, excepting in occasional instances when the sun shines with peculiar warmth, and is not seen again until the beginning of spring. The young, to the number of four or five, are produced in May, and there is generally a second brood some time in August. A rather pugnacious animal is the male Hackee, and during the combats which are frequently waged when several males meet, their tails have been known to snap asunder from the violence of their movements, for these members, it is undoubtedly true, are wonderfully brittle in their structure.

Figure 62. Hackee, Or Chipping Squirrel. Laying up food for the famine of winter.

Pretty as he is, and graceful as are his movements, it hardly pays to keep the animal in a domesticated state, for his temper is very uncertain, and he is generally sullen even towards his keeper. But could he be induced to take to the life of a captive kindly and pleasantly, he would, by his cunning little ways, prove a most agreeable companion.

Some years ago an American writer of note had a pair of these animals which made their home in the foundation wall of her house. A row of wild cherry trees stood near the lawn in the rear of the building, which the little fellows were wont to visit many times daily, carrying off in their pouches quite a number at a time of the numerous cherry pits that lay scattered over the ground.

The season being dry, one morning early the person to whom reference has been made repaired to the lawn and poured a pitcher of water over some plants that grew near her porch, when one of these squirrels was observed to pass among them on his way to the trees. He paused from his journey, sat up on his haunches, took one of the wet leaves in his hands, pressed the sides together for a trough for the moisture, and holding it to his mouth drank from it the water in the most comical fashion imaginable. He then went to another and another, drinking from five or six leaves in all, while she stood watching near by. A large saucer filled with water was placed near the plants, which the little fellows quickly discovered, and both thereafter drank and washed regularly at the dish.

A practice of testing their knowledge of nuts was then made. When cracked hickory nuts were given them, they at once sat down and picked out of them the meats, which they eagerly devoured. Cracked nuts, it would seem, were deemed worthless for storage. But, on the contrary, when whole nuts were given, they tested them, evidently by weight, to see if they were sound. Sound nuts were promptly transported to their burrow, but the poor ones were dropped. They were never known to be mistaken in their judgment, for the rejected nuts on being cracked were always found to be worthless.

Although the food of the Hackee is mostly vegetal in character, yet, like his English relative, he is occasionally carnivorous in his appetite, for he has been detected in the cruel act of robbing birds' nests and devouring their callow young.

Some Squirrels are remarkable for their extreme agility in climbing trees, and in making extraordinary leaps from one bough to another or from some elevated spot to the earth. The Ground Squirrels, however, are intended to abide on the earth, and are seldom known to ascend trees to any great height. As they possess cheek-pouches, they are placed in a separate genus under the name of Tamias, which is a Greek word, signifying a store-keeper, and are distinct from the others in being furnished with these appendages. *Tamias striatus* is the appellation by which the subject of our sketch is known to the books.

Chapter 38

CANINE SAGACITY

Many years ago, two decades or more, the writer was the possessor of a little dog—a French poodle by breed. A more knowing animal of his kind never lived. He was a pretty creature, with hair as white as driven snow, and manners the most agreeable. Great pride was taken in his appearance. That his dress should maintain its natural purity, he was weekly subjected to a warm-water bath. This task devolved upon a little brunette, for whom the canine had contracted a strong affection.

Frisky, for such was our pet's name, had never before coming into the family known what it was to receive a good washing. His first experience was as uninteresting as it was novel and strange. It was anything but pleasant to him, but the little fellow bore it like a martyr.

Such treatment, by the ordinary cur, would have been resented with snaps and snarls, but his was a gentle nature that knew no such untoward manifestations. But there was, all the same, an aversion to the bath, as looks only too plainly indicated. So pronounced was the dislike, that the very sight of water caused his delicate frame to shake like a child's with the cold.

Had not the greatest care been taken in the preparation of the bath, it might have been thought that the tremors that shook his by no means robust frame were induced by the water's chilliness or by its undue warmth. But this could not be the case, as the fluid was always tempered to the most sensitive touch.

But there came a time, however, when Frisky was determined to evade these kindnesses upon the part of his mistress. He had pleaded immunity from them in pitying glances, but without avail. Something must be done, his looks would seem to say, as he lay cuddled up by the cosy kitchen fire. One could almost read the thoughts that were shaping themselves in his mind.

For three long years Frisky, who had been allowed to sleep at nights in the sitting-room, was accustomed, when morning broke, to visit the different members of the family in their respective dormitories, and have a lively, rollicking time. These visits were always looked forward to, and in no instance, during the whole of that period, were they ever intermitted. To have missed one of these exciting romps, would have been a keenly-felt deprivation. But that we were to be doomed to such disappointment and affliction, subsequent events only too clearly showed.

One Saturday morning, for it was always on the Jewish Sabbath that the bath was given, Frisky failed to make his accustomed calls. This was noticed by everyone, and no amount of comment was provoked. Loudly his name was spoken, but no response was elicited, and it soon became evident that the cunning little elf was beyond the reach of calling. Search was instituted, but to no effect. His absence was lamented, and it was feared some calamity had befallen him. A silence, like unto death, filled the house. There was weeping and wailing about, for Frisky was not.

But just as the shadows of night were deepening, and hope was dying out of the bosoms of all, the patter of little feet was heard upon the pavement leading to the back-door. The sounds were too familiar to be those of a stranger. All listened with breathless silence. "'Tis Frisky, 'tis Frisky," went up a chorus of voices, and we all rushed to the door to welcome the runaway back to the fold. Not a chiding word was spoken, not a look of reproof given, as with outstretched arms the culprit was received to our hearts. A more crestfallen, conscience-stricken being, however, could hardly be conceived to exist.

Things resumed their wonted sway. Happiness reigned once more in the family. Frisky's matutinal visits were as though they had not been interrupted. His frolics had all their former vivacity. The sin committed had been condoned, and he in splendid repute again.

Figure 63. My Dog Frisky. How he greeted his master.

A week since his first wrong-doing had elapsed. Would he repeat his plan of getting rid of the obnoxious bath?—had never entered our minds. The day dawned bright and lovely. All was bustle outside, and the slamming of shutters told that the servant was astir in the kitchen. As was her usual custom, the entry door was left open for Frisky. All ears were on the stretch. There were no familiar signs. The sharp, glad bark that always heralded his coming was wanting, and so, too, the timing of little feet upon the stairs. Not a sound of breathing, not a rustle of counterpane, was heard. Still and motionless we all lay, till the minutes seemed hours, and then came the thought that it was Saturday and Frisky had again disappeared. Search was everywhere made, but the missing one was nowhere to be found. That he had slipped out when the door was opened, was now most obvious. No effort was made to find his hiding-place, for we all knew that he would come back with the shadows.

His coming was later this time than before. The sun had long gone to rest. It was pitch dark when the pawing of little feet against the door announced his return.

This second offence was passed over as the first had been, and Frisky was his jolly, frolicsome self once more. A score of Saturdays was thus managed and the hateful bath escaped, for well this cunning bit of flesh and fur knew that the seventh was the only day of the week when it was convenient for his mistress to attend to his ablutions.

That Frisky was able to count, or had some means of determining the coming of the day he so thoroughly detested, there can be no question. But the exceeding cuteness of his nature not only showed itself in his manner of getting rid of the hateful bath, but in various other ways. He seemed equal to every emergency that could arise. Oftentimes I have watched him, as he lay upon a rug by the kitchen-hearth, or upon the pillow of a new-made bed, for he was at liberty to go where he pleased about the house, and I have fancied that I could see him thinking, or read the train of thoughts passing through his mind, so human-like seemed he in these reflective moments.

When scolded for some trifling misdoing, or threatened with denial of some expected pleasure, no so-called brute could show more pitying glances. His grief was often heart-rending to behold. Prostrate upon the ground or carpet, or in what place soever he chanced to be, he would moan and moan for hours together, and only consent to be comforted when the burden was lifted from off his soul by a kind word spoken, a smiling look given, or a quick, hearty shake of his delicate paw. When happy, and it did not take much to make him happy, he was full of life and vivacity, capering and prancing about with the utmost *abandon*, and doing his very best to show off his happiness and pleasure. His eyes seemed kindled with a holy affection, and a blaze of heavenly sunshine would appear to play over his features. I have seen him, when in such mental agony, to actually shed tears, a sight that never could fail to reach and melt the flintiest hearts. He knew and understood every word that was spoken to him, and responded by a shake of the head, or a low, soft bark. A conscience within told him the right from the wrong, and though he sometimes knowingly erred, yet he was always truly sorrowful for his sins afterwards.

There came a day, however, when the idol of the household went out and never returned. Some unlucky event had doubtless befallen him, or he had been spirited away to parts unknown. If living, I trust he is being cared for as he richly deserves. He was a kind, gentle, loving being, and I cannot help thinking that some day I shall meet him in the beautiful world beyond the grave.

Chapter 39

Feline Intelligence

Probably no creature has been more calumniated by man than the Domestic Cat. While wonderful intellectual powers, as well as the most amiable traits of character, have been accredited to the dog, and rightly so, it seems rather strange that so little of good has been found to exist in the subject of our sketch. She has been held up to reprobation as a thoroughly selfish animal, seeking her own comfort rather than that of others, and manifesting a stronger attachment to places than to owners. Sly and treacherous as her untamed kindred of the forests and jungles are known to be, she receives no higher commendation, and is even accused of concealing her talons in her velvety paws when matters go pleasantly with her, and ready to use them even upon her best friends when crossed in her purposes.

Whatever may have been the experience of those who have so grossly libelled the Cat, my own large acquaintance with the animal has led to different conclusions. Nearly all the Cats with which I have been most familiar have been as docile, tractable and affectionate as any dog could be, and have exhibited an amount of intellectual ability unsurpassed by few dogs. There is as much to be said about the good and bad temper of the Cat as of the dog, while, as to her mental capacities, the advantage is not so decidedly upon the side of the dog as is generally supposed. Nor is my own experience exceptional, for in all instances where friends have possessed favorite Cats their experiences have been similar to my own.

Self is not always paramount to everything else with Cats. Some are generous to a fault. Mothers have been known, whose devotion to their young has been so strong that they have hunted all day for their benefit, even when the latter were full-grown, scarcely taking any nourishment for themselves. But such feelings are perfectly natural. When, however, we see a Cat that is willing to share its food with a stranger, one cannot resist the thought that here is a case of real generosity. A friend once possessed a fine black Cat. He was dainty in his eating, scrupulously exact in his dress, and well-mannered in his deportment. No Cat ever received better training. Unlike the average Cat, he could be trusted in the presence of tempting viands, and was never known to abuse the confidence reposed in him. Beauty, for so he was called, was a model fellow, and well deserved the name. The education he received, while it made him gentle, kind and affectionate, and gave him reliability of character, did still more, for it endowed him with a soul that was not a stranger to the noblest impulses. Life had few luxuries that he did not enjoy; but a sprig of catnip was more to him than the choicest steak or raciest tidbit, and to this luxury he was weekly treated. Notwithstanding his fondness for the herb, he was never reluctant to share it with another, whom Fortune had less favored.

Cats, at least such as are well circumstanced, possess some knowledge of the uses of things. We once knew a Cat that would, when out of doors, make its presence known by a few loud raps upon the closed door, administered by its right front paw. If the call was not immediately answered, a few more raps, louder than before, would be given, and then the Cat, unable to restrain its impatience, would spring up to the latch, striking it a downward blow, as though endeavoring, human-like, to effect an entrance.

But quite as interesting as any of the foregoing cases is that of a female Cat that had run a spine into one of her hind feet. Limping upon three legs she made her way to her mistress, and, raising her foot, implored with a piteous look and sad, distressing cries the removal of the offensive spine. A child could not have made its suffering better understood, nor supplicated the needed relief more intelligently, than did this poor creature, which thoughtless man in his self-glorification is so prone to regard as a senseless, unintelligent and unreasoning being, which has no existence beyond this sublunary sphere.

While Cats are useful in the destruction of vermin, and afford man no little amusement by their wonderful antics, yet they seldom put themselves to any practical use. The Cat, about which we shall now have something to say, is an exception to the rule, and quite a marvel in his ways. He is a resident of a far-away town in New Jersey, and came to his present quarters a long, gaunt, wild-eyed, unfed creature. But something in his looks told of a soul within that fore-shadowed a great deal of good, and so the Cat, which at first seemed an unwelcome guest, began to be looked upon in an appreciative manner. And now Tom, as the Cat is called, is a fixture in the household.

Almost from his advent into the family Tom began to give an exhibition of his common-sense. This first remarkable show of intelligence was on the Sunday succeeding his adoption. The family had repaired to church, leaving Tom contentedly snoozing in a corner of the kitchen. But their surprise can hardly be pictured when in the midst of the sermon Tom came flying down the aisle to the place where his master was seated, and clawing the legs of the trousers of the latter, began yelling at the top of his voice. The minister stopped in the midst of his talking, and everybody got up to see what the trouble was, but Tom, utterly oblivious of them all, continued his strange behavior.

Figure 64. Tom on Duty. Guarding his master's cows.

Convinced that the actions of the Cat were not the result of an epileptic fit, but foreboded something wrong at the house, the male portion of the congregation started thither, and when the house was reached a dense column of smoke was seen pouring from the kitchen window. The door was thrown open, and the carpet on the floor was found burned to a cinder. A coal of fire had evidently fallen from the stove-grate and started the fire. That Tom had understood the danger, was shown by his actions.

One day, a horse, belonging to a neighboring farmer, ran away, and tore down the road past Tom's home at a thundering gait. Tom was sauntering around the yard, and his attention being drawn to the rattling of the wagon, he was soon in the road to see what the trouble was, and observing that the team was unaccompanied by a driver, he leaped upon the head of the runaway horse and hung on with teeth and claws until the animal was secured. On another occasion a tramp, happening along the road, descried a bicycle that belonged to one of the inmates of the house. He was soon astride the wheel, and might have made his escape had not he fallen under the eyes of Tom, who, as quick as a flash, was after the thief. Leaping into the air, he fell on the man's shoulders and set his teeth firmly into his neck. There was a howl and a crash that brought the family to the yard, and there they found the tramp rolling on the ground and making desperate efforts to get away from Tom's rigid jaws. Finally the Cat was induced to relax his hold, and the wounds of the tramp being cared for, he was allowed to proceed on his way.

More wonderful still is what follows: When the master wants to bait his cows and keep them within a certain area he instructs Tom to watch them, and the allotted task is performed with all the faithfulness and wisdom of a shepherd's dog. Any disposition to stray outside the limits is checked, the erring animal being hustled back by Tom, who, attaching himself to her caudal extremity, remains there until she is brought back to where she belongs.

No animals seem to require human sympathy so much as Cats, or to be so capable of giving sympathy in return. Where they have formed a strong attachment to a person they are loath to be away from his society and follow him wheresoever he goes, giving caresses and expecting a liberal share of the same in return. I have been upon a bed of sickness and a favorite Cat, which I always addressed as Puss, would, whenever the opportunity occurred, make her way into my room, and, jumping upon the bed, lay her head against my face in the most endearing manner, and purr her sweetest and gentlest, ever and anon stopping to express her sympathy by licking my forehead or uplifted hand. Even when Puss has been suffering from maladies to which all flesh of her kind is heir, I have sat by her side and stroked her head, and have read in the look which she gave me that she felt my sympathy and appreciated it beyond any power of expression of hers to declare. She seemed to think at all times that I was wholly her own, and no other Cat, not even one of her own offspring, would be allowed such familiarities, as any attempt was sure to provoke the most intense jealousy. Nor was I permitted to lavish attentions upon any of her kith, for she would soon become wrought up to a high pitch of excitement, and instant vengeance would be wreaked upon the recipient of my favors.

Much more might be said about the Cat. It has its good qualities and its bad qualities. There is hardly a trait of character which the human animal possesses that it does not possess. Of course I now speak of our Domestic Cat. In the long-past times, when the Egyptian nation was at the head of the civilized world, *Felis maniculata*, which is the reputed origin of our Domestic Cat, was universally domesticated in their homes, and it is not unknown the very high position it held in the love and esteem of the people, for it was deified and worshipped as a god. Even in England, still later down in time, the Domestic Cat was so scarce that royal edicts were issued for its preservation. Yet in those days, A.

Chapter 40

BRIGHT LITTLE CEBIDAE

Next to man, in descending the scale of animal life, come the Quadrumana, or Four-handed Animals. They are represented by the Apes, Baboons, Monkeys and Lemurs. Excepting the last, and a few other species, these animals are not very pleasing in aspect or habits, some of them, the larger apes and baboons, being positively disgusting. The air of grotesque humanity that characterizes them is horribly suggestive of human idiocy, and we approach an imprisoned gorilla or baboon with much the same repugnance that we do a debased and brutal maniac. This aversion seems not to be produced so much by the resemblance that the ape bears to man, as by the horror felt lest man should degenerate to the condition, character and physiognomy of the ape. But to the naturalist, who sees wonder or beauty in all things animate, these creatures are no less pleasing than others that are not so repugnant.

Were we to take a survey of the varied forms which the Quadrumana of the Old World assume, we would find that the forms would show such diversification that there would hardly seem scope for further modifications. Yet the prolific power of nature is so inexhaustible, that the depth of our researches would only bring to light objects of such infinite variety of form as to overwhelm the mind with surprise and admiration. Thus it would be found to be with the Cebidæ, or American Monkeys. While they would be shown to possess the chief characteristics of the monkey nature, thus establishing their close relationship with the Old World monkeys, yet they would be seen to exhibit the strangest modification of details. Their four hand-like paws, and other quadrumanous peculiarities, would indicate their *status* in the animal kingdom, while sundry differences of conformation would show that they were intended to live under conditions that would ill suit their relatives on the other side of the globe. Curious it is to observe how the same idea of animal life is repeated in various lands and climates, even though seas, impassable to creatures unaided by the light of civilized reason, intervene. So we have the Simiadæ of Asia and Africa represented by the Cebidæ of America. Nor is this wonderful idea restricted exclusively to the man-like animals. The lion, tiger and other feline races of the

Eastern Continent find Western representatives in the puma and jaguar, and the same circumstance may be observed of nearly all the mammalia, the birds, the reptiles, the fishes, and, in short, through the entire animal kingdom.

But of all the monkeys of the New World, and they are numbered by hundreds included in several genera and species, there are none that deserve more consideration than the Capuchin Monkeys. They are active, little animals, lively and playful. So similar are all the species in general habits, that a description of one will equally serve for any other. Their youthfulness and sportive manners make them very desirable companions, and hence we frequently find them domesticated by the native Indians and European settlers. Like other small monkeys, the Capuchin often strikes up a friendship for other animals that may happen to live in or near its home, the cat being one of the most favored of its allies. It is sometimes the case this familiarity is carried so far that the cat is turned into a horse by the monkey, who, seated upon her back, perambulates the premises. More unpromising subjects, we are told, have been pressed into similar service. Humboldt cites the case of one that was accustomed to catch a pig every morning, and, mounted upon its back, was known to retain its seat during the entire day. Even when the pig was feeding in the savannas its rider remained firm, and bestrode the animal with as much pertinacity as one skilled in equestrianism would the most rampant steed.

No little difficulty is experienced in settling the species of the Capuchins, for their fur is rather variable in tint, and some individuals differing so greatly as to cause them to look like another species. The general color of the Capuchin—*Cebus apella*—is a golden olive, a white fur bordering the face in some, though not in all individuals. *Cebus fatuellus*, commonly called the Horned Sapajou or Capuchin, is much more conspicuous than the last, as the erect fringe of hair that projects so prominently from the forehead indicates it at once: hence from the front, the hair assumes the appearance of two tufts or horns, from which peculiarity the animal derives its name. These horns are not completely developed until the monkey has attained maturity. There is also a manifest difference in color of hair, the Sapajou having a constant tinge of red in its fur. It is usually of a deep brown color, but in some individuals there is a marked resemblance to that peculiar purple-black which is obtainable by diluting common black ink with water, while in others the ruddy hue is so pronounced as to impart a chestnut tint to the animal's hair. The fringed crest is tipped with gray.

Perhaps no more interesting form of the Capuchins exists than the Weeper Monkey, or Sai, or, as it is called in the books, *Cebus capucinus*. As in the case of the two preceding species, it is an inhabitant of Venezuela and Brazil, and as lively as any of its congeners. Like its brethren, its tail is invested with a dense growth of hair, but this does not interfere with its prehensile powers. The Sai is possessed of a large amount of intelligence, and its quaint little ways make it a great favorite with those who delight to watch its quick and agile movements. While things of a vegetable character constitute the chief part of its food, yet it manifests a fondness for various kinds of insects, and is sometimes known to ascend to higher prey, for it has been observed to feed upon birds, which it devours with avidity, not even waiting to pluck off the feathers. Eggs are also thought to form a no inconsiderable part of this Capuchin's diet.

Some few years ago, Prof. Cope had in his possession a tame Sai, which was kept in a cage, or, rather, was supposed to be kept in it, for the animal had a strong aversion to confinement, and was sure to break loose therefrom sooner or later. When in durance vile, and wishing to break prison, he always directed his attention to the hinges, and no matter how firmly they were fixed, he was sure before long to extract the staples, pull out the nails, and so open the door at the hinges, and not at the latch.

Finding that the cage could not hold him, his master had him confined by a strap fastened around the waist, after the fashion of monkeys. The strap, however, proved to be of no more use than the cage, for the crafty animal soon contrived to open it, and this he did by ingeniously picking out the threads by which the strap was sewn to the buckle, and so rendering the fastenings useless.

Again he was confined to the cage and carefully watched. Having rid himself of the strap, he began to consider how he might apply it to some useful purpose. So, having perceived that some food had fallen beyond his reach, he took one end of the strap in his paw, flung the other over the morsel of food, and so drew it toward him. In this feat he displayed great accuracy of aim, seldom missing the object which he desired. Once or twice, when he had to make a longer throw than usual, he loosened his hold of the strap. The first time that this happened, some one handed him the poker. He took it, drew the strap toward him, and resumed its use as before.

No intelligent person can deny that these acts were prompted by reason. So far from even being aided by instinct, the animal was certainly acting in direct opposition to it. The instinct of an animal when confined or tethered in any way is to break loose by main strength, and the instinct of the monkey would have impelled him to force his way through the bars of the cage or to strain at the strap until he had broken it in two. But it was his reason that taught him to look for the weak part in both cage and strap, and, having found it, to devote his energies to that part until he had succeeded in his object.

Was it possible for instinct to teach him that the hinges were the weak part of his cage, and that, if he could only remove the staples or nails, the door would open and he would be free? Could instinct teach him that the stitches of the strap-buckle were to the strap what the staples and nails were to the hinges, and that if he could but pick out the threads, the fastenings of the strap would be rendered of no effect? Neither could instinct teach him to use the strap after the manner of a lasso, nor to employ the poker in regaining his lost weapon.

Figure 65. Jack At Dinner. Showing his use of table implements.

Not only did he thus show his ability to deal with the obstacles that stood in the way of his freedom, and without even the slightest suggestion from the mind of his master, but he also gave evidence that he had the capacity to profit by many of the civilities by which he found himself surrounded in the life in which he was placed. Monkeys are remarkable for their power of imitation, and Jack, as this Capuchin was called, proved himself to be no ordinary fellow in this respect. He had seen his master eat out of a dish, using knife, fork and spoon when occasion demanded, and nothing would do but an abandonment of his old habits—the using of his fingers, which his ancestors were wont to do—and the assumption of civilized practices. In time he became quite skilful in the use of these table implements and showed greater dexterity in handling them than many a man has shown. Accustomed to their use, he would never have things any other way. The writer has repeatedly been present when he was taking his meals. Seated upon the ground, his head and body slightly bent forward, with his plate of food before him, the ground serving him as a table, Jack would help himself in a quiet, cool and deliberate manner, all the while evincing in movement and look an air of the most consequential importance. To say that he was proud of the success which he had achieved in the correct use of table implements but tamely expresses the feeling which would dominate his bosom at such times. No human individual who had accomplished some wonderful discovery or striking feat at arms that had caused the earth to resound with his praises, could have felt more of the emotion than Jack. Indeed, it was a remarkable feat for Jack, and he had a right to feel vain over its accomplishment. All the while he was eating he would chatter in his uncouth guttural tongue, as though he had learned, like his human brethren, that conversation gave relish to a meal and was a powerful aid to digestion.

While Jack was a very useful fellow to have about, especially where cats without owners abounded, for he was a terror upon these feline nuisances, yet he had a few faults which detracted very much from his otherwise good character. Like some boys, he was addicted to the habit of throwing stones, but I am more than half disposed to believe that this was an acquired propensity, which he had learned by seeing his master engaged in a similar diversion, or perhaps, which is not at all unlikely, he had been trained to such exercise and pastime by his master. Well, he could throw stones with considerable force, and with as much precision as any well-trained lad of fourteen summers could do. Let the master but give him a stone, and say, "Now, Jack, hit that fellow," and Jack needed no second telling. Throwing his right arm back, just as a boy would do, in order to give the necessary impetus to the missile, he would send the stone flying in the right direction. It required no little skill and celerity of movement to dodge the projectile, as the writer had more than once learned by painful experience, for Jack's wonderful and well-directed aim seldom went astray of its purpose.

Towards his master Jack showed great deference and attention, and was ever ready to obey his slightest wish. No one's society he enjoyed better. It was always a pleasure to be near him, but strangers he seemed to despise and treat as enemies. He would always eye them with a suspicious look, and could never tolerate their presence for any considerable length of time without giving vent to his annoyance by the most angry vociferations and hideous grimaces. Should this not have the effect of causing them to retire, he would emphasize his objection to their presence by pelting them with stones and such other missiles as were convenient to hand. That he had a considerable affection for his master, and respected him, no stronger evidence could be given than what has already been adduced.

After all that has been said concerning Jack, yet the world is full of people, educated and intelligent as they consider themselves to be, who cannot see that this bit of flesh and spirit has been endowed by the same wise Creator with the same traits of character, but differing in degree, that they themselves possess. Going back to the ingenuity which Jack displayed in the cases of the cage and the strap referred to, it may be said to his credit that even Baron Trench himself could not have shown greater skill in the discovery of the weak parts of his prison and bonds than did this so-called brute, nor could he have exhibited more patience and perseverance in working at them. Indeed, there are many human beings that would not have been half so sensible as Jack, but still we must believe that such high intelligence, comparatively speaking, must inevitably perish with the body, through which as a vehicle it was made to manifest itself. All intelligence is an emanation from the Divine Intelligence, and, when the life has gone out of the body from which it was made to shine forth, then it, instead of perishing with the material, returns to the Source of all intelligence, not to be re-absorbed, but, as I think, to continue as a separate intelligence, drawing its life and light from the great Central Head, like as the planets derive theirs from the centre of our material universe—the Sun.

Chapter 41

UNTUTORED MAN

Strange and unique as are the plants and animals of Australia, yet nothing definite can be affirmed of its native human inhabitants. They are a peculiar people, separated by a wide remove from the Papuans, the Malays and the Negro. Of a dark, coffee-brown complexion, rather than actually black, the Australian is but little inferior to the average European in height, but is altogether of a much slimmer and feebler build, his limbs, particularly, being very lean and destitute of calves, a defect which is a peculiarity of the darker races of man. His head is long and narrow, dolichocephalic in type, with a low brow, prominent just above the orbital regions, but receding thence in a very marked degree. The nose, proceeding from a comparatively narrow base, broadens outwardly to a somewhat squat end, the eyes on each side of its attenuated root appearing drawn together. His face bulges into high cheek bones; his mouth is large and grotesque, the jaw-bone contracted, the upper jaw projecting over the lower, but with fine, white teeth; the chin cut away, and his ears slightly pricked forward. Not only the head and face, but the entire body as well, is covered with a profusion of hair, which, when freed of its enclogging dirt and oil, is soft and glossy. Like most savage peoples, the effluvium of his skin, offensive as it naturally is, is very much exaggerated by the fish-oil he uses in the anointment of his person.

Almost exclusively directed on the means of procuring sustenance, the intellect of the Australian operates wholly within the range of the rudest bodily senses. But inside that simple, elementary sphere he displays no little nimbleness and dexterity. In tracking and running down his prey he is unsurpassed. His weapons, though of the most primitive forms, are well adapted for the purposes of the chase. Rude and uncouth as his culinary and domestic apparatus appear, yet they serve equally well the objects for which they were designed. Some imitative facility, or rude sense of elementary art, is possessed by him, as is evidenced by the crude figures of sharks, lizards and other animals that may be seen carved in caves in the north-east of Australia, and on the rocks of New South Wales. That he has some exuberance of rude sense is still further shown in his language, which, within its very circumscribed sensuous sphere, is fairly expressive and complete, and likewise in the ease with which he learns to chatter the languages of peoples with whom he has been thrown into contact.

Outside the circle described, all is blank to the Australian. He has no architecture, no pottery and almost no weaving, and may be said to have no religion. His sensations may scarcely, if at all, be said to have attained the dignity of sentiments, much less that of sentimentalities. The man domineers over the woman, who is as much his property as his boomerang or dingo. Male offspring are held in considerable estimation, and a father will bewail the death of a son for months, and even for years. Old men and old, infirm women, on the other hand, are cruelly abandoned, and left to starve to death, for they are considered worthless and a burden, and consumers of the food that should go to the support of the young and physically strong. During the summer they roam about naked, utterly strangers to shame, which seems not to be innate to their natures. Wives are accounted an item in a man's chattels, the stealing of which being met with some definite punishment. Caves, where they abound, afford shelter and security for some of the tribes, but where these are not found, screens of twigs and bushes covered with leaves or turf, or logs of wood and turf, serve for protection and cover for a few days or weeks, till the pursuit of food calls them elsewhere.

Thrift is unknown to the Australian. His life alternates between satiety and semi-starvation. In summer he goes naked, but in winter he wraps himself in kangaroo skins. A girdle of hair bound about his loins holds his dowak, as his digging-stick is called, and an apron of skins suspended from the girdle affords a protection from shrubs. His food consists largely of animals, which he devours alive, and includes lizards, snakes, the heads being rejected, frogs, white ants, larvæ and moths. Other animals are roasted, showing that the Australian knows, contrary to an opinion that once prevailed, the method of kindling a fire. In seasons of dearth, when there is a paucity of food-material, cannibalism is general. He then makes an attack upon a neighboring tribe who is his enemy, and if he cannot obtain food in this manner, he scruples not to fall back upon his wife and his children. One obligation of the wife is to keep her husband supplied with vegetable food, such as the roots of the wild yam, seeds of the acacia, sophoræ, leaves of the grass-tree, etc. Failing to produce a sufficiency, she is liberally treated with maulings and spearings, so that a wife generally appears bruised and gashed all over her body.

Among the different tribes of Australians, the boomerang is the principal weapon. This is a flat stick, three feet in length, and curves at the centre. It is thrown into the air among birds, jerks in a zigzag, spiral or circular fashion, and when thrown by a person skilled in its use is sure to bring down a few individuals at every throwing. Besides this weapon they have the throwing-stick, flint-pointed spears, shields, stone-hatchets, digging-sticks, netting-needles, nets of sinews, fibres or hairs, water-skins and canoes.

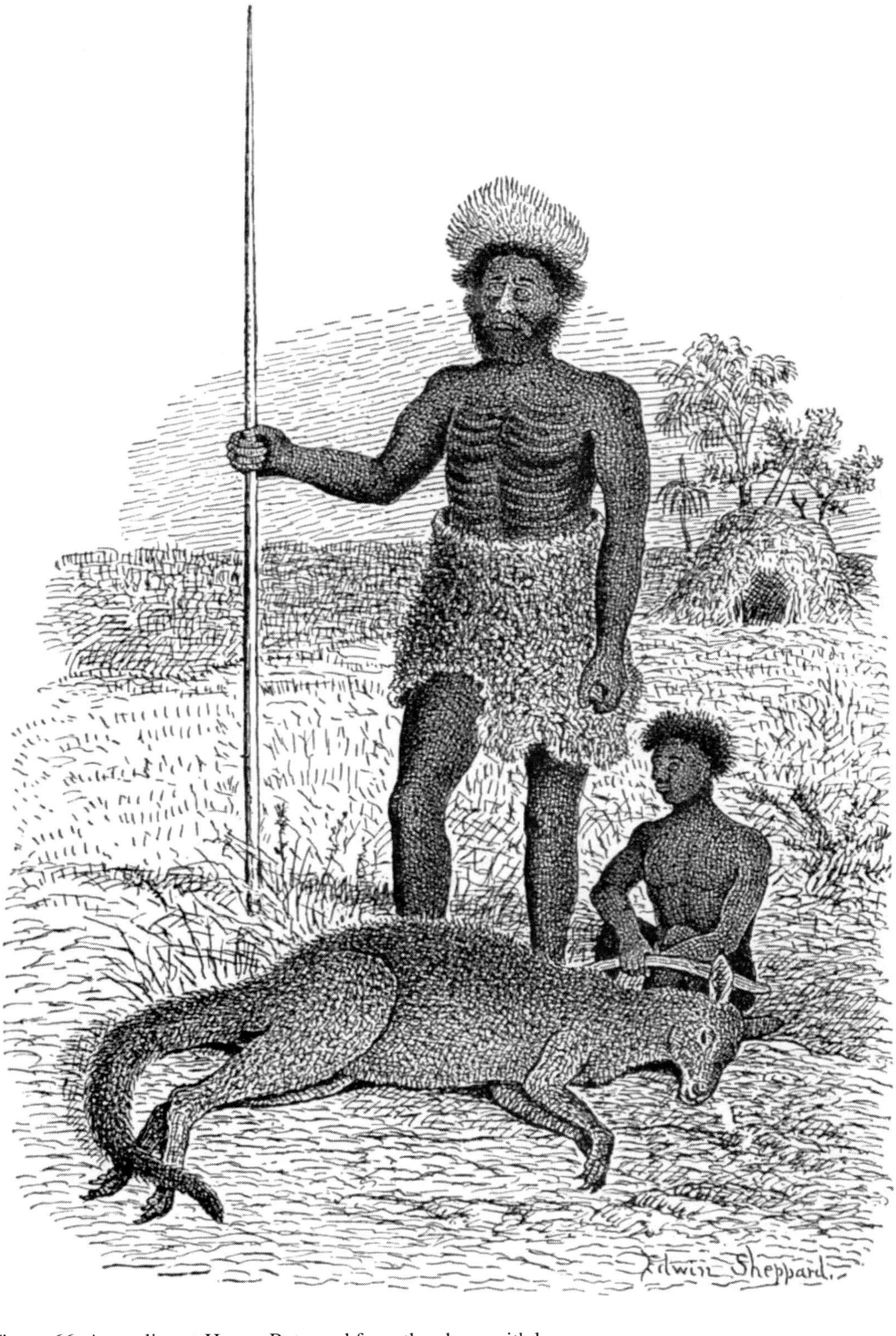

Figure 66. Australian at Home. Returned from the chase with kangaroo.

No government exists among this people outside that of the family, and no laws except certain traditionary rules about property. As for their religion, they have little save their terror of ghosts and demons, and certain superstitious traditional rites applicable to epochs in a man's life, but more especially so at the time of his burial. At ten years of age, a boy is covered with blood; at ten to fourteen, he is circumcised in the north and south of Australia, but not in the west or on the Murray River; and at twenty, he is tattooed or scarred. Felicity after death is the reward of proper burial, but a man dying in battle or rotting in a field becomes an evil genius.

No more perfect example of tribal organization exists than that of the tribes of Australasia. In a very large proportion of existing tribes, the tribe is an aggregate of several stocks or distinct bodies of kindred, the persons composing the tribes being included in stocks which are, or are accounted, distinct from each other. Two tribal customs, namely, the prohibition of marriage between persons of the same stock, and the reckoning of kinship through females only, so that children are accounted of the stock of their mother, sustain this organization. Persons of the same stock, too, owe duties to each other, and are to some extent participants in each other's liabilities. An injury done by a man is an injury done by his stock, which may be avenged upon any member thereof; or an injury done to a man is an injury done by his stock, for which every member of it is bound to seek vengeance. As a consequence of these customs, a husband must be of a different stock from his wife or wives, and therefore must be accounted of a different stock from his children; and if he has wives of different stocks, then their respective children are accounted of different stocks. More than one stock, it will thus be perceived, is represented in every household. And since a man owes duties to his stock—the duties of acknowledged blood-relationship—while to those of his family who are not of his stock, there being nothing but the accident of birth to unite him, it necessarily follows that the family among these tribes has very little cohesion.

Wholly sensuous is the language of the Australian, their abstraction tending only in the way of arithmetic as far as the number five, and that itself being quite an unusual stretch. Polysyllabic as it is in formation, and having the accent on the penultimate, it is not at all inharmonious. Though it comprehends many divergent forms, yet they seem to be all fundamentally connected, constituting a group entirely isolated from any of the linguistic families of the other parts of the world. Within its narrow confines the language is well developed and sensuously copious and expressive.

Like almost all other savages, the native Australians are rapidly disappearing before the spread of civilization. The European settlers crowd them out of all the more fertile and habitable lands, pressing them more and more into the desert of the interior, where they find it exceedingly hard to obtain in their roving, unsettled lives the necessary means of subsistence. Great numbers are thus forced to succumb to deprivations not of their own bringing, and not a few to the diseases and vices brought among them by the new possessors of their domains. The lowest estimate of their number, prior to the settlement of Europeans among them, gives over 150,000, but the natives still surviving scarcely figure one-half of that population. It is only a question of a decade or two when the Australian, like the Tasmanian, who was once his near neighbor, will have vanished from off the face of the country, leaving behind him his implements of war and the chase, his culinary and domestic apparatus, and the rude carvings of his hands in caves and in rocks, as the principal evidences of his earthly existence.

By competent critics the Australian is pronounced to be the most degraded of human beings, and the lowest type of man. In reason, love, generosity, conscience and mere responsibility he is the inferior of many of the lower animals, and in the erection of a house for comfort, shelter and security he is surpassed by creatures even as low in the scale as the worms and insects. It is true, when hunger has to be met, that he has shown some skill in the manufacture of implements necessary to the obtainment of his food, and also in resisting the attacks of his own kind and of the natural enemies by which he is surrounded. There is no doubt that he is well satisfied with his condition in life, and could hardly be induced to exchange it for another. He has doubtless fulfilled the purpose of his being in the world, and unable to cope in the struggle for existence with a superior civilization must succumb to the latter which is better fitted to endure, a sad but impressive lesson which is the teaching of every chapter of the world's geologic story.

Chapter 42

LIVING SOULS

All things were made by the Word of God. In this Word was life, spirit or energy. Without it was not anything made that was made. Hence, says Elihu, "the *Spirit* of God hath made me, and the *breath* of the Almighty hath given me *life*;" or, as Moses testifies, "the Lord God formed man, the dust of the ground, and breathed into his nostrils the breath of lives; and *man* became *a* Living Soul."

Now, if it be asked what the Scriptures define a living soul to be, the answer is a living natural, or animal body, whether of beasts, birds, fish or men. The phrase living creature is the exact synonyme of living soul. The words *nephesh chayiah* are in Hebrew the signs of the ideas expressed by Moses, *nephesh* signifying *creature*, *life*, *soul*, or *breathing frame* from the verb *breathe*, and *chayiah*, a noun from the verb *to live*, *of life*. *Nephesh chayiah* is the genus which includes all species of living creatures. In the common version of the Scriptures, it is rendered *living soul*, and, therefore, under this form of expression they speak of all flesh which breathes in air, earth and sea.

From the evidence adduced a man then is merely a body of life in the sense of his being an animal or living creature—*nephesh chayiah adam*. Therefore, as a natural man, he has no preëminence over the creatures God has made. Moses makes no distinction between him and them, for he calls them all living souls, breathing the breath of lives. His language, literally rendered, says, "and God said, the waters shall produce abundantly *sheretz chayiah nephesh* the *reptile living soul*;" and again, "*kal nephesh chayiah erameshat* every living soul creeping." In another verse, "let the earth bring forth *nephesh chayiah* the living soul after its kind, cattle, and creeping thing, and beast of the earth after its kind," and "*lekol rumesh ol earetz asher bu nephesh chayiah* to everything creeping upon the earth which has in it living breath," that is, the breath of lives. And lastly, "whatsoever Adam called *nephesh chayiah* the living soul that was the name thereof."

Not even are quadrupeds and men living souls, but they are vivified by the same breath and spirit. *Neshemet chayim*, or the *breath of lives*, and not the *breath* of *life* as the text of the common version has it, is said to be in the inferior creatures as well as in man. *Chayim* in the Hebrew is in the plural number, and therefore the words *neshemet chayim* should be rendered as above. Thus, God said, "I bring a flood of waters upon the earth to destroy *all flesh* wherein is *ruach chayim* spirit of lives." And in another place, "they went in to Noah into the ark, two and two of *all flesh*, in which is *ruach chayim* spirit of lives." And *all flesh* died that moved upon the earth, both of fowl, and of cattle, and of beast, and of every creeping thing, and every man; all in whose nostrils was *neshemet ruach chayim*, BREATH OF SPIRIT OF LIVES. Now, as has been previously affirmed, it was the *neshemet chayim* with which God, according to the testimony of Moses, inflated the nostrils of Adam. If, therefore, this were a particle of the divine essence, as it is declared, which became the immortal soul in man, then all other animals have likewise immortal souls, for they all received breath of spirit of lives in common with him. Begotten of the same Invisible Power, and formed from the substance of a common earth mother, man and beasts were animated by the same spirit, and constituted to be *living breathing frames*, though of different species, and in God they lived, and moved, and had their continued being.

Returning to the philology of our subject, it is to be remarked that by a metonymy, or a figure of speech where the container is put for the thing contained, and conversely, *nephesh*, *breathing frame*, is put for *neshemet ruach chayim*, which, when in motion, causeth the frame to respire. Hence *nephesh* signifies not only *breath* and *soul*, but also *life*, or those mutually affective, positive and negative principles in all living creatures, whose closed circuits cause motion of and in their frames. By Moses these principles, or qualities of the same thing, are apparently styled the *Ruach Elohim*, or by Timothy the Spirit of Him "who only hath immortality, dwelling in the light which no man can approach unto, whom no man hath seen, nor can see," and which, when the word was spoken, first moved upon the face of the waters, and afterwards disengaged the light, evolved the expanse, gathered the waters together, brought forth the green vegetation, manifested the celestial universe, vitalized the breathing frames of the dry land, the firmament and the seas, and formed man in His own image and likeness. This *ruach*, or spirit, was the instrumental principle commissioned by the glorious Increate for the elaboration of the natural world, the erection of this earthly house, and its equipment with living souls of every species; and it is this same instrumentally formative power that, together with the *neshemeh*, or breath, that keeps them from perishing, or returning to the dust. "If God set his heart against man, He will withdraw to himself *ruachu veneshemetu*, that is, *His spirit and His breath*; all flesh shall "perish together, and man shall turn again to dust." "By the *neshemet el*," or breath of God, "frost is given." Speaking of reptiles and beasts, David saith, "thou withdrawest *ruachem—their spirit*—they die; and to their dust they return. Thou sendest forth *ruheck*—thy spirit—they are created."

Figure 67. Representative Life of Western Asia. Illustrating the scriptural idea of living souls.

From this cumulative evidence it is manifest that the ruach is all-pervading. It is in heaven, in sheol, or in the dust of the deepest hollow; in the uttermost depths of the sea; in the darkness as well as in the light; in all things animate and inanimate. In the broadest, or I may say, in an illimitable sense, it is a universal principle. It is the substratum of all motion, whether manifested in the revolutions of the planets, in the ebb and flow of the sea, in winds and storms and tempests, or in the organisms of plants and animals. The atmospheric expanse is charged with it; but it is not the air. Animals and plants breathe it, but it is not their breath; yet without it, though filled with air, they would die. Neshemet el, or atmospheric air, is the breath of God, as Job puts it, or the mighty expanse, as affirmed by Moses. What the ruach, or spirit, is, none with certainty can say. Extending from the centre of the earth, and thence in all directions through the immensity of space, is the Ruach Elohim, whose existence is demonstrable from the phenomena of the natural order of things. It penetrates where neshemet el cannot penetrate, but when speaking of the motivity and sustentation of organized dust, or souls, they co-exist with them, the Ruach Elohim becoming the ruach chayim, or spirit of lives; the neshemet el, the neshemet chayim, or breath of lives, and both together in the elaboration and support of life, the neshemet ruach chayim, or breath of the spirit of lives. Living creatures, or souls, are not animated, as is erroneously supposed, by a vital principle which is capable of disembodied existence. On the contrary, souls are made living by the coetaneous operation of the ruach chayim and the neshemet chayim upon their organized tissues according to certain fixed laws, called natural laws. When the as yet occult laws of the all-pervading ruach, or spirit, shall be made known, men will be astonished at their ignorance respecting living souls, as we are at the notion of the ancients that their immortal gods resided in the stocks and the stones they so ignorantly worshipped.

Though lent to the creatures of the natural world for the allotted period of their living existence, yet the ruach chayim and neshemet chayim are still God's breath and God's spirit, and to distinguish them from the expanse of air and spirit in their totality, they are sometimes specifically styled "the spirit of man" and "the spirit of the beast," or collectively "the spirits of all flesh," and "their breath." Thus it is written in Ecclesiastes, "they have all one ruach, or spirit, so that man hath no preëminence over a beast; for all is vanity or vapor." "All go to one place; all are of the dust, and all turn to dust again." And in the sense of supplying to every living creature, or soul, spirit and breath, Jehovah is styled by Moses in the book of Numbers,—"God of the spirits of all flesh."

Enough has been advanced to show the Scriptural import of the text already quoted, that "the Lord God formed man, the dust of the ground, and breathed into his nostrils the breath of lives; and man became a living soul." The simple, obvious and undogmatic meaning of this is, that the dust being animalized, and then organized, was next set in motion by the inrush of the air through his nostrils into his lungs according to natural laws. This phenomenon was the neshemet el, or "breath of God," breathing into him; and as it was the pabulum of life to all creatures constituted of dust, it was very expressively styled the "breath of lives," and not the "breath of life." God breathes into every man at his birth the breath of lives to this day, and there can be no reason, Scriptural or otherwise, why we should deny that He breathed it into Adam as He hath done into the nostrils of his posterity by the operation of natural laws. Man, as soon as he began to respire, like the embryo passing from fœtal to infant life, "became a living soul," that is, nephesh chayiah, a living, breathing frame, or body of life. All kinds of flesh, whether of man, beast, fowl and creeping thing, are made alive by the same breath and spirit. They all become, in consequence, living souls, so that, having an oneness of spirit, a man hath no superiority over a beast.

Having now proved, as we think, beyond the possibility of a doubt, that men and beasts "have all one ruach, or spirit," and hence are all living souls, we now approach a form of life, termed vegetable life, about which the Scriptures have little to say. Neshemet el, or atmospheric air, is just as essential to plants as to animals. Deprived of it they wither and die. No less necessary is the all-pervading ruach, or spirit. It is in the air, though not of the air. Plants, equally with animals, breathe it, but it is not their breath. Without it, even though filled with air, they would perish. Perhaps it is the base of each of the elementary constituents of the air. Uncombined, may it not be that wonderful fluid whose explosions are heard in the thunder, whose fiery bolts overthrow the loftiest towers and rive the sturdy monarchs of the woods, and whose influence, though in less intensity, gives polarity to light, the needle, and the brain?

Living plants are a part and parcel of the life of our globe. They preceded in the grand scheme of creation animal existences. Low down in the scale of life are forms about which it cannot be predicated these are plants and these are animals. Scientists are unable to say where plant-life ends and animal-life begins. No hard-and-fast line can be drawn between the two vast kingdoms of life, and it is often wholly impossible to decide whether we are dealing with an animal or a plant. There can be no question that the earliest life was vegetable by nature, and that its habitat was the primeval ocean. This is no less the teaching of science than that of the Scriptures. From some such life, originating de novo as the Spirit of God passed over the waters, the two great branches of animate nature may have taken their rise. What the form of this life may have been, whether cellular or a mere mass of formless protoplasm, the mind of man cannot asseverate. It is a mystery, and will doubtless ever remain as such to finite intelligence. That this life, no matter how apparently insignificant it must have been, breathed in its own simple fashion, that is, by the coetaneous operation of the ruach chayim and the neshemet chayim upon its simple substance in accordance with natural law, there can be no dispute. Breathing is not always conditioned by the existence of nostrils. Plants respire, or, in other words, take in carbonic acid from the air through their stomata, or mouths, which they separate into its components of carbon and oxygen, appropriating the former, which they build into solid matter, but usually throwing off the latter into the great receptacle of atmosphere from which it was extracted. Even a moner, which has no distinction of parts, may be said to breathe, but it breathes by means of its whole external surface, for neshemeh and ruach are as necessary to it as to man himself. It will thus be obvious that plants are living, breathing frames, or bodies of life, and hence are as much entitled to be considered as living souls as animals are. Let but God withdraw his ruach, or spirit, from them, and they die and to their dust return. Surely no more could be predicated of animals.

Chapter 43

CONSCIOUSNESS IN PLANTS

Plants, it has been vaguely asserted, differ from animals by not having the power of movement. Rather should it be stated that plants acquire and display this power when it is to their advantage. This will be found to be of comparatively rare occurrence, as they are affixed to the ground, and food is brought to them by the air and rain. Evidence of the very high position a plant may attain in the scale of organization may be seen when we look at one of the more perfect tendril-bearers. As a polypus adjusts its tentacula for action, so a plant places its tendrils. If the tendril be displaced, it sets to work to right itself. Acted on by the light, it bends towards or from it, or disregards it altogether, whichever course may be the most advantageous. For several days the tendrils or internodes of the plant, or both, spontaneously or otherwise revolve with a steady motion. But should they strike some object, they curl quickly around it, grasp it with wonderful firmness, and in the course of a few hours contract into spirals, dragging up the stems, and forming most excellent springs. All external movements now cease, and by growth the tissues soon become surprisingly strong and durable.

Such a movement, as has just been considered, is a widely prevalent one in plants, and is essentially of the same nature as that of the stem of a climbing plant, which successively bends to all points of the compass, so that the tip is made to revolve. This movement has been called *revolving nutation* by some writers, and *circumnutation* by others. In the case of the circumnutating movement of the tip of the radicle of some plants, there can be no doubt that it is it that affords the radicle some slight assistance in penetrating the ground. But whether or not a radicle, when surrounded by softened earth, is aided in making a passage for itself by circumnutating, one thing is certain, that is, that this movement, by guiding the radicle along a line of least resistance, can hardly fail to be of high importance. Should, however, a radicle in its downward growth break obliquely into any crevice, or an opening left by a decayed root, or one made by the larva of an insect, and more especially by worms, the circumnutating movement of the tip will materially aid it in following such open passages. Not only our own observation, but also those of such eminent authorities as Darwin and Hensen, conclusively show that roots commonly run down the old burrows of worms.

But radicles of seedlings, as well as those of more vigorous plants, would pass over stones, roots and other obstacles, which they must necessarily encounter in the soil. This they are abundantly able to do, for they are exceedingly sensitive just above their apices, and bend like a tendril *towards* the touching object. When, however, one side of the apex is pressed by any object, the growing part bends *away* from that object, and this seems a beautiful adaptation for avoiding obstacles in the soil, and for following the lines of least resistance.

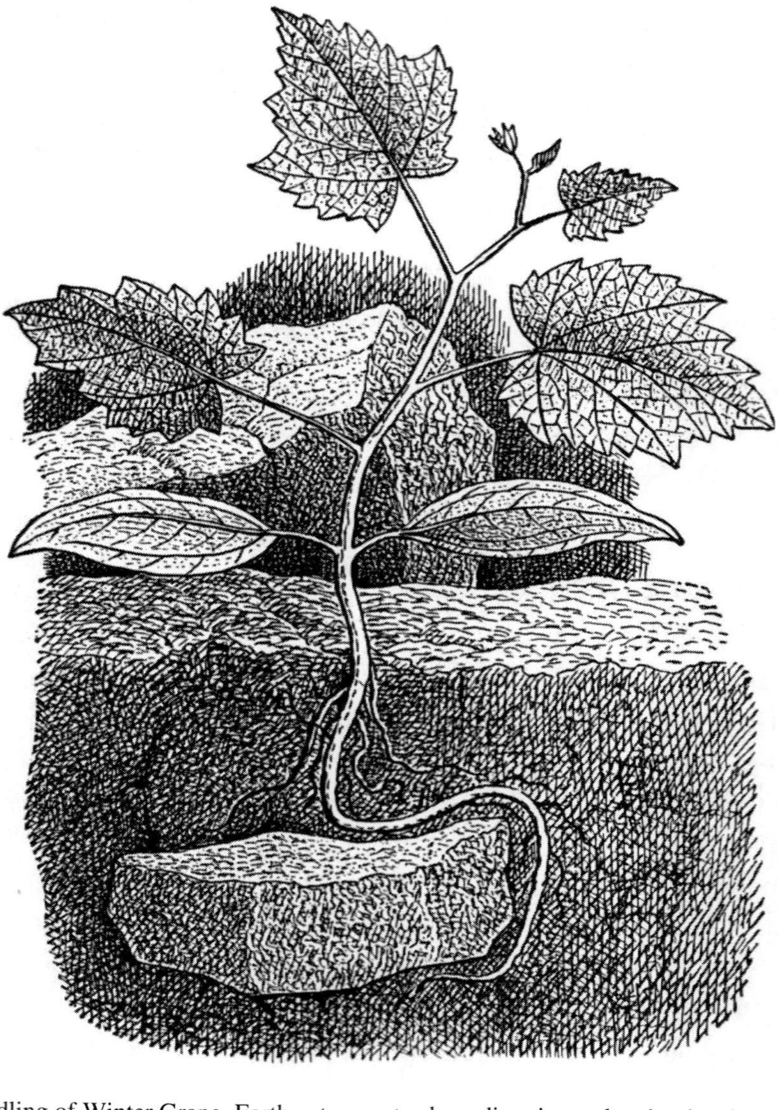

Figure 68. Seedling of Winter Grape. Earth cut away to show directions taken by tip of radicle in avoiding a stone.

that the knowledge of human language does extend to the signification of particular words. Parrots, as is well known, are well acquainted with the meanings of the words which they speak. Examples have been known to the writer of parrots that were able to speak in two languages, and, when addressed, always replied in the language used by their interlocutors, speaking English or Spanish, as the case might be. "Go, bring up the cows," was an order that was daily given to Lion, a large black dog, with a shaggy head, that belonged to my maternal grandfather, an old-time farmer who lived way back in the fifties. So well did he understand the significance of these words, and the labor, worry and responsibility which they implied, that he did not have to be told a second time, nor have to have their import conveyed to him by sign or by action of the farm lad whose business it was to see that the animals were brought to the barn-yard at milking time. Obedient to orders, he would trot to the pasture-ground, nearly a quarter-mile distant, open the bars between the lane and the field with his mouth, and then start on his business with a full sense of its requirements. His coming was well known to the cattle. While the most of them would take their way in a quiet, orderly manner to the lane, yet there were some unruly ones among them who gave Lion a great deal of trouble, but he always succeeded in overruling their contrary tendencies. When there was a tumult in the hennery, accompanied by loud noises, the command, "Go, see what the trouble is!" was performed to the very letter, and the trouble, if any, was speedily announced by a series of loud, sharp, quick barks, which soon brought some one or more members of the family to the scene of disorder. If nothing unusual was happening, Lion would return to the house in a slow, leisurely way, and by his looks convey, as clearly as man could do it, the utter needlessness of the command.

Not only is the dog capable of understanding many things that are said to him, but is even capable of forestalling one's wishes. Part of one of the writer's vacations was spent in a small country town not very remote from Philadelphia. There was in the family with whom he boarded a dog called Prince. He was a very great favorite, and was once noted for his lively, vivacious disposition and jolly manners. But at the time of my introduction to him, he seemed to be suffering from some bodily affliction, which had not only taken away his appetite for food, but the very *animus* of his being. Upon inquiry I learned that the master of the house, to whom Prince was so deeply attached, had died the year before, and that the dog had taken his death so completely to heart that he had lost all of his former vivacity. He refused all food, often going for days without taking a single mouthful. Life seemed to have lost for him all its charms. Sad and dejected he would lie upon the porch-floor or ground, seemingly unconscious of everything and everybody. That he was slowly dying seemed evident to all. But a change from our first interview appeared to come over the animal. From some cause or other, he had taken quite a fancy to me. He would greet me with considerable friendliness when I would come down in the morning, and always seemed glad to be in my presence. My first business, on coming downstairs, was to go for the newspaper, which was always to be found inside the yard, some thirty steps from the house. I would then sit down upon the porch and read it, but Prince was always close-by,

So feeble is the circumnutating movement of the terminal growing part, both of the primary and secondary radicles, that it can assist them but little in penetrating the ground, excepting when the superficial layer is very soft and moist. But it must aid them materially when they chance to break obliquely into cracks, or into burrows that have been made by earth-worms or larvæ. Moreover, combined as it is with the sensitiveness of the tip of the radicle to contact, it can hardly fail to be of the highest importance, for as the tip is always endeavoring to bend to all sides, it will press on all sides, and will thus be able to discriminate between the harder and softer adjoining surfaces. Consequently, it will tend to bend from the harder soil, and will thus take the directions of the least resistance. So it will act if it meet with a stone or the root of another plant in the soil, as must incessantly occur. If the tip were not sensitive, and did not excite the upper part of the radicle to bend away, whenever obstacles were encountered at right angles to its growing direction, it would undoubtedly be liable to be doubled up into a contorted mass. But with radicles growing down inclined plates of glass, as shown by experiment, it has been observed that as soon as the tip merely touched a slip of wood cemented across the plate, the entire terminal growing point curved away, so that the tip soon stood at right angles to its former direction; and thus, as far as the pressure of the surrounding soil would permit, would it be with an obstacle encountered in the ground. Thick and strong radicles, like those of the horse-chestnut, are endowed with less sensitiveness than more delicate ones, and would therefore be the better able by the force of their growth to overcome any slight impediment to their progress. Further, as radicles perceive an excess of moisture in the air on one side and bend towards this side, it is reasonable to infer that they will act in a similar manner with respect to moisture in the earth, for the sensitiveness of moisture resides in the tip, which determines the bending of the upper part. May not this capacity partly account for the extent to which drain-pipes often become choked with roots? The direction which the apex takes at each successive period of the growth of a root, ultimately determines its whole course. It is therefore very important that the apex should follow from the first the most advantageous direction. We can thus understand why sensitiveness to geotropism, contact and moisture should all reside in the tip, and why it should determine the upper growing part to bend either from or to the exciting cause. Darwin has compared a radicle with a burrowing animal, such as a mole, which wishes to penetrate vertically into the ground. By a process of circumnutation, or the movement of his head from side to side, he is enabled to feel any stone or other obstacle, as well as any difference in hardness of soil that may exist, and will therefore turn from that side; but if damper on one side than on the other, will turn thither as a more suitable hunting-ground. Nevertheless, after each interruption, he, guided by the sense of gravity, will be able to recover his downward direction and to reach to a greater depth.

Destruction of the tip of a radicle does not prevent the adjoining part from bending, if this part has already received some influence from the tip. As with a horizontally extended radicle, whose tip has been cut off or destroyed, the part which should bend most remains motionless for many days or hours, even though exposed at right angles to the full influence of gravity, we cannot do otherwise than conclude that the tip alone is sensitive to this power, and transmits some stimulus to the neighboring parts, thereby causing them to bend. Direct evidence of such transmission has been obtained. When a radicle was left extended horizontally for an hour or an hour and a half, by which time the supposed influence will have travelled some distance from the tip, and the tip was then cut off, the radicle subsequently became bent, although it was placed in a perpendicular position. Terminal portions of several radicles thus treated continued for some time to grow in the direction of their newly-acquired curvature, for being destitute of tips they were no longer acted upon by the power of gravity. New vegetative points, however, appeared, and being acted on by this influence coursed themselves perpendicularly downward as was their custom.

Investigation having shown that it is the tip of the radicle that is sensitive to geotropism in the members of such distinct families as the Leguminosæ, Malvaceæ, Cucurbitaceæ and Gramineæ, which may be represented by the Clover, Mallow, Gourd and Rye, we may justly infer that this character is common to the roots of most seedling-plants. Whilst a root is penetrating the ground, the tip must take the incipient step, as it has to determine the direction of the entire root. When, however, it is deflected by any subterranean obstacle, it is essential that a considerable length of the root should be able to bend, particularly as the tip itself grows slowly and bends but little, so that the proper downward course should be recovered. Immaterial as it would seem whether the entire growing part should be so sensitive to geotropism as to effect this movement, or that it should be brought about by an influence transmitted exclusively from the tip, we should, however, remember that it is the tip that is sensitive to the contact of hard objects, causing the radicle to bend away from them, thus directing it along certain lines in the soil where the least opposition interposes. It is again the tip that is alone sensitive, at least in some instances, to moisture, causing the radicle to bend towards its source. These last two kinds of sensitiveness conquer for a time the sensitiveness to geotropism, which, however, ultimately prevails. But the three kinds most often come into antagonism, first one prevailing, and then the other. It would, therefore, be an advantage, perhaps a necessity, for the interweighing and reconciling of these different kinds of sensitiveness, that they should all be localized in the same group of cells which have to transmit the command to the adjoining parts of the radicle, necessitating it to bend to or from the source of the irritation.

Though generally believed by authors that the modification of the upper or lower surfaces of a radicle, whereby curvature is induced in the proper direction, is the direct result of gravitation, yet there can be no question from all that has been said that it is the tip alone that is acted on and that transmits some influence to the adjoining parts, causing them to curve in a downward manner. Gravity, it would seem, does not act in a more direct way on a radicle than it does on any lowly-organized animal, which moves away when it feels some weight or pressure.

When we consider what we have written, it is impossible not to be impressed with the resemblance between the movements of plants and many of the actions performed by the lower animals. With plants an astonishingly small stimulus suffices. One plant may be highly sensitive to the slightest continued pressure, while a closely-allied form just as highly sensitive to a slight momentary touch. The habit of moving at certain periods is inherited both by plants and animals; and other points of similitude have been specified. But the most striking resemblance is the localization of their sensitiveness, and the transmission of a stimulus from the exciting point to another, which consequently moves. Yet plants do not, of course, possess nerves or a central nervous system. May we not therefore infer, and wisely so, too, that with animals such structures but serve for the more perfect transmission of impressions, and for the more complete intercommunication of their several parts?

No structure in plants seems more wonderful, as far as its functions are concerned, than the tip of the radicle. Lightly pressed or burnt or cut, it transmits an influence to the upper adjoining part, causing it to bend away from the affected side. But more surprising, however, is the fact that the tip can distinguish between a slightly harder and softer object, by which it is simultaneously pressed on opposite sides. Let the radicle be pressed by a similar object a little above the tip, and it will be noticed that the pressed part does not transmit any influence to the more distant parts, but bends abruptly towards the object. Perceiving the air to be moister on one side than the other, it likewise sends out an influence to the upper adjoining part, which deflects towards the source of the moisture. When excited by light, the neighboring part bends from the light; but when excited by gravitation, the same part bends towards the centre of gravity. In almost every instance the ultimate purpose or advantage of the several movements can be clearly perceived. Two, or perhaps more, of the exciting causes often act simultaneously on the tip, and one conquers the other, doubtless in accordance with its importance for the life of the plant. The course pursued by the radicle in penetrating the ground being determined by the tip, has acquired for it the diverse kinds of sensitiveness which it possesses; and it is hardly an exaggeration to assert that the tip of the radicle thus endowed, and having the power to direct the movements of the adjoining parts, acts like the brain of one of the lower animals, which organ, seated within the anterior end of the body, receives impressions from the sense-organs, and directs their several movements.

In animals possessed of a nervous system, contractions only follow stimuli, which are carried to the contractile elements by nervous threads, the internal energy representing the external stimulus being called nervous energy or neurism. But where a nervous system does not exist, as is the case in some low animals and in all plants, external stimuli must be justly supposed to be converted into the same form of energy, which in such organisms has a general circulation throughout the contractile protoplasm. The attainment of some position, favorable for the procurement of relief from some unpleasant sensation, or the acquisition of some agreeable one, or for both, is the important thing directly subserved by such movements in the generality of animals. While we have the best of reasons for believing this to be true in the vast majority of animals, because fundamentally their structure is similar to our own, yet the inference that the same is true of the lowest forms of life is justifiable until it is proved to be mistaken.

Whatever be the nature of any movement, whether the projecting of portions of its own body-substance as pseudopodia in the primitive animal, the movement of flagella or cilia in more specialized forms, or the turning of the radicle of a plant-seedling in overcoming some obstacle, there is no resisting the conclusion that the functions of these organs, when once called into existence, are due to stimuli not unlike those which affect the motions of the limbs of the higher animals, and that the preliminary to all such movements, which are not automatic, is an effort. And as no adaptive movement is automatic the first time it is performed, effort, therefore, may be regarded as the immediate source of all movement. Now, effort is a conscious state, and implies a sense of resistance to be overcome. But when an act is performed without effort, resistance has been overcome, and the mechanism requisite for its performance has been completed. Automatism has now been reached. New movements, in their incipiency, necessarily meet with resistance. How this resistance is overcome, there seems to be some diversity of opinion among physiologists and metaphysicians, but it is generally believed that some such mental state as a sensation or a desire, which may or may not stimulate a natural process as an intervening element in the circuit, is concerned in its subduement. That sense-perceptions are stimuli to the immediate appearance of structural changes or movements is shown by the production of color-changes in animals through changes in the condition of the organs of sight and in the bending of the radicle of a seedling-plant a short distance above its tip in obedience to a communication from the tip of a sensation of hardness, caused by contact with a stone experienced in its downward progress in the ground.

New conditions bring forth new acts in animals. No one can deny this statement, as instances of its truth are too frequent to believe otherwise. That such may be predicated of plants, which have not the ability, as a rule, to meet with new conditions by reason of their being affixed to the soil, very few persons are willing to admit; but there is no getting away from the fact. The tip of the radicle of a plant not only has the power, acting as a brain, as it would seem, of guiding the root out of the reach of an obstacle that would be injurious, or in the direction of water when it would be an advantage, but a tendril has also the ability, in obedience to some inherent force, of making its way to a support that has been purposely placed in the near distance for its especial benefit. No external agencies, which the materialistic naturalist has devised for accounting for the movements of plants and low types of animal existences that are devoid of a visible nervous system, can possibly explain these movements, which are only explicable on the theory that nervous energy may be elaborated and be distributed without such a system by and through the general mass of the plant or animal, or by and through such parts as may be necessary to its good.

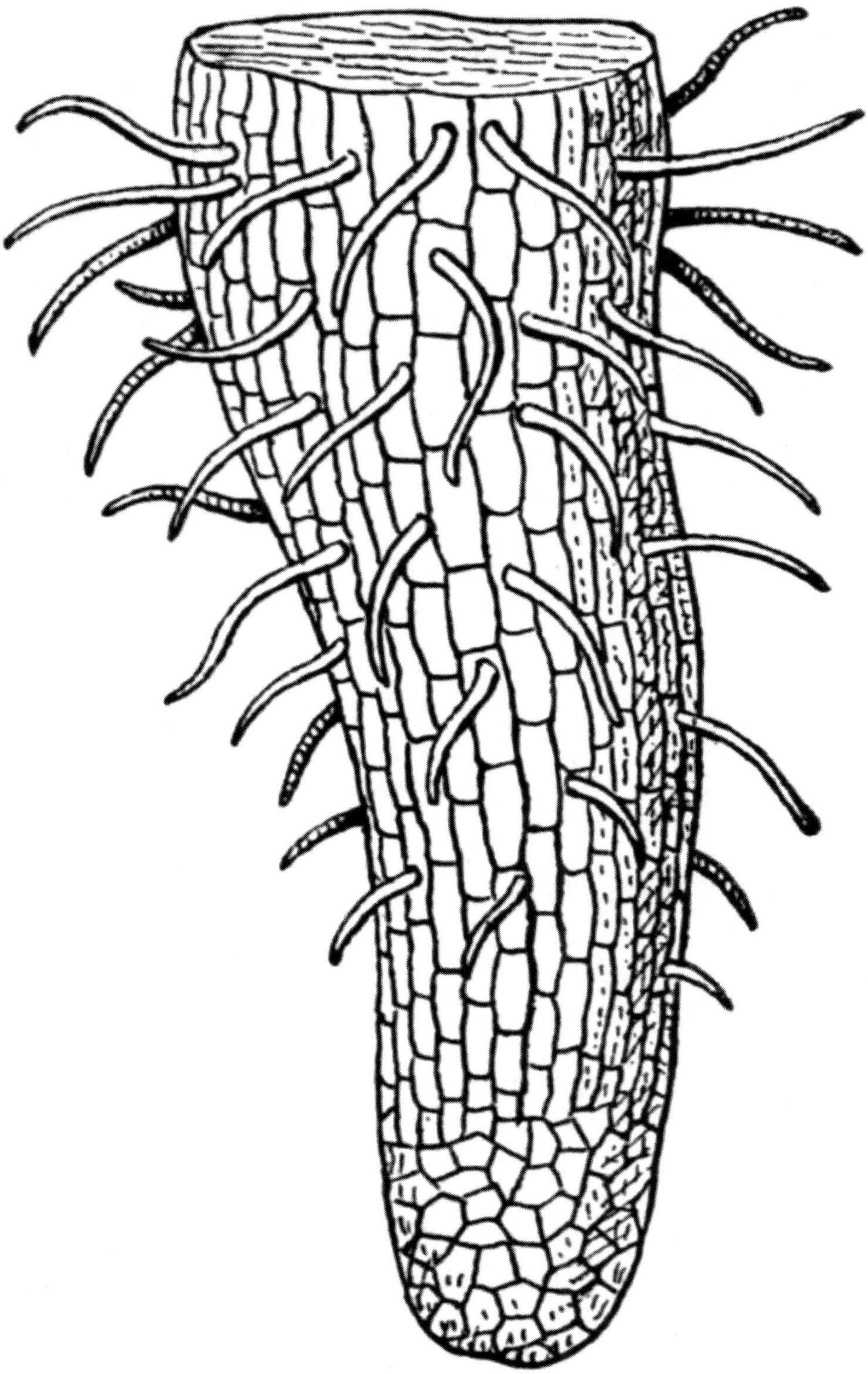

Figure 69. Tip of Radicle of Seedling Maple. Lower cells show where consciousness is supposed to reside.

No one who has experimented with the Droseras or Sundews, can have failed to observe the extreme sensitiveness which resides in their leaves. That these plants manifest a comparatively high order of consciousness, there can be no question. Try them with insects, or rare bits of meat, as articles of diet, and in a few hours, if vigorous leaves have been experimented with, the leaves will have folded around the food and commenced their curious process of assimilation. Mineral substances, such as bits of chalk, magnesia and small pebbles, have no such effect. They seem to ignore these things, just as an intelligent animal would if they were placed by its side. Some experiments made by Mrs. Treat, several summers ago, go far to confirm the statement that plants are endowed with some sort of consciousness. *Drosera filiformis* was the species used in her experiments. Some living flies were pinned one-half an inch from the leaves, but near their apical extremities. In forty minutes the leaves had perceptibly bent toward the flies, and in little more than an hour had reached the prey, the legs of the latter being entangled and held fast by the tentacles of the leaves. Next, the flies were removed three-quarters of an inch further from the leaves, but the latter, even though bent away from the direction of the light, failed to reach them at this distance. What was it that induced the leaves to stretch in the direction of the flies? Had the sun been shining from that side, it might be said that the movement of the leaves was influenced by its light and heat, for plants as a general rule turn toward that part of the heavens where these energies are the most effective. It cannot be that they were produced by some emanation of moisture from the bodies of the flies, or by any influence that might be exercised by the vibratory movements of their wings. No vain imaginings of such character will suffice for their explanation. The energy necessary to explain this phenomenon must come from within the leaves themselves. There was felt within them a desire for food, and it was this desire that led the leaves to bend away from the light and in the direction of the objects whose presence created in them that sensation. But how they were able, in the absence of any visible sense-organs, to determine the presence of these objects, is difficult to surmise. That they are sensitive to contact is generally conceded. And in them, no doubt, the sense of touch is keenly developed. Granting this to be the truth, then they see, as a blind man sees, by the sense of feeling. Currents of air, established by the vibration of the insect's wings, impinging upon the epidermis of the leaves, affect the cells beneath, and a nervous influence is started, guided by some central agency, of which we know nothing, causing the leaves to bend in the proper direction. But why the leaves do not thus bend when impinged upon by currents other than those produced by insects, I am unable to say. Even as a blind man, though deaf, is able through the sense of touch to discriminate moving objects by the currents of air they excite, so it may be presumed that the leaves of Drosera are endowed with the same wonderful and intelligent capacity. Such a feeling once experienced would be apt to be known again, for it would become fixed in consciousness by a process of memory. That Drosera, whose habits are more animal-like than plant-like, must occupy a high position in the scale of vegetable life, there can be no reason to doubt from what has been said, and this assumption receives a most remarkable confirmation from the fact that there are evidences, not apparent however, of a sort of nervous system in its make-up, as shown by the discovery of Darwin that by pricking a certain point in a leaf one-half of its substance becomes paralyzed.

Wonderful as these facts are, yet they are not more so than some recent discoveries made by Stahl while studying the simple movements and physical conditions of certain low plants called Myxomycetes. In their young stages these plants wander from the parts of the deposit on which they are creeping, and which are gradually drying up, toward those which are more moist. It is possible, by bringing moist bodies in proximity to any ramifications, to produce pseudopodia, which lift themselves from the deposit, and soon come into contact with the moist object, so as to enable the whole mass of the plasmodium, that is, the large, motile, membranous protoplasmic body formed by the coalescence of the swarm-spores of the Myxomycetes, to migrate thereon. But on the entrance of the plasmodia into the fructifying condition, the Myxomycete quits the moist deposit, technically called the substratum, and creeps upwards on to the surface of dry objects. Unequal distribution of warmth in the substratum and unequal supplies of oxygen and chemical substances soluble in water also cause locomotion in these strange organisms. Let the plasmodia come into contact on one side with solutions of saltpetre, carbonate of potash or common salt, and they at once withdraw from the dangerous spot; but an infusion of tan, or a dilute solution of sugar, causes a flow of the protoplasm and an ultimate translocation of the entire plasmodial mass towards the source of nourishment. Some solutions have an attractive or repulsive effect, but this is in accordance with the degree of their concentration. Unlike what is so natural to plants in general, the Myxomycetes seem to have an aversion to light, as shown by their disposition to withdraw from its presence.

How such tender structures as the Myxomycetes, which are destitute of every kind of external protection, are enabled to carry on their existence, the knowledge of the remarkably delicate reaction of their plasmodia under external influences prepares us to understand. Plasmodia, which are not yet ripe for reproduction, are kept in the moist substratum by their peculiar affection for moisture and utter dislike of the light. But within the darkness and moisture of the substratum the plasmodia do not necessarily remain in one place, for the differences in the chemical composition of the substratum cause continual migrations. Nothing more remarkable can be said of the plasmodia than that they have a wonderful faculty of avoiding harmful substances, and, traversing the substratum in all directions, of taking up the materials they require for food and growth. When, however, their internal changes have advanced so far that the plasmodia are approaching the fructifying condition, they are brought by their dislike for moisture, which now sets in, from the moist ground of forest or wood which they affect to the surface, where they creep up various upright objects, frequently not doing more than forming rigid reproductive capsules at some height from the ground. If, however, the substratum becomes gradually colder, as is the case in autumn, a change which sets in at the surface moving downwards, then the plasmodia migrate into deeper regions still having a higher temperature; but when the cooling proceeds very gradually, which especially happens in large tan-heaps, the plasmodia may in their migration attain considerable depths, where they then change into sclerotia, which are hard tuberous substances, resembling the tubers and bulbs of flowering plants. If, however, the temperature begins to ascend, the sclerotia again germinate, and movement takes place from the deeper and cooler parts to the upper already named.

Thus we see, in the locomotion of the Myxomycetes, extremely interesting cases of movements due to stimulation. Light, heat, moisture and gravitation are, in general, stimulus-movements, and ultimately all growth depends on stimulus-movement, the most primitive kind of protoplasmic movement. No causes other than those which actuate higher organisms can be discerned to account for this lowest type of organic movement. What form of inorganic energy can be cited of sufficient potency to cause the organism to change, and without regard to gravitation or any known form of attraction or repulsion, its position in obedience to stimuli acting for its self-preservation? There is none. In the Fuligo, or Tan Flower, a most remarkable example of designed movement has been observed. This form will, according to H. J. Carter, in its early amœbula stage, when isolated from the sawdust and chips of wood among which it has been living, adapt itself to the water of a watch-glass, or any other shallow vessel, in which it may happen to be placed. But, if the watch-glass be placed upon the sawdust, then it will make its way over the side of the glass to get to the sawdust. Here is probably shown a sense-perception of the presence and position of the tan-bark, as well as a feeling of desire to go to it. May not this desire have been due to a sense of discomfort induced by the surrounding water, or to the calling up in memory of some superior comfort associated with the tan-bark?

Man in his self-complacency thinks that he knows the plants about him. It is true that he has noted their form, their anatomy, their color and their resemblances and differences, but how few have studied them in meadow and woods by the light of a lantern at night or by the silver rays of the moon. One feels on such an occasion as though he had stepped from his threshold upon a foreign soil. Folded leaves and strange sleeping forms will be found to confront you in every direction. Of the nature of the nocturnal movements of plants, as well as their varied and curious attitudes, both in leaves and flowers, much speculation has been rife among botanists. In many flowers the night attitudes have been conclusively shown to have relation solely to their fertilization by insects; but the drooping night attitudes of the leaves were supposed to indicate an aversion to moisture, many plants seemingly verifying the conjecture by the assumption of the same position during rain as in the dew. But when the same pranks were played on a cloudy day or a dewless night, the explanation had to be abandoned. With the clovers, the nocturnal positions of the heads seem to be assumed only in the darkness, and this invariably, dew or no dew, while the leaves appear to revel in the rain, remaining freely open, their chief concern being the protection of the young blossom-clusters.

Were our eyes sharp enough we might discern a certain strangeness in the nocturnal expression of every plant and tree. But in no tree is this expression so remarkably emphasized as in the locust, a member of the same leguminous order of plants with the clover. These trees are especially noted for the pronounced irritability of their leaves, and odd nocturnal capers, whose seeming vital consciousness has induced some authorities to place them at the extremity of their system, in contact with the limits of the animal kingdom. How strange the pigweeds look at night! Their upper leaves, which during the day had extended wide on their long stems, now incline upward against the stalk, enclosing the tops of the younger branches, but still older plants are seen with leaves extended much as at mid-day, but nearly all turned edgewise by a twist in the stem. Circling in a close curve, the creeping-mallow blossom now ignores her proud array of cheeses, and the oxalis flower has forgotten her shooting pods to keep the vigil, closed and nodding upon her stem, while her leaves masquerade in one of the oddest disguises, their three heart-shaped leaflets being seen reflexed and adjusting themselves back to back around the stem with many contortions. Whatever the function of this strange nocturnal movement may be, and it is still a matter of dispute with botanists, one thing we are certain about, that is, its essential condition to the life of the plant, careful experiment having demonstrated, according to one authority, that "if the leaves are prevented from so regulating their surface, they lose their color and die in a few days"—a fact which Darwin has just as conclusively shown to be the case with other plants.

Flowers that bloom by night could hardly be suspected of that vanity which Rhodora has been made to confess by Emerson in his beautiful lines to this flower. Our evening primrose does not bloom in the dark hours for mere sentiment or moonshine, but from a nature which lies, figuratively speaking, much nearer her heart. "Often when the nights are very dark," says an old writer, "her petals emit a mild phosphorescent light, and look as if illuminated for a holiday. And he who does not fear to be out in her mild and lovely haunt may see a variety of nocturnal ephemeræ hovering around the lighted petals, or sipping at the flowery fountains, while others rest among the branches or hurry up the stems as if fearing to be too late." From the first moment of her wooing welcome it would seem that our evening primrose listens for murmuring wings, and awaits that supreme fulfilment with joyous expectancy, for it will invariably be found that these blossoms, which open in the twilight, have adapted themselves to crepuscular moths and other nocturnal insects, a fact which finds a striking illustration in the instances of very long tubular-shaped night-blooming flowers, like the honeysuckle and divers orchids, whose nectar is beyond the ability of any insect but a night-flying hawk-moth to attain. True, it is, that in other less deep nocturnal flowers the sweets could be reached by butterflies or bees if the blossoms were left open. But the night-murmurers receive the first invitation, which, if accepted, leaves but a wilted, half-hearted blossom to welcome the sipper of the sunshine. This beautiful expectancy, somehow or other, determines the limit of its bloom. However, in the event of rain or other causes preventive of insect visits, the evening primrose will remain open for the attention of the butterflies during the ensuing day, when otherwise it would have perceptibly drooped, and extended to them but a listless welcome. Most strikingly may this fact be seen illustrated in a spray of mountain-laurel. For nearly a week have I observed in my house these blossoms lingering in patient expectancy, when the flowers on the parent shrub in the woods had fallen several days before, their mission in life having been fulfilled. In the house specimens the radiating stamens, which are naturally dependent upon insects for their release, and the consequent discharge of the pollen, remained in their pockets on the side of the blossom-cup, a support, as it seemed, for the bracing up of the corolla upon its receptacle. But when the operation of releasing the stamens was artificially consummated, the flower-cup soon dropped off or withered upon the peduncle.

Not mainly has the writer, in attributing a phosphorescent quality to the evening primrose, followed the license of fancy, for, if scientists are to be believed, the regular luminous glow of this and other nocturnal flowers has long attracted the attention of the curious, and positive qualities of inherent light have been accorded in many instances. It is true, as one authority asserts, that "the evening primrose is perfectly visible in the darkest night," from which fact phosphorescent properties have been ascribed to it. Many well-authenticated cases are on record of luminous, electrical, lightning-like phosphorescence playing about flowers, the daughter of Linnæus having been the first one to note such an interesting phenomenon. Similar flashes or corona have been observed in nasturtiums, double marigold, geraniums, red poppy, tuberose, sunflower and evening primrose. According to various authorities, and it would be a rash and presumptuous commentator who would dare to challenge such an array of competence, many beautiful surprises await the traveller among the dewy shadows. Whoever has made such a journey will not only return with the consciousness that he has doubled his possessions, but that he has also explored a new world—a realm which he can look in the face on the morrow with an exchange of recognition that was truly impossible yesterday.

Whether or not all the facts that have been adduced show that plants are conscious organisms in the particulars for which it is claimed, it matters not, for enough have been set forth to demonstrate beyond the shadow of a doubt the position that they are endowed with a consciousness, no matter how infinitesimally small a part it plays in nature. Everyday observation of the botanist teaches the fact. Sensation, which is consciousness, has preceded in time and in history the evolution of the greater part of plants and animals, unicellular and multicellular, and, therefore, if kinetogenesis, or the doctrine of the effects of molar motion, be true, "consciousness," as Cope alleges, "has been essential to a rising scale of organic evolution." Animals which do not perform simple acts of self-preservation must necessarily, sooner or later, perish. Impossible it is to understand how the lowest forms of life, wholly dependent as they are on physical conditions of many kinds, should to-day exist if they were not possessed of some degree of consciousness under stimuli at least. We have but to picture to ourselves the condition of a vertebrate, without general or special sensation, would we obtain a clear perception of the essentiality of consciousness to its existence. If now use, as has been maintained, has modified structure, and so, in coöperation with the environment, has directed evolution, we can understand the origin and development of useful organs, and also how, by parasitism, or some other mode of gaining a livelihood without exertion, the adoption of new and skilful movements would be unnecessary, and consciousness itself seldom aroused, for continual repose would be followed by sub-consciousness, and later by unconsciousness. Such appears to be largely the history of degeneracy everywhere, and such is, perhaps, in a great measure the history of the entire vegetable kingdom, for plants, from their ability to manufacture protoplasm from inorganic substances, do not bodily move about in quest of food as animals generally do, and therefore require no conscious conditions, it would seem, to guide their movements. They become fixed, and their entire organization, except in specialized instances, becomes monopolized by the functions of nutrition and reproduction. Their movements are mostly rhythmic or rotary, but that they exhibit the quality of impromptu design more frequently than scientists are willing to allow must be admitted, or facts and the conclusions which naturally flow therefrom constitute no criteria of judging. Too much stress, I fear, is placed in these days upon the action of certain supposed forces that are resident in the plant's or animal's environment in accounting for its behavior, to the utter exclusion of any energy that may be acting from within the organism itself. "That consciousness as well as life preceded organism, and has been the *primum mobile* in the creation of organic structure," as Cope assumes, there is no doubt; but that it early abandoned the vegetable world, and also that all the energies of vegetable protoplasm soon became automatic, causing plants in general to become sessile, and therefore parasitic and in one sense degenerate, I cannot wholly accept. That insects have, in the matter of evolution of plant-types, exerted considerable influence on the conditions of almost all of their organs, the forms of the organs of fructification and especially of the flowers, through certain stimuli and strains to which they have become subjected by reason of these insects and their occupancy of parts as dwelling-places, there can be no doubt; and it is probable also, as has been maintained, that we owe to insects, directly or indirectly, not only the forms, but also the colors of the flowers, and their odors and peculiar markings as well. And thus while degeneracy, as observed in the abortion of ovules, carpels and perianth, may be seen everywhere, which the influences that have acted upon them have induced, yet it is the height of presumption to assert that consciousness has entirely abandoned the members of the vegetable kingdom, and that they are reduced to the condition of mere automata. It is true, as has been claimed, that the permanent and the successful forms of organization have ever been those in which motion and sensibility have been preserved, as well as the most highly developed; and just as true it is that plants, even though fixed to the soil and unable to effect a change of environment in consequence, are not so incapable of conscious actions as not to be able to meet any changes, and these changes do very often occur, that climate, new conditions of soil, helps or hindrances to growth and wear, may bring about. That they must adapt themselves to such changes, or perish in their struggle to exist, none can question. It is not enough to say that natural selection affords an explanation of every phenomenon that they may exhibit. There is an energy within the plant, think and write as we will, and it is this that comes to its aid and directs the movement that will be productive of the most good.

Concluding, then, let me aver that no plant can exist or fulfil its allotted part in the drama of life without the possession of some form or degree of consciousness. If it be true that life and consciousness preceded organization, and the statement can hardly be disputed, and have been the *primum mobile* in the creation of organic structure, what reason, seeing that life necessarily persists in vegetable organism, can be given for their dissociation in existing forms of plants, as seems to be the tendency of modern scientific thought? That plants once possessed consciousness, there can be no difference of opinion. Well, then, what has become of this consciousness? It could not have been destroyed, for energy or force, and consciousness certainly must be placed under this category, can never be destroyed. I repeat the question. What has become of it? Either it exists in the plant in a dormant condition, awaiting opportunities to call it into existence, or it has returned to the great Source of all consciousness, whence each individual organism, whether of plant or animal, obtained its *quantum*. It still exists, but how or under what conditions, I cannot affirm, and is to plants what mind is to man and animals, controlling their actions when such are for their well-being and good. If mind persists in a future state, then consciousness, which may be considered as mind in plants, must also persist, for it is not at all likely that the Source of all consciousness, which we worship as God, the Creator of all things, could be unmindful of the least of His children.

Chapter 44

Mind in Animals

That the lower animals are in possession of all the characters of the mind or soul that are either the inherited or acquired properties of man, some evidence will now be adduced. Foremost among these qualities is Reason. Much vagueness of idea exists as to what constitutes reason, the general tendency being to confound it with instinct, and to wonder where the one ends and the other begins. Hundreds of anecdotes, too familiar for mention, might be instanced, which have been described as wonderful examples of instinct, but which, upon careful examination, have been shown to be undoubted proofs of reason. That disposition of mind by which, independent of all instruction or experience, animals are unerringly directed to do spontaneously whatever is necessary for the preservation of the individual or the continuation of the species, is instinct. It is instinct that teaches the newly-born child to breathe, or to seek its mother's breast and obtain its nourishment by suction. Instinct teaches the bird how to make its nest after the manner of its kind, but it is reason that leads it to construct a fabric radically different from the typical form. Taking the case of insects, there can be no doubt that it is instinct that teaches the caterpillar to make its cocoon, to remain there until it has developed into an imago, and then to force its entrance into the world. Ducks, though hatched under a hen, instinctively make their way to the water, while chickens, though hatched under a duck, instinctively keep away from it. Man, as well as the lower animals, has his instincts, but very few of them are apparent, for he is able to bring the most of them under subjection by the power of his reason. Some, however, remain and assert themselves throughout the entire period of his life.

There is the widest possible difference between reason and instinct, the former being an exercise of the will, while the latter is independent thereof. Instinct comes in at birth, but reason is an after-growth of the mind. No exercise of thought does instinct require, but when the mind reasons some conclusion is deduced from the premises which it has assumed. All animals, in common with ourselves, possess the power of reasoning, although in a less degree. It is by the superiority of our reason over theirs that we maintain our supremacy. False premises often lead to wrong deductions, but their process is still one of pure reason. With them, as well as with ourselves, reason, especially in the case of domestic animals, often conquers instinct, and so by contact with a higher order of reason, that of man's, their own is more fully developed. They, in a sense, become civilized. Let a hungry dog and a cat be left in a room where food is unguarded, and their instincts will urge them to jump upon the table and help themselves. But if they have been trained, their reason restrains their instinct, and, no matter how hungry they may be, they will not touch the food until it is given to them. Some few years ago a matronly lady and her dog, a beautiful pug, were accustomed to take their dinner at a saloon which the writer daily visited. The dog was given a chair on the side opposite his mistress. He was a well-mannered animal, and never during his many visits to the place did he ever violate the laws of good manners. Patiently he would wait until the food was put upon his plate, and not even then would he take it, for he had been taught that it was something that should not be hastily seized and eaten. The idea that food cost money was distinctly impressed upon his mind, and this the owner did by thrice repeating, "This cost money." It was evident that the dog understood what was said from the thoughtful look he gave her. In a little while he was given the command to eat, but, like the cultured he was, everything was done orderly and decently. Almost any animal can be thus trained to subject its natural instincts to its reason.

Fishes are not known to possess much reason. There is not an angler, nevertheless, that will not tell you that he has had the powers of his mind taxed to the utmost in his efforts to induce an old and wary trout to take the bait, and even when he has succeeded in hooking him, it has greatly tried his genius for planning to prevent the fish from breaking his line. Natural instinct teaches a fish to fly from man, and even one's shadow on the water will frighten away the fish and destroy an angler's hopes of success. Yet we have seen a pond full of gold-fish which were quite tame, and which, when they saw a human being at the side of the pond, would come forward instead of showing alarm. They were so perfectly confiding that they would take a piece of bread or biscuit out of his hand. Here, then, is an example of the instinct, which urges them to flee from man, being overcome by the reason, which tells them to approach him.

Animals of burden may often be seen attending to prescribed work without any supervision. Dray-horses, as is well known, sometimes take pleasure in their work. I knew of a horse of the kind that was as much interested, apparently, his work as his owner. He never had to be told when to move, for all the while the dray was loading he was observant of everything, and, knowing the capacity thereof, was ready when the look from the master told him to proceed. Horses have sometimes shown a knowledge of the amount of work they are supposed to perform in a day. A case has been cited of a horse by Mr. Wood that was capable of doing his work without a driver. He belonged to the owner of an American mine. As soon as his cart was filled with ore, at a given signal he went off to the spot where the ore was to be dumped, waited until the cart was unloaded, and then returned for another load. So many loads had to be carried daily, and, strange to relate, the animal knew when his task was finished as well as any of the men. When the last load for the day was deposited, he could be seen trotting off in the direction of home, where he knew he would receive a kind reception from his mistress.

Figure 70. Wonderful Equine Intelligence. A horse that knew when his day's work was done.

Enough has been said to show that animals have and do exercise powers of reason. That they have the means of transmitting ideas to their fellows is not to be questioned. Language is the means of transmission. Not only are they able to interchange thoughts with each other, but with man also when they are brought into contact with him. They must possess a language of some kind, whereby they can understand each other, can comprehend human language, and make themselves intelligible to man. All these conditions are fulfilled in the lower animals, but there is one distinction between the capability of understanding their own language and that of man, and that is, that they are born with the one and have to learn the other. Newly-hatched chickens, although they have only entered the world an hour or so ago, understand perfectly well their mother. They know what to do when she calls them to find what food she has unearthed, and they know what to do when she warns them of danger. Who has not heard them talk to her? But how different are their tones under various circumstances. The little piping notes of content when all is going on well can never be confounded with the cry of alarm when they have lost their way or are otherwise frightened.

Wasps, as everybody knows who has studied these insects, carry out one of the first principles of military art. They always have the gate of their fortress guarded by a sentinel. Should danger be imminent, the alarm is given by the sentinel, and out rush the inhabitants to wreak vengeance upon the offender. Out of a full-sized nest, consisting of many hundred wasps, it is evident that the individual who is to act as sentinel must be selected, and its task appointed. How the selection is made, no one knows. But that such is done, there can be no question, for the rest of the community acknowledge their sentinel, trust to it for guarding the approaches of the nest, while they busy themselves with the usual task of collecting food for the young and new material for the nest.

Nearly related to wasps are the ants. Some of their performances are truly astonishing. They have armies commanded by officers, who issue orders, insist on obedience, and will not permit, while on the march, any of the privates to stray from the ranks. There are other ants which till the ground, weed it, plant the particular grain on which they feed, cut it when ripe, and store it in their subterranean granaries. Arrant slaveholders are others, who make systematic raids upon neighboring species, carry off their yet unhatched cocoons, and rear them in their own nests to be their servants. Somewhat recent discoveries show that there are ants which bury their dead. Two pairs of bearers are chosen to carry the corpse, one pair relieving the other when tired, while the main body, often several hundred in number, follow behind. So much could be said about ants, so closely do their performances resemble the customs of human civilization, that the subject could never grow uninteresting, but we must, for the present, forbear. All these various performances could not be possible were there not some way by which communication, or interchange of ideas, could be carried on among the individual members of the same community. Sometimes one species of ant is capable of carrying on a conversation, so to speak, with another. Bees, wasps and ants are the best linguists of the insect race, their language being chiefly conducted by means of their antennæ.

Who has not often observed two dogs, members of the same household, holding sweet converse with each other? Pug and Gyp were two animals that belonged to the family where

I spent a summer vacation. They thought much of each other when romping together in the yard, or in foraging the neighboring woods and fields for rabbits and ground-hogs. Never would they start out on an expedition for game without having previously laid their plans. It was interesting and amusing to watch them. They would bring their heads into close contiguity, remaining in this position for two or three minutes, when, by mutual consent, they would separate, look each other in the eyes, and then start off in different directions for the scene of their projected enterprise. Times out of number I have observed such behavior and have always discovered that they meant something of the kind. There were no audible utterances, no visible gestures, yet there was an interchange of ideas. Through the medium of the eye were the thoughts conveyed. It was spirit speaking directly to spirit, conveying by a single glance of the eye thoughts which whole volumes would fail to express.

Each species of animal has its own dialect. Yet there is another language, a sort of animal *lingua franca*, which is common to all. A cry of warning, no matter from what bird or animal it emanates, is understood by them all, as is well known to many a sportsman who has lost his only chance of a shot by reason of an impertinent crow, jay or magpie which has espied him, and has given its cry of alarm. There is not a bird of garden or orchard, or a fowl of the barnyard or doorside, that does not understand the peculiar cry of the rooster when a hawk is seen careering overhead, or perched upon the summit of a near-by tree. With one accord they flee to their coverts, and there remain until the danger is past.

No more quarrelsome and pugnacious species of bird exists than the English sparrow. He appropriates every available locality for nesting purposes, and our native species are driven to the necessity of fighting for their rights, or of seeking quarters in the rural districts which these birds do not infect. Thus it is that many a useful robin, bluebird or martin is driven from our midst. Many have witnessed encounters between these birds and the robins. The author once saw a contest between a pair of sparrows and a pair of robins for the possession of a certain tree that grew in his yard. Now the robin, single-handed, is more than a match for a sparrow. In the engagement referred to, the robins were getting the better of the sparrows, which the latter were not slow in perceiving. Instantly the sparrows set up the wild, ear-piercing harangue for which they are peculiarly noted, when more than a score of friends from the immediate vicinity gathered to their assistance. But the war-cry which they sounded not only summoned help to their standard, but it was equally understood by all the other birds of the neighborhood, who flocked to the defence of their brethren against the alien. The battle waged warm and fiercely for some minutes, when the sparrows were forced to seek safety in retreat.

Not only can crows and rooks assemble, hold council and agree to act on the result of their deliberations, but other birds are known to do the same things. Birds are able to communicate their thoughts to each other by means of a language, but it is not likely that in their language, or the language of animals in general, there are any principles of construction such as are possessed by all human languages. But the same effect may be produced by different means, and the reader will see that in the above instance no human language, however perfect its construction, could have served its purpose better than did the inarticulate language of the sparrows. They told their friends that their territory was usurped by an intruder too strong to be ejected by them, and implored their assistance. But while it told them this, it did still more, for it conveyed the report to their numerous foes, who winged their way to the support of their opponents. In fact, whenever animals of any kind form alliances and act simultaneously for one common purpose, it is evident that language of some sort must be employed.

That beasts possess a language, which enables them to communicate their ideas to each other, has been clearly shown. It is just as apparent that they can act upon the ideas so conveyed. We have now to see whether they can convey their ideas to man, and so bridge over the gulf between the higher and the lower beings. Were there no means of communicating ideas between man and animals, domestication, it is true, would be impossible. Everyone who has possessed and cared for some favorite animal must have observed that they can do so. Their own language becomes in many instances intelligible to man. Just as a child, that is unable to pronounce words, can express its meaning by intimation, so a dog can do the same by its different modes of barking. There is the bark of joy or welcome, when the animal sees its master, or anticipates a walk with him; the furious bark of anger, if the dog suspects that anyone is likely to injure himself or his master, and the bark of terror when the dog is suddenly frightened at something which it cannot understand. Supposing, now, that its master could not see the dog, but could only hear its bark, would he not know perfectly well the ideas which were passing through the animal's mind? Most certainly he would. There is a difference between the mew of distress and the ordinary conversation, the purr of pleasure, of a cat. A pet canary always knows how to call its mistress, and when it sees her will give a glad chirrup of recognition quite distinct from its ordinary call. Bees and wasps have quite a different sound in their wings when angry than when in the discharge of their ordinary work. Any one conversant with their ways understands the expression of anger and makes the best of his way off.

All the foregoing are but examples of sound-language. The gesture-language of animals, however, is wonderfully extensive and expressive. A cat, could it say in plain words, "Please open the door for me," could not convey its ideas more intelligently than it does by going to the door, uttering a plaintive mew to show that it wants help, and then patting the door. Dogs, or, in fact, all animals that are accustomed to live in the house, will act after a similar fashion. There, then, we perceive that the lower animals can form connected ideas, and can convey them to man, so that the same ideas are passing at the same moment through the minds of man and beast, evidencing that they possess the same faculties, though of different extent.

Figure 71. Papier-Maché Palace of the Hornet. Sentinel guarding the entrance to the palace.

Some few examples must suffice to show the power of gesture-language in the lower animals. I once owned a dog, a variety of hound, which was as companionable as any animal could possibly be. He was never happy unless he was on the go. So fond was he of travel and sight-seeing, that I gave him the name of Rover. My occupation calling me from home every day of the week, except Saturday and Sunday, but giving me a few hours of each day before the shadows began to settle round, Rover was forced to spend his time during my absence as best he could. He was no ordinary dog. Little he cared for the dogs of the neighborhood. His was a superior nature, and rather than associate with his neighbors when my companionship could not be had, he would perform his journeys alone, sometimes being gone nearly the entire day. But he managed to keep a pretty fair record of the time, for he was always on hand to greet me on my return home. His joy at my coming knew no bounds. He would rub up against my side, caper around me, assuming a hundred different attitudes, leap up into my face, which he would caress with his tongue. I shall never forget the barks of delight, nor the smile, as I would call it, for it verily seemed a smile to me, which lit up his intelligent face. Then he would slowly meander his way to the gate. Reaching it, he would place his right front paw upon the latch, spring it, and, taking hold of the top with his mouth, fling it wide open. He was then a very happy fellow. That he appreciated the favor I was about to show him, there could be no question, as he plainly showed it in his look, gesture and speech. Sometimes it was not convenient for me to take a walk with him, or I was not in the physical or mental condition to do so. It was not necessary for me to tell him in so many words that the pleasure would have to be foregone for the present, for his keen, discerning mind could read it in my looks. I never liked to disappoint him, for the grief which he manifested was piteous in the extreme. He would prostrate himself to the ground, place his head between his front paws, and look the very picture of inconsolable distress. The low, sorrowful moan which he would emit, when the disappointment was the keenest, was so heart-rending, that many a time I would reverse my purpose and say, "Come, Rover, master will not deny so good a creature the pleasure of his company for an hour or so in the woods." Instantly his whole expression would change, and there would be exhibited a joy as intense as the grief which had depressed him to the earth. Rover was no hypocrite. His sorrow was not assumed, but as real and poignant a sorrow as ever possessed a human breast. I have known him to grieve for hours, and even to refuse the daintiest food when he has been disappointed. Were he dissembling, seeing that it availed him not, he would not be likely to have kept it up so long, and to his sore discomfort and detriment. Examples of animals making their language intelligible to man could be multiplied *ad infinitum*, but we must pass on to say something about their capability of understanding the language of man.

That many of the lower animals understand something of human language is a familiar fact. All the domesticated animals, notably the dog and the horse, can comprehend an order that is given to them, though, perhaps, they may not be able in all instances to understand the precise words which are used. There are many occasions, however, when it is evident

a willing spectator. One morning, however, instead of going to the gate for the paper as was my custom, I stood debating in my mind whether to go or not, when, to my utmost surprise, the dog, after watching me for a while, walked very soberly down to the gate, picked up the paper in his mouth, and brought it to me, not laying it down at my feet, but placing it in my hands. I thanked him for his kindness, gave him a few gentle pats upon the head, and he walked away as pleased as a child would have been who had received a few pennies for a similar service. The dog had evidently read in my looks the debate that was going on in my mind, and knowing that I always read the paper when I came down from my room, anticipated my wishes by bringing it to me.

Figure 72. Unsolicited and Unlooked-For Kindness. How prince forestalled my wishes by bringing me the morning newspaper.

There is in the two interesting stories just related a singular aggregation of faculties which are held in man to belong to the immortal, and not to the mortal part of his being. Reason, or the deduction of a conclusion from premises, is strikingly exhibited. Then there is the power of forming ideas and communicating them to man, and the capability of understanding man's language, and even of anticipating the wishes of human friends. And lastly, there is the intense love for the master, combined with the power of self-sacrifice, which enabled Lion and Prince to act as they did, while instinct was urging them to take their exercise in the open air, or in the enjoyment of luxurious ease.

No faculty of the mind gives greater trouble to materialists than Memory. It is that which survives when every particle of the material brain has been repeatedly changed. It is that which more or less deeply receives impressions and retains them through a long series of years. And even when they are apparently forgotten, hidden as it were behind a temporary veil, a passing odor, a dimly-heard sound or a nodding flower may rend the veil asunder in the twinkling of an eye, and scenes long forgotten are reproduced before the memory as vividly as though time had been annihilated. Nothing is omitted. There comes up to view a minute and instantaneous insight into every detail, and for a moment we break loose from our fleshy tabernacle, and see and hear with our spiritual and not with our material eyes and ears. Man expects that he shall retain his memory and carry it into the next world. He also expects to recognize in the spiritual world those whom he has loved in this temporal sphere. Memory, therefore, must be spiritual and eternal; and wherever it can be found, there exists an immortal spirit. No stronger evidence, apart from Revelation, exists of a future life of man than memory. And if we apply this proof to ourselves, then, in pure justice, we should apply it wherever memory is found.

But some have claimed that memory is a mere emanation from the brain. That an inferior brain is coupled with an inferior intellect, and that if the brain be slightly or seriously injured, the powers of thought will be weakened or utterly held in abeyance, are arguments that have been made to prove that thought is the creation of the brain. The facts in themselves are true, but the conclusion is false. The brain is but the organ or instrument of the thought-power, and stands in the same relation to it that a tool does to a carpenter. However good an artisan a carpenter may be, it is but common-sense to say that he cannot turn out good work with a blunt instrument, or any work at all with a broken one. So it is with the brain. It is but the tool of the spirit, and, if it be damaged in any way, the keenest intellect will not be able to work with it. Memory, moreover, exists in creatures which are devoid of brain. No real brain, but only a succession of nervous ganglia running the entire length of the body, is found in insects, and indeed in many of them the faculty of memory is very strongly developed.

Then there is the moner, a mere speck of formless protoplasm, that has not the slightest trace of a specialized nervous system, yet it has the power of throwing out arms and of retracting them into the general body-mass, of opening out mouths where a particle of food strikes it, of digesting its food, and of circulating its fluid without the necessity of canals. But how are these movements effected? Certainly a nervous influence is the prime mover of all its actions. Nerve-matter, mayhap, constitutes its entire body-mass, or it may be all brain as well as all muscle. Though the lowest and simplest of all animal life, yet it possesses an innate consciousness and intelligence. Memory is not wanting as a faculty of the mind of this all-brain animal, which I have thought fit to characterize it, as some actions of it already described under the head of "Slime Animals" seem very clearly to indicate.

Some fifteen years ago I mentioned in an article, entitled "Insect Pets," a pair of flies, the common *Musca domestica* of our houses, which had been closely observed by Mr. Forestel, the gentleman who at that time had charge of the distributing department of the Philadelphia *Record*. This position necessitates nocturnal employment. While taking his midnight lunch, Mr. Forestel's attention was directed to a pair of these insects that had located themselves upon his plate. Had it been in the summer when flies were plentiful, the event would hardly have been noticed; but being in the winter, a season notable for their great scarcity, they could not but impress his mind with something out of the ordinary. Night after night these self-invited and curious guests presented themselves at the same place, and it was a long time before he observed the regularity of their visits. At first he was disposed to view the alighting of two flies upon his plate as a mere coincidence, but he at length became so deeply interested in the affair, that he resolved to watch their actions very closely. It was not long before he became convinced that they always waited for the commencement of the meal, when they would deliberately fly down for their regular lunch. So closely did he watch them, that he was soon able to discriminate between the two, and to discover beyond a doubt that it was not a series of two flies, but always the same pair. As time progressed, Mr. Forestel and the flies grew to be famous friends. They in time became so friendly, that they would permit themselves to be handled. Although at first they would only appear when Mr. Forestel was alone, yet they soon became accustomed to strangers. On the nights when their friend was not on duty, others have spread their lunches on the table used by him, but the flies were not slow in making the discovery, and, instead of alighting, would quickly hasten away without their accustomed meal. Who can deny the possession of memory to these two flies? Had the discovery of the food been an accidental occurrence the first time, could it have been so the second and all the succeeding times? Then, again, the flies always came at the right time, showing that they had some idea of the passing moments. Even admitting that this latter thought is out of the range of probability, there can be no doubt that they were not observant creatures, else how would they know when to come, or whether or not the man that sat at the table was the same that had shown them so much kindness on their previous visits. That they did know these things, there cannot be the slightest doubt. But how did they know them? There is only one answer to the query. They knew them through the exercise of memory, these creatures impressing on their minds the appearance of the objects near the table, the form and color of the table itself, the look, manner and dress of the man who sat by it, and acting on the result of these impressions. Human beings act in just the same way in traversing for the first time a locality through which they will have to return. And yet, as has already been stated, these insects have no true brains.

Considerably removed from insects are the vermes, or worms. Man, in his overweening opinion of self, would hardly credit the earth-worm with the possession of any mental qualities; yet it has been shown that it can reason, and can communicate after its fashion with its fellows. It is now my intention to prove that it has the power of memory. Has the reader ever seen an earth-worm trying to carry into its burrow a pair of pine-needles joined at their bases? It knows just where to seize the pair. This it determines by feeling, or moving its head along the needles, the sense of touch being very acute in this portion of its body. Hardly ever is a mistake made by seizing the free or apical extremities. Once it has discovered where to act, this position is fixed in memory, and the animal exercises the latter power in dealing with objects of the kind in all subsequent operations.

Almost any living being can by means of the faculty of memory be taught by man. But were it absent, no teaching would be of the slightest avail. In most cases where an animal is ferocious, I firmly believe that fear, and not ill-temper, is the real cause of its conduct. Let a little kindness be shown, and the animal will never forget it. Such acts, repeatedly performed, assure it that your intentions are well-meant, and it soon learns to recognize in you a friend. The memory of your goodness will often be recollected after long years of separation, and the most joyous feelings be manifested at the sight of your presence upon returning home. Everyone who has had personal experience of domesticated animals must have remarked the great strength and endurance of their powers of memory. The dog, the cat, the horse and the ass afford so many familiar anecdotes in point, that I shall be obliged to pass them over and restrict my illustrations to a few animals about which little has been said.

For obstinacy of opinion no animal can excel the pig. He is a creature whom few, on account of his uncleanly person and disgusting habits, would care to caress. Yet there is no animal under man's care that enjoys such treatment better than he does. He will stand for hours while you rub his head and back, the very impersonation of contentment, never failing to express his thanks and appreciation by occasional monosyllabic grunts. A friend of ours, living in Northern Indiana, had a fine fellow, whom he had raised from infancy. When he was quite young, he began to show him considerable attention, picking him up in his arms, and fondling him in the most affectionate manner. The choicest food was always reserved for him, and the cosiest bed of straw provided for his nightly rest. In process of time the animal grew to great size, but he never forgot these early attentions. He expected them all the same. When denied what he deemed were his lawful rights, he would set up an unearthly squealing, enough to split the ears of the groundlings, and refuse to be comforted until his demands were satisfied. Never was the master, when out of the house, safe from his intrusions. He would besiege him in the presence of company, command his attention, and cry in his own peculiar fashion if he thought himself ignored. Many a rough-and-tumble game, which reminded me of boys in my childhood days, would they have together, and it was really amusing to see them. They enjoyed these tussles, which were always of the most friendly character.

Stupid as the life of a cow may seem to be, yet there has been known to the writer some cows which were far from being dull and prosaic. Our same Hoosier friend had such an animal, whom he called Daisy. She was very docile and affectionate, and would come, even when grazing in the most delightful pasture of clover, whenever her name happened to be mentioned. Daisy was a pretty creature, and very exemplary in her conduct. When her companions would break into a field of corn, where they had no right to be, she would not follow their wicked example, but remained where her master had placed her and the rest of the herd, showing them, as it were, that she did not approve of such wilful waywardness. No member of the bovine family of animals ever showed a greater fondness for love than Daisy. The master could put his arms around her neck, and lay his face against the side of her own. That she approved of such familiarity was evident, for she would show that she did by placing her lips against his in true lover-like fashion. But there came a time when this attachment to the master became dissolved. On account of the bad behavior of the herd in general, and to make it a law-abiding community, it was resolved that each member should have its horns sawn off close up to the skull. This, it was thought, would improve the temper of the herd, and make it less troublesome to manage. No fear was entertained, however, for Daisy, who was already as good as she could be, but Daisy must undergo the same cruel punishment for the sake of uniformity in this particular in the herd. It had, however, the opposite effect upon Daisy from what it had upon the rest of the herd, for it made her sullen and morose, and from that time she resented all familiarity upon the part of the master. She seemed to view him as her worst enemy. All attempts to settle her grievances were viewed in a suspicious manner, and the matter of reconciliation had at length to be abandoned.

Beasts, there is no doubt, were intended to be the servants of man, and there is nothing in his hands half so powerful in the accomplishment of this end as thoughtful kindness. Inflexible decision, combined with gentleness and sympathy, are irresistible weapons in his power, and no animal exists, I firmly believe, which cannot be subdued if the right man undertakes the task. By this mixture of firmness and kindness many a wild beast of a horse has been in a half-hour rendered gentle and subservient by Rarey, obeying the least sign of his conqueror, and permitting himself to be freely handled without displaying the slightest resentment.

That there is something more in memory than a mere production of a material brain must seem probable from the examples given. In several cases the animals were without any brains at all, but in others, where a brain did exist, its material particles must have been repeatedly changed, while the ideas impressed upon the memory still remained in full force.

Perhaps no attribute of the mind is better fitted to follow that which has just been treated than Generosity. But whether we accept it in the sense of liberality or magnanimity, it is certainly a very lofty quality, and one which infinitely ennobles the character of those who possess it. Taken in the former sense, it is an attribute of Deity, who gives us freely all that we have, and so sets us an example of generosity to our fellow-creatures. Now, if it be admitted that the possession of generosity ennobles man's character, while the lack of that quality debases it, then the inference is undeniable that when we find a beast possessing generosity, and a man devoid of it, the beast is in that particular the superior of the man. And that generosity, being a divine attribute, belongs to the spirit and not to the body, no believer in Christianity is likely to deny. Therefore, wherever we find this characteristic developed, we must admit the presence of an immortal spirit.

That the lower animals do possess generosity in the sense of Liberality will now be proved from circumstances that have occurred within my own observation. My first proof is a very interesting one, and is drawn from the life of a dog that was the companion of my school-boy days. Sport was the name of the animal. He was not a greedy, selfish creature, but a generous, noble fellow. Many an act of self-sacrifice had he been known to perform, and he was never happier than when he was doing some good to his fellows. It was not unlike him, when he would meet a poor, strange and hungry animal of his own kind by the roadway, to bring him to his master's house, and at the meal-hour divide with the unfortunate his noon-day allowance. Between him and a certain cat, called Blackey, which was also a member of the same household, there existed a very strong friendship. Any injury done the cat was most summarily resented by Sport. He would share his meals with her, and never seemed satisfied unless she would consent to take the choicest bits. But the generosity was not all on his side, for the cat certainly rivalled him in the exercise of this noble trait, which all acknowledge to be one of the noblest characteristics of the human mind. When Blackey was sick, and unable to be around, much of the time of the dog would be spent in her presence. He would caress her with his paw, smooth her silken, jet-black fur with his tongue, and seek by every means in his power to raise her drooping spirits and alleviate her miseries. No animal, not even man himself, could show more real sympathy for a fellow in distress than Sport did for Blackey.

No bird, it would seem, could be expected to manifest so little of generosity as the sparrow. As a rule, sparrows are remarkable for their ability to take care of themselves. Theirs is a nature which is based upon self. They are an avaricious species, and little they reck for their neighbors. As the eagle is known to treat the osprey, and the skua-gull its weaker brethren, so the sparrow has been known to act towards its neighbors. But exceptions exist to every rule, and we are pleased to record an honorable one in the case of this most detested species. Close by a maple-tree, which a pair of sparrows had appropriated and made the support for their home, dwelt a sturdy robin with his mate. Their home, a mud-lined domicile, was placed in the crotch of a small tree. Three children appeared in process of time to bless the happy couple. Everything went along smoothly and pleasantly with the robins, the sparrows being too much engrossed with their own affairs to think of giving them any trouble. But a tragedy soon happened which, sad to relate, foreboded evil and consequent death to the nest-full of young robins. Father and mother had, while searching for food for the little ones, been cruelly killed by a conscienceless sportsman. But the fledglings, which seemed doomed to die the death of starvation, were spared by some good genius who put it into the heart of the sparrows to pass that way, and thus was their sad and pitiable condition brought to the light of day.

Their heart-rending appeals for food, combined with their orphaned situation, struck a sympathetic chord in the breast of the sparrows, and day after day these birds, whose chief concern naturally seems for self, might be seen acting the part of the good Samaritan towards these unfortunate of God's children.

But let us now pass to that form of generosity which has been called Magnanimity. Few qualities in human nature are more noble than the capability of foregoing revenge when the offender is powerless to resist. This unwillingness to resent an injury, even though the power to do so is present in the individual, is what is implied by magnanimity. When we find those beings whom we designate brutes rising to a moral grandeur which few men can attain, disdaining to avail themselves of the opportunity of vengeance, and even rewarding evil with good, it does seem an utter absurdity to affirm that they are not acting under the inspiration of Him who gave us the celestial maxim, "Love your enemies." By their actions they show themselves worthy of everlasting life, and what they deserve they will assuredly receive at the hands of Him who is Justice and Truth. Consciously, or unconsciously, the feeling of magnanimity is acknowledged among mankind. Even in the lowest stratum of society it is recognized. As with man, so with the lower animals; and there are many instances on record where the strong have disdained, no matter what the offence had been, to make reprisals upon the weak.

Bus and Jack are two dogs whose acquaintance I made three years ago. The one is a beagle, and the other a pug. No one that has seen these animals in their frolics and plays, would ever suspect that any differences could arise between them. But when such disagreements do occur, and there is hardly a day that does not witness a dozen or more, it is always Bus that is the instigator. The most trifling act upon the part of Jack will be made the cause of offence, and an excuse for the precipitation of a quarrel. In a rage, Bus will fly into the face of Jack, but the latter will coolly shake him off and walk leisurely away. No provocation will induce him to resent an insult or an injury, especially where Pug or a dog smaller than himself is concerned. It is not that he is afraid of Pug, for, when once aroused in the presence of equal or even superior strength, he becomes a terror. He is too magnanimous to avenge a wrong done him by one less powerful than himself. The look which he would give Pug, after one of these attacks had been made, was one of pure contempt, and said, as plainly as words could have said, "Your assaults are mere child's play, and are unworthy of recognition by one who is so much your superior in feats of valor." That Pug felt the meaning and force of the look was apparent, for he would always slink away abashed to some corner, where he would remain for an hour or two without showing himself. Over and over again has Jack allowed little dogs to bite him without troubling himself to retaliate; but if a big dog ventured upon an insult, that dog had to run or pay the penalty for his temerity. No dog could give a more disdainful look than Jack, and that look always gave him an easy and uninterrupted passage wherever he chose to go.

Other anecdotes of a similar nature might be given to show that animals can act magnanimously towards each other. That they are as capable of displaying the magnanimity of their nature towards men whom they hated has frequently been observed. The manager of a mill in Fifeshire, Scotland, was, according to Rev. J. G. Wood, very much disliked by the watch-dog, probably from some harsh treatment which the animal had received from his hands. One very dark night the manager had strayed from his path and fell over the dog. Seeing the mistake he had made, and finding that he could not recover himself, he gave himself up as lost, for the dog was a very powerful animal. But the dog was magnanimous enough to spare a helpless enemy, and to lay aside old grievances. Instead of seizing the prostrate man by the throat, as a brute would be expected to do, the dog only licked his face and exhibited his sympathy. Ever afterward the man and the dog were fast friends.

Just as there are animals capable of exercising great self-denial by giving to others what belongs to themselves, and even manifesting a generosity which would put human nature to the blush, so there are animals which can cheat like accomplished swindlers. As all Cheatery requires the use of the intellect, it is therefore evident that the most intellectual animals will be the most accomplished cheats. Dogs have shown themselves to be considerable adepts in cheating, and this we would naturally expect. Some curious and rather ludicrous instances of cheatery upon the part of the dog are noticed. We once knew a pair of dogs, a spaniel and a pug, that were inmates of the same house. They were very jealous of each other so far as the master was concerned, and neither could endure to see the other caressed. It happened that the spaniel was taken quite ill, and was in consequence very much cared for and petted. His companion, seeing the attention and sympathy that were bestowed upon him, pretended to be sick herself, and, going to a corner of the room, lay down upon the floor and looked the very picture of misery and distress. A cat and a dog, that for many years were members of the writer's family, had taken a fancy to the same spot, a soft cushion at the head of a sofa. While they were the best of friends, yet a difference of opinion would occasionally arise, and a slight loss of temper would be the result. When the cat would be in the possession of the cushion, the dog would torment her in every possible way with the view of causing her to abandon the pet spot. He would pull at the cushion, seeking to drag it to the floor, or, seizing the occupant by the ear or tail, endeavor to dislodge her by force. But the cat, seemingly unmindful of what was going on, and the very impersonation of patience all the while, would refuse to give up so comfortable a couch. At last the dog hit upon a *ruse* which he knew would bring the cat down from the sofa. He rushed out into the kitchen, and began acting as though in pursuit of a mouse. He and puss had often engaged in such diverting business. This was more than the latter could stand. She was down from her cozy bed in an instant, and was soon by the side of the dog. But as soon as puss, all ablaze with excitement, had her head in a corner and was straining her eyes to get a glimpse of the supposed mouse, the dog ran to the sofa at full speed, jumped on the cushion, curled himself round, and was happy. Poor puss, perceiving that the dog had left her, was not slow to discern that she had been imposed upon by the latter, and that it was only a trick that had been played upon her by her shrewd companion, that he might get possession of the soft spot upon the sofa. She, however, bore it good-naturedly and decorously, and was ever afterward on the alert for these little tricks of her canine friend.

Birds can be as capable of cheating, not only each other, but other animals. A crow, belonging to John Smedley, a resident of Lima, Pa., was an adept in the business. When dinner was preparing, he would fly around the corner of the house, set up a terrific cawing as though in great distress, and when the mistress of the house, with whom he was a great favorite, would come out on a tour of investigation, the rascally bird would elude her and manage to steal round to the table in the opposite direction and seize what food suited him the best, which he would carry to the top of the house, where he would eat it at his leisure. No persuasions would induce him to come down, for he knew that such action was a breach of the peace, and he was fearful of the punishment, that of confinement to a cage, which would follow. When, however, he felt assured that his mistress had forgiven the wrong-doing, he would fly down to the porch, and do his utmost to convince her that he was a well-meaning bird, and that he was thoroughly ashamed of his actions. But there was one member of the family that utterly detested the bird. It was the dog Rover. Many a trick had the bird practised upon the latter, especially at meal time. Poor Rover was not allowed to eat in peace. When he would be wholly absorbed in his dinner, the crow would approach him in the rear, give him a severe twirl of the tail, and then in a twinkling fly to one side, looking the very picture of innocence. But ere the dog had recovered his self-possession and was ready to resume his feeding again, the bird had captured the daintiest morsel, and was off to the tree-top. Discomfited and outwitted, the dog would rush to the base of the tree, bark his growls of anger and defiance, while the crow would look quizzically down from above, and chuckle with delight.

Many of my readers may, perhaps, remember the story of the two dogs that used to hunt the hare in concert, the one starting the hare and driving it toward the spot where his accomplice lay concealed. I recall an instance where a somewhat similar arrangement was made, only the two contracting parties, instead of being two dogs, were a dog and a hawk, the latter making use of his wings in driving the prey out of the copse into the open ground. Innumerous examples of such alliances are known, and in all of them there is manifest the curious fact that two animals can arrange a mode of cheating a third. One of the principal stratagems used in war, that is the ambuscade, whereby the enemy is induced to believe that danger is imminent in one direction, when it really lies in the opposite and unsuspecting direction, is employed. No one would admit that a general who contrived to draw the enemy into an ambuscade acted by instinct. The act would be construed as proof of the possession of reasoning powers surpassing those of the adversary. And if this be the case with the man, why not with the dog, or with the raven or hawk, when the deception is carried out by precisely the same line of reasoning?

Beasts possess, in common with man, the sense of Humor. This is developed in many ways. Generally it assumes the phase of teasing or annoying others, and thus deriving pleasure or amusement from their discomfort. Sometimes, both with man and beast, it takes the form of bodily torture, the struggles of the victim being highly amusing to the torturer. Civilized man has now learned to regard the infliction of pain upon a fellow as anything but an amusement, and would rather suffer the agony than inflict it upon another. But with the savage it is otherwise, for there is no entertainment so fascinating as the infliction of bodily pain upon a human being. Among our Indian tribes, torture is a solemn usage of war, which every warrior expects for himself if captured, and which he is certain to inflict upon any prisoner whom he may happen to take. The tortures which he inflicts are absolutely fiendish, and yet a whole tribe will assemble around the stake, and gloat upon the agonies which are being borne by a fellow-creature. Similarly the African savage inflicts the most excruciating sufferings upon the man or woman accused of witchcraft, employing means too horrible to be mentioned. But in all these cases the cruelty seems to be in a great measure owing to obtuseness of perception. Yet the savage who binds his victim to a stake, and perforates the sensitive parts of his body with burning pine-splinters, behaves very much like a child who amuses itself by catching flies, pulling off their wings and legs, and watching their unavailing efforts to escape.

Many years ago cockchafers were publicly sold in Paris for children to torture to death. The amusement consisted in running a hooked pin through the insect's tail, fastening a thread thereto, and watching the poor creature spin in the air. After the poor beetle was too enfeebled to expand its wings, it was slowly dismembered, the child being greatly amused at its endeavors to crawl, as leg after leg was pulled from the body. A similar custom, though in a more cruel form, prevails in Italy, the creatures which are tortured by way of sport being more capable of feeling pain than are insects. Birds are employed in this country for the amusement of children. A string is tied to the leg of the bird, and the unfortunate creature, after its powers of flight are exhausted, is generally plucked alive and dismembered. The idea of cruelty does not seem to enter at all in these practices, but they are done from the sheer incapacity of understanding that a bird or a beast can be a fellow-creature. Italians are notorious for their cruel treatment of animals, and if remonstrated with become very much astonished and reply, "Non è Cristiano," that is to say, "It is not a Christian." Englishmen have little to boast of on this score. Bear-baiting was abolished by the Puritans, not because it gave pain to the bear, but because it gave pleasure to the spectators. Even at the present day, both in England and in this country, there is a latent hankering after similar scenes, and dog-fighting, rat-killing and cock-fighting, even though they are now contrary to law, are still practised in secret. Similarly the sense of humor is developed in the lower animals by causing pain or annoyance to some other creature, and the animal acts in precisely the same manner as a savage or a child.

Sparrows, as might be expected from their character, will gratify their feelings of aversion by banding together for the purpose of mobbing some creature to which they have an objection. In Hardwicke's *Science Gossip* for December, 1872, there is a short account of a number of sparrows mobbing a cat. Evidently the cat had intended making a meal on one of the birds, but was greatly mistaken, for the sparrows dashed upon him so fiercely, that he soon turned tail and ran into the house, one of the sparrows actually pursuing him into the house. The poor cat ran up-stairs, and was found crouching in terror under one of the beds. This happened in London, where the sparrows are less numerous now than they used to be.

No bird of my knowledge possesses a larger amount of humor than the crow. I have known him to feign an attack upon a distant part of a field of newly-sprouted corn, which was being guarded by a farmer with his gun. When the latter would be drawn to that part of the field where the attack was to be made, the sagacious bird would manage to outwit him, slip around to the other side, drop down into the field and obtain a few tender sprouts before the farmer hardly knew what was going on. But he was always up and away at the opportune moment, and, perched upon a fence-rail, beyond the range of the gun, would enjoy one of his rollicking cawing laughs at the farmer's expense. Crows that are tame have the sense of humor more keenly developed than their wild brethren of the fields and the woods. I once knew a tame crow that took great pleasure in annoying a dog that lived in the same family. Carlo, as the dog was called, was never so contented as when allowed to sleep the hours of the morning away, after a night's carousal, in a quiet, sunny spot in the backyard. When the dog had become fast wrapped in the arms of the god of slumber, the crow would steal to his side, give his ear a sharp pull, and when the dog would awake and look around the crow would be busy in gleaning, the most unconcerned creature in the whole yard. Again and again would she annoy the poor animal, and always with the same evident sense of delight, which I could always read in the mischievous twinkle that lurked in her eyes, till the dog, bewildered and unable to account for such mysterious actions, would silently skulk away to other parts, where he hoped to be free from all intrusion. Even the mistress of the house was not exempt from her annoyance. She would carry off everything she could lay hold of, and always hid them away in one place, that is, in a large crevice on the top of the house between the peak of the roof and the chimney. One day the mistress's spectacles disappeared. Search was instituted everywhere, but without effect. None knew better than the bird what the trouble was. While the search was going on, she busied herself in looking around, and seemed as desirous of finding the missing glasses as any member of the household. The look which the bird gave showed that she enjoyed the situation of affairs immensely, and considered it a fine joke that she had played upon her mistress. After a few days the lost spectacles were restored to their accustomed place, but no one ever positively knew how they came thither.

Domestic birds, as a rule, are remarkable for the generosity which the master-bird shows to his inferiors. He will scratch the ground, unearth some food, and then, instead of eating it himself, will call some of his favorites, and give them the delicacy for which he labored. But I have met with a few cases where the cock scratched as usual, called his wives, and, when they had gathered round him, ate the morsel himself. It was but a practical joke that he had perpetrated upon them, and that they felt it as such their looks only too strongly testified. There was a relish of delight in it for the cock, for the cackle, which he immediately gave, assured me of this fact as much as the laugh of a man could have done who had played such a joke upon one of his fellows.

Parrots are much given to practical joking, after the ways of mankind. A parrot, belonging to an aunt, had a bad habit of whistling for a dog, and then enjoying the animal's bewilderment and discomfiture. She would call the cat, as her mistress was accustomed to do, and when puss would come, expecting some dainty article of food, she would call out in her severest tone, "Be off, you hussy!" and the cat would make all possible speed for a place of security, greatly to the amusement of the parrot from her perch in the cage. There have been known parrots that would play practical jokes upon human beings, but dogs and cats seem to be the principal victims of the parrot's sense of humor.

Animals not only show their playfulness in such tricks as have been mentioned, but many of them are able to appreciate and take part in the games played by children. When I was a boy I knew a dog, a species of greyhound, which was an accomplished player at the well-known game called tag, or touch. Quite as much enthusiasm was displayed by the animal as by any of the human players. He would dart away from the boy who happened to be "touch" with an anxiety that almost appeared terror. It was an impossibility to touch the clever canine player; but he was a generous creature, with a strong sense of justice, and so, when he thought that his turn ought to come, he would stand still and wait quietly to be touched. His manner of touching his play-fellows was always by grasping the end of their trousers with his teeth, and as it was impossible for the boy to stop when so seized in full course, the dog was often jerked along the ground for some little distance.

Hide-and-seek is a game which is often learned and enjoyed by many animals. I have often been an interested spectator of the play in which two dogs were the participants. It was as exciting as such a diversion could possibly be between two children. For an hour at a time I have watched the fun, and the players seemed not to abate the least jot or tittle from their ardor and enthusiasm. They were apparently as fresh then as at the beginning. In due time the game ceased as if by mutual consent, but the animals did not seek some cool, quiet spot for comfort and rest, but started off to the woods for some further diversion, from which their voices were soon heard, telling that they were in pursuit of a rabbit or the ignoble ground-hog.

We have far from exhausted the list of examples at hand to show that the lower animals possess a sense of humor. But what use, it may be asked, can the capacity of humor subserve in the next world? Much the same, I presume, that it subserves in this. There are some in this world in whom the sense of humor is absolutely wanting. Estimable as they may be in character, they are just solemn prigs, and I should be very sorry to resemble them in the world, whither, it is hoped, all life tendeth.

Pride, Jealousy, Anger, Revenge and Tyranny, while not very pleasing characteristics, belong, as such, to the immaterial, and not to the material, part of man. That the lower animals possess these qualities will be seen from what follows. Hence the inference to be drawn from that fact must be quite obvious.

Taking these characteristics in order, Pride, or Self-esteem, is developed as fully in many animals as in the proudest of the human race. Most conspicuously is this shown in animals which herd together. There is always one leader at the head, who will not permit any movement to be made without his order, and who resents the least interference with his authority. This is particularly the case with the deer, the horse and the ox. Even when these animals are domesticated, and the habits of their feral life have materially changed, the feeling of pride exists to the fullest extent.

Whoever has carefully watched and studied the inhabitants of a farm-yard cannot fail to have observed that the cows have their laws of precedence and etiquette as clearly defined as those of any European Court. Every cow knows her own place and keeps it. She will never condescend to take a lower, nor would she be allowed to assume a higher. A new-comer in a farm-yard has about as much chance of approaching the rack at feeding-time as a new boy at school has of getting near the fire on a cold winter day. But as the young calf increases in growth, and is nearing maturity, she is allowed to mingle with her companions on tolerably equal terms. Should, however, a younger animal than herself be admitted, it is amusing to see with what gratification she bullies the new-comer, and how much higher she ranks in her own estimation when she finds she is no longer the junior.

But should the fates be propitious, and she should arrive at the dignity of being senior cow, she never fails to assert that dignity on every occasion. When the cattle are taken out of the yard to their pasture in the morning, and when they are returned to it in the evening, she will not allow any except herself to take the lead. An instance is recorded where the man in charge of a herd of cows would not permit the "ganger," as the head cow is often called, to go out first. The result was that she refused to go out at all. Therefore, to get her to go out of the yard, the man had to drive all the other cows back again, so that she might take her proper place at the head of the herd.

Few people know much about the real disposition of the mule. Judging from popular ideas respecting the animal, one would think that it had no pride in its composition. It is in reality a very proud animal, and fond of good society. One of his most striking characteristics is his aversion to the ass, and the pride which he takes in his relation to the horse. An ass would be hardly safe in a drove of mules, for he would, in all probability, be kicked and lamed by his proud relatives; whereas a horse, on the contrary, takes a distinguished position, the mules not only crowding around him and following his movements, but exhibiting a violent jealousy, each striving to get the nearest to their distinguished relative.

Figure 73. Exhibition of Grandeur. Male peacock in presence of some barn-yard fowls.

We have seen the pride of rank and love of precedence in cows, and the pride of ancestry in mules. There is, however, a pride that takes the form of sensitiveness to ridicule. Nothing is so galling to a proud man as to find himself the object of ridicule. The same trait of character is to be found in many animals, and especially in those that have been domesticated, for it is in these that we have the most opportunities for observation. All high-bred dogs are exceedingly sensitive to ridicule. We knew of a cat that was quite conscious if spoken of in a disparaging manner, and testified his disapprobation by arching his tail, holding himself very stiff indeed, and marching slowly out of the room.

There is, however, another form of pride which is often to be seen among the lower animals, but more especially among birds notable for gaudy or abundant plumage. This is the pride which manifests itself in personal appearance. Vanity is the name which is currently applied to this form of pride. Those who have observed a peacock in all the glory of his starry train will recognize the intense pride he feels at his own splendor. This display of his magnificent train is not for the purpose of attracting the homage of his plainly-attired mates solely, but seems to be intended to evoke the admiration of human beings as well. Not even the homage of birds, whom he regards his inferiors, is to be despised.

We have seen him, with his train fully spread, displaying his grandeur around a dozen or more barn-yard fowls, and apparently as satisfied with the effect he produced, as he stalked majestically among them, as if he had been surrounded by his own kith and kin. Then there is the turkey. No movements are more grotesque than his. See him as he struts about in his nuptial plumage, and yet no bird, notwithstanding the ludicrousness of his behavior, surveys himself with greater complacency. The whidah-bird, or widow-bird, as it is often called, exhibits this trait of character in its highest development. He is wonderfully proud of his beautiful tail, and, as long as he wears it, loses no opportunity of displaying it to every person who visits his cage. But when the moulting season has arrived, and he has taken on the plain, tailless attire of his mate, a change as great has come over his manner, and, instead of exhibiting himself in all his pride and glory, he mopes listlessly and stupidly about, and seemingly ashamed of his mean condition. In all these instances the character of pride in personal appearance is as strongly developed as it is possible for it to be in any human being.

That peculiar uneasiness, which arises from the fear that a rival will dispossess us of the affection of one whom we love, or the suspicion that he has already done so, is termed jealousy. There are two forms of this passion, one connected with the love of some other being, and the other dependent on the love of self. But it is the former, whose definition begins the present paragraph, with which we shall exclusively deal. It is evident from the meaning of jealousy, as given above, that the power of reasoning is implied, and that any creature by which it is manifested must be able to deduce a conclusion from premises. No matter if the conclusion drawn by the animal be wrong, the process, however incorrect it may be, is, it cannot be denied, still one of reasoning. All who have possessed pet animals must be familiar with the exceeding jealousy displayed by most of them. Most strongly is this feeling manifested when an animal has been the only pet and another is introduced into the house. Where there are two or more dogs in the same family, one is often amused at the boundless jealousy displayed toward each other while engaged in the service of the master, although at other times they were on the most excellent terms. Bus is the name of a favorite dog belonging to a friend. No more affectionate dog ever lived. Beagle was his companion. When they were by themselves, life was a round of frolics and rambles. No matter how rough and exciting their plays were, they never got cross, but endured everything with patience and forgiveness of spirit. Beagle was a clever animal, and very fond of the chase. Many a ground-hog would he dislodge from its burrow and fight to the death, while Bus would look on with wonder and admiration. But let the slightest attention be shown by the master to Beagle, and Bus's jealousy and anger became unbounded. He would fly at his friend in the most infuriated manner, rending him with tooth and claw, while Beagle would quietly slip around the corner of the house to get out of the reach of his companion's temper. Beagle, being a large and powerful dog, had in him the ability to give Bus a very sound whipping, but he was too noble and magnanimous a creature to take advantage of one younger and smaller than himself. He would always allow Bus to have his own way, knowing that the passion which was lacerating the bosom of his young companion and playmate would soon spend itself, and the latter, ashamed and abashed, would be soon seeking forgiveness and reconciliation.

Even in such rarely tamed animals as the common mouse the feeling of jealousy has been known to be so intense as to lead to murder. A young lady, one of Rev. J. G. Wood's correspondents, had succeeded in taming a common brown mouse so completely that it would eat out of her hand and suffer itself to be taken off the floor. She had also a tame white mouse in a cage. One morning when she went to feed the white mouse, as was her usual custom, she found it lying dead on the bottom of the cage, and beside it was its murderer, the brown mouse. The cage being opened, the latter made its escape, as though fearful of the consequences that might ensue, but how it had managed to gain admission was always a mystery.

Instances are on record where the jealousy of a rival has been restrained for long years through fear, and has ultimately broken out when the cause of the fear has been removed. A case of the kind came under our notice some few years ago. There were two cocks, belonging to different breeds, whom fate had placed as denizens of the same family. One was a magnificent dunghill cock, and the other a Malay, a cowardly caitiff, that was kept in fear and subjection by the former. In the course of events the dunghill cock suddenly died. His rival, coming by chance on his dead body, and perceiving that the time had come to wreak out the mixture of hatred and revenge that had lain smouldering in his bosom for years, instantly sprang upon it, kicked, spurred and trampled upon the lifeless bird, and, standing upon the corpse, flapped his wings in triumph, as it were, and crowed himself hoarse with the most disgusting energy. He immediately took possession of the harem, but he was far from being the noble, generous and unselfish creature that his predecessor had been. Again, comparing man with beast, it is at once apparent that the bird in this instance acted exactly as a savage does when his enemy has fallen, for the savage not only exults over the dead body of an enemy, especially if the latter has been very formidable in life, but also mutilates in futile and silly revenge the form which he feared when alive.

Tyranny, or the oppression of the weak by the strong, is another of the many traits of character common to man and the lower animals. But whether or not that strength belongs to the body or the mind, it is tyranny all the same. Taken in its most obvious form, it not only manifests itself in many of the animals in the oppression of the weak by the strong, but also in the killing and the eating of the same, even though they be of the same species. Human cannibals act in just the same manner, eating their enemies after they have killed them. There is hardly an animal in which the milder forms of tyranny may not be found. Insects, especially, manifest it in a light manner when they drive away their fellows from some morsel of food which they desire to keep to themselves. Among gregarious animals, the herd or flock is always under the command of an individual who has fought his way to the front, and who will rule with imperious sway until he has become old and in turn has been supplanted by a younger and more vigorous rival. In the poultry-yards the same form of tyranny is manifest, one cock invariably assuming the leadership, no matter how many may be the number of birds.

There is a curious analogy between these birds and human beings, especially those of the East, whether at the present day or in more ancient times. Many petty chieftains are found in Eastern countries, but there is always to be met with one among them who is more mighty than the rest, and who holds his place by superior force, either of intellect or military power. Challenged by one of the inferior chiefs and victorious, he retains his post, but if vanquished, his conqueror takes his place, his property and his wives. But curious to relate, with men as with birds, the members of the harem seem to trouble themselves very little, if any, about the change of master. The Scriptures are full of allusions to the invariable custom that the conqueror takes the possession of the harem belonging to the vanquished. David did so with regard to the women of Saul's household, and when Nabal died, who had defied the authority of David, so the latter, as a matter of course, took possession of his wife, together with the rest of his property. And when Absalom rebelled against David, he publicly took possession of his father's harem, which was a sign that he had assumed the kingdom.

Where a number of creatures are confined in the same place, a very curious sort of tyranny is sometimes manifested. Mandarin ducks, according to Mr. Bennett, when confined to an aviary, show a very querulous disposition at feeding-time. The males of one and the same kind of a different species endeavor to grasp all the nourishment for themselves, unmindful of the wants of others, and will not even permit their companions to perform their ablutions without molestation, although they may themselves have completed what they required. Often the mandarin ducks have been observed to excite the drakes to assail other males or females of the same species, and other kinds of birds in the aviary, against whom the ladies, from some cause or other, have taken a dislike. One pair of these ducks are always to be noticed that exercise a tyranny over the others, not allowing them to wash, eat or drink, unless at their pleasure and approval.

But, of all tyrants, none can be compared to a spoiled dog, who is even worse than a spoiled child. Obedience is a stranger to his nature. Does his master want him to go out for a walk, and he prefers to stay at home, he stays at home, and his master is compelled to go out without him. But if he wants to go for a walk, he makes his master go with him, and even to take the direction he prefers. Duchie is the name of a Skye terrier whose history is given in a work on the latter breed of dogs by Dr. J. Brown. So completely had this little animal domineered over her mistress, that the latter could not even choose her own dinner, but was obliged to have whatever the dog preferred. It is related that for a half of a winter's night she was kept out of bed, because Duchie had got into the middle and refused to move. Certainly, no better example of tyranny could be adduced.

That so-called brutes possess, in common with ourselves, a Conscience, that is, a sense of Moral Responsibility, and a capability of distinguishing between right and wrong, may seem a very strange assertion to be made, especially to those who have never studied the ways of the lower animals. Animals which are placed under the rule of man, and those, like the dog, which belong to his household and are made his companions more particularly, would naturally be expected to show the strongest development of the principle. Conscience, in their dealings with man, constitutes their religion, and they often exercise it in a way which would put many a human being to the blush. This feeling it is that induces the dog to make himself the guardian of his master's property, and often to defend that property at the risk of his life. However hungry may be the dog that is placed in charge of his master's dinner, nothing would, as a rule, tempt him to touch a morsel of the food, for he would rather die of starvation than eat the food which belongs to his master. Often have we seen field-laborers at work at one end of a large field, while their coats and their dinner were at the other end, guarded by a dog. Not the least uneasiness did they seem to manifest about the safety of their property, for well they knew that the faithful animal would never allow anyone to touch either the clothes or the provisions.

There could hardly be a stronger instance of moral responsibility than the one which I shall now relate, which is substantially the same as appears in Wood's "Man and Beasts Here and Hereafter." Living in an unprotected part of Scotland was a poor woman, who unexpectedly became possessed of a large sum of money. She would have taken it to the bank, could she have left the house, but lack of bodily health prevented her from so doing. At last she asked the advice of a butcher of her acquaintance, telling him that she was afraid to live in the house with so much money about her. "Never fear," said the butcher, "I will leave my dog with you, and I'll warrant you that no one will dare to enter your house." Towards the close of the day the dog was brought, and chained up close to the place where the money was deposited. That very night a robber made his way into the house and was proceeding to carry off the money, when he was seized by the dog, who held him a prisoner until assistance arrived. The thief turned out to be the butcher himself, who thought he had made sure of the money, but he had not considered that his dog was a better moralist than himself, for who would, rather than betray a defenceless woman, take her part against his own master. Kindly pardoned by the woman, the intending robber made his way home, and it is to be hoped that for the future he learned a lesson from his own dog and amended the evil of his ways.

Not only does the dog guard the property which is intrusted to its charge, but frequently goes a little further and assumes a charge on its own account. When the writer was a boy living in the country, where much of the spring and summer of the year was spent in working upon a farm, he became on very excellent terms with a little bull-terrier, named Tip, that belonged to a certain farmer by whom he was employed. Upon my first introduction to Tip, I felt a sort of aversion towards him. This grew out of the mysterious actions of the animal. He was always around when I was busy at work and seemed to be eying me in a suspicious sort of manner, which at times made me feel very unpleasant. After the lapse of a few days I discovered that I was not so closely watched as before, and that I was treated by him as he was accustomed to treat the other members of the family. Upon inquiry I learned that he always acted in this way toward people whom he did not know intimately, and that, after a time, he had confidence in their honesty and left them alone. While in many instances Tip was entirely wrong in his surmises, yet cases are recalled where the dog was right and acted in a manner that would have been creditable to a human being. One of the men employed upon the place, presuming upon the friendship of the dog, sought to carry away under cover of darkness something belonging to the farmer, but he was immediately beset by the animal, who was an eye-witness of the proceeding, and compelled to desist from the intended theft. From that time the man was under the closest surveillance by the dog. Unable to effect a reconciliation, and chafing under the look of suspicion with which he was always greeted, the man soon took his departure, much to the delight and satisfaction of the faithful canine, and was never afterwards seen.

Quite a common form of conscience among the lower animals is that which may be defined as a recognition of having done wrong, and acknowledgment that punishment is deserved. Animals have in their way very pronounced ideas as to right and wrong. When they have committed an act which they know will offend their master, they display as keen a conscience as any human being self-convicted of sin could exhibit. In many instances, the offence in not merely acknowledged, but the creature remains miserable until forgiveness has been granted. This condition of mind, if manifested by man, is called Penitence, and, assuredly, it cannot be known by any other name when manifested by animals that are lower down in the scale of life. My little dog Frisky, about whom mention has already been made, affords a very fine illustration of this phase of conscience. Whenever he did wrong, the severest punishment that could be meted out to him was to ignore his presence and decline his offered paw. For hours the poor fellow would moan and cry, and even refuse food, when he thought I was angry with him. But a word or a look of forgiveness was sufficient to change his sadness into joy. A shaking of hands, so to speak, would then follow, and master and dog would be good friends again. No love could be more intense than his, and this was especially shown when I would return from a short absence, when the little fellow would almost overwhelm me by his affectionate caresses.

No loftier characteristic adorns humanity than Love. But how far it is shared by the lower animals it is now our purpose to inquire. That there are many phases of development cannot be doubted. Sympathy, or that capacity of feeling for the sufferings of another, is the first phase. Many, and perhaps all, living creatures possess the capacity of sympathy. In the majority of cases it is not restricted to their own species, but is extended to those beings which appear to have very little in common with each other. Ordinarily, however, it is exhibited between animals of the same species, and it is often seen in the dog, as, for example, where a dog, having been cured of an injury, has been observed to take a fellow-sufferer to his benefactor. Such sympathy, it need hardly be remarked, could not be carried out unless the animals possessed a language adequately defined to enable them to transmit ideas from one to the other. Cats are often kind to each other, sympathizing under difficulties, and helping their friends who require assistance. A cat, belonging to a friend, has been known, when oppressed with the cares of a family, to employ a half-grown kitten to take charge of the young while she went for a ramble. Between the cat and the dog an enmity exists that is hereditary, and yet, when in good hands, they are sure to become very loving friends, and even to show considerable sympathy towards each other. Such an exhibition of good feeling was observed by the writer a few years ago. The dog, a large black Newfoundland, had contracted a warm and devoted friendship for a gray cat that was an inmate of the same family. When the cat was assailed by one of her kind, or by a strange dog, the Newfoundland would pick her up in his mouth and carry her to the house out of reach of danger, the cat maintaining all the while the most perfect serenity of composure, knowing that she was in the care of one who meant her no ill. When the same cat would become sick, the Newfoundland would lie down by her side, caress her with his tongue, and show in every way possible that he was sorry that she was sick.

Many examples are recorded of birds feeling sympathy with the lost or deserted young of other species, and that have taken upon themselves the task of feeding the starving children. A pair of robins had constructed a nest near to the writer's home in the country, where in due season a family of four children was raised. Disaster soon came to the little ones, for both parents were slain by some wicked boys of the neighborhood. There dwelt in the same locality a pair of bluebirds, but between the two families there had never been apparent the least interchange of friendship. Each family kept to itself, and attended to its own business. But when the cry of the young robins in their piteous demands for food rent the air, the bluebirds came over to their home to discover what the trouble was. They were not slow to perceive the sad state of things. Their sympathies were at once aroused, and their energies soon bent in the direction of relieving the sufferings of the little orphaned robins. For the next two weeks they had all they could do in providing meat for their own and the robins' young.

Figure 74. Four Orphaned Robins. Kind-hearted bluebirds assuming the role of parents.

While capable of showing sympathy for near as well as distant kin, the lower animals have also the capacity to sympathize with human beings in distress. Cats occasionally manifest a sympathy for suffering humanity. As for sympathy displayed by dogs, there is no need to cite examples. No human being, I am safe in saying, was ever free from troubles of some kind, and I am equally sure that no one who had a companionable dog felt that he was without sympathy. Full well does the dog know when his master is suffering pain or sorrow, and his nose pushed into his master's hand, or laid affectionately upon his knee, is a sign of sympathy worth possessing, even though it exists only in the heart of a dog. From that moment there has been established a bond between the soul of the master and the dog, and certainly no one can believe that the bond can ever be severed by the death of the material body, whether of the man or the animal.

That Friendship, which is another branch of love, exists among animals, is a well-known fact. But it is among the domesticated animals that it most frequently exhibits itself. Horses, as everyone knows, which have been accustomed to draw the same carriage are usually sure to be great friends, and if one be exchanged the other becomes quite miserable for want of his companion and seems unable to throw any spirit into his work. Dogs, too, are very apt to strike up friendships with each other. Among animals it is not confined to one species, but is occasionally found to exhibit itself in those which might be supposed to be peculiarly incongruous in their nature. That cows and sheep live, as a rule, on good terms with each other in the same pasture is a familiar experience, though sometimes the former are a little prone to domineer over the latter. But a very strong affection sometimes exists between animals so different, and when once they have accustomed themselves to each other's society neither can be happy without the other. The goat and the horse frequently become friends, and a peculiarly vicious horse has been known to allow a goat to take undue liberties with him without the least manifestation of resentment. In many places the stable-cat is quite an institution. Its usual place of repose is upon the back of the horse, and the latter has been known to grow very uneasy if left for any length of time without the companionship of his little friend. A very singular instance of friendship occurred at the rural home of a near relative. He had a fine mastiff which had taken a fancy to a brood of young chickens, and which acted as their protector. They were not at all unwilling to accept him in this capacity, as they followed him about just as though he had been their mother. Quite an interesting sight it was to watch the dog and the chickens as they would take their *siesta*. The dog used to lie on his side, and the chickens would nestle all about him, though one chicken in particular would invariably scramble upon the dog's head, and another just over his eye, but both parties appeared equally satisfied with this remarkable arrangement.

Already have we referred to the intense yearning which is felt by many of the lower animals for human society. This yearning is indeed but the aspiration of the lower spirit developed by contact with the higher in domesticated animals or those which are in perpetual contact with man. This feeling is a matter of no great surprise. But that it should be exhibited in feral animals and birds, and even in insects, is a fact well worth considering, as it furnishes a clew to some of the many problems of life which are as yet unsolved. That power of attraction exercised by the spirit of man upon that of the lower creation is well exemplified in many wild animals, who are known to forsake the society of their own kind for the companionship of the being whom they feel to be higher than themselves.

Perhaps one of the wariest of wild animals is the squirrel. He is horribly afraid of human beings, and if a man, woman or child come to the windward of him, the little animal is sure to scamper off at his fleetest pace, scuttle up the nearest tree, and conceal himself behind some branch. Yet, wild as he may be, he is peculiarly susceptible to the influence of the human spirit, and for the sake of human society will utterly abandon that of his own kind. I once knew a pet gray squirrel by the name of Charley. He had been taken from the nest when very young. His home for awhile was one of those whirl-about cages. Charley did not like his cage, but preferred to be outside in the unrestrained enjoyment of the dictates of his own free will. So it was difficult to keep him behind the bars. When awake he loved to follow his own devices; but when tired he usually slept on a soft cushion on the sofa, or found his way into some bed-room where he would nestle under a pillow. Nothing was more to his satisfaction and pleasure than a share of the bed of his mistress, but he was always a troublesome nest-fellow. Charley had, as must be obvious, perfect freedom. He was allowed to go as he pleased. There was no coercion in his case. Had he wished to escape, there was nothing to prevent, and nothing bound him to his mistress but an "ever-lengthening chain" of love and aspirations which none but a human being could satisfy. The sparrow, one of the most independent and self-reliant of birds, has been known to abandon its kind for the sake of human beings. Wood cites a case of a bird of this species that had been rescued from some boys who had been robbing the nest. The bird was brought home, but was never confined in a cage, but was permitted to fly freely about the house. As there was a cat about the house, she had to be closely watched lest she might do the bird some injury. On Sundays, when the family went to church and no one remained to keep an eye on the cat, the sparrow was turned into the garden, where it flew about until the family's return. The opening of the dining-room window by its mistress, and the display of her ungloved hands, was the signal for its entry. But if the mistress stood by the window with her gloves on, then the bird showed not the slightest disposition to enter.

Such is the intensity of the love which the lower animals sometimes entertain toward man that they have been known to grieve themselves to death on account of his loss. A dog by the name of Prince, who lived in the family where the writer spent a few weeks of a summer, is a case in point. He had a good master, and one to whom he was strongly attached. The year before the master sickened and died, and Prince felt the loss so keenly that he refused to take any food, and even to notice the surviving members of the family. He was pitiable to behold. Life had lost all attractions to him, and he showed that he was slowly but surely grieving his life away. Some few weeks after the writer's departure, the poor animal breathed his last, and his spirit, it is to be hoped, went to join that of his master, while his ashes became mingled with the dust of the earth as his master's had been.

What a wonderful power do some animals have of returning to their beloved master, even though they have been conveyed to a considerable distance. This is especially true of the dog. So many examples of such feats are on record that I refrain from mentioning them, but will give but a single example. Rover, a pet greyhound that belonged to the writer, had become such an annoyance to the neighborhood where he lived, that the master determined to provide him a home in the country some fifty miles away. He was conveyed to his destination in a covered wagon, and after his new master had reached home, the poor animal was placed in a stable for several days, where he was daily visited and fed, and every effort possible made to attach him to the place and family. On the fourth day of his arrival he was given his freedom. With a long, loud wail he saluted the neighborhood, and the next moment was off at full speed across the country, all efforts to stop him being unavailing. In less than a week from his leaving he was at home again, hungry and jaded out with fatigue and travel, but not too tired nor too hungry to express the great joy he felt for the old master. How he ever accomplished the journey, and what vicissitudes and difficulties he encountered on the way, no one will ever know. After this I had not the heart to send him away again, but put up with his capers and tricks as best I could, and when complaints were preferred against him endeavored to excuse them as a parent is prone to do in the case of a spoiled and wayward child. But a day arrived when Rover to me was no more. What had become of him I was never able to discover, but I always blamed a near-by neighbor, a man who had neither love nor charity in his soul, for his sudden disappearance.

That cats are selfish animals, attaching themselves to localities and not to individuals, I do not believe. This idea has, perhaps, some ground of truth, for the nature of a cat is not so easy to understand as that of a dog. But when a cat is not understood, it is very probable that she cares less for the inhabitants of the house than for the house itself. Frequent instances are known by the writer where cats have been in the habit of moving about with their owners, and have been as much unconcerned as dogs would have been. True they have, like women, a curious and prying disposition. I have seen them in new and strange quarters go sniffing about every room of a house, and at last settle down in some cozy, comfortable place, well satisfied with their tour of investigation. Where the house fell short of their expectations, if they have been cats that have received due consideration from their mistresses or masters, they have tried to live down their objections and to learn to be happy and contented with their lot. Only cats that have not been much thought of are inclined to show their disapproval to changes of residence which they deemed unsuitable by refusing to stay with their masters. Blackie, a favorite cat of ours, never seemed to care where her home was, so long as her friends were there to pet, caress and pamper her with choice dainties.

All animals, so far as can be learned, have not only a capacity for the society of man, but an absolute yearning for it. This feeling may be in abeyance, from not having received any development at the hands of man, but it nevertheless exists, and only awaits to be educed by someone capable of appreciating the character of the animal. Tigers, as is well known, are not generally considered the friends of mankind, and yet the Indian fakirs will travel over the country with tame tigers, which they simply lead about with a slight string, and which will permit small children to caress them with their hands without evincing the least disposition to hurt them.

When we survey the examples of love displayed by animals towards human beings, which we have just detailed, and recall the hundreds that we know and have read about, is it possible to believe that such love can perish? We apprehend not. Unselfish love as this, which survives ingratitude and ill-treatment, belongs to the spirit and not to the body, and all beings capable of feeling such love must possess immortal spirits. All may not have an opportunity of manifesting it, but all possess the capacity and would, were the conditions favorable, manifest it openly.

Few animals, as may easily be imagined, manifest Conjugal Love. Most species have no particular mates, but merely meet by chance, and seemingly never trouble themselves about each other again. No real conjugal love, therefore, can exist, and it is rather curious that in such animals a durable friendship is frequently formed between two individuals of the same sex. But when we come to polygamous animals, such as the stag among mammals and the domestic poultry among birds, we meet with a decided advance towards conjugal love, although as in the case of polygamous man, that love must necessarily be of an inferior character. There is seen, at all events, a sense of appropriation on either side. Take the example of the barn-yard fowl, as has already been mentioned in that part of the chapter which deals with jealousy, where it is shown that the proprietor of the harem resents any attempt on the part of another male to infringe on his privileges.

This brings us to the consideration of birds, where the many are mated for the nesting-season, but subsequently do not seem to care more for each other than they do for their broods of children. If one of the pair be killed at the nesting-time the survivor, after a brief lamentation, consoles itself in a few hours or days with another partner, for there really appears to be a supply of spare partners of both sexes always at hand. And now we come to those creatures which are mated for life, and often we find among them a conjugal love as strong and as sincere as among monogamous mankind. Prominent among them are the eagle, the raven and the dove. And while we praise the turtle-dove for its conjugal fidelity, and credit it with the possession of all that is sweet, and good, and gentle, how remarkable is it that we forget to accredit with the same virtue the eagle and the raven, that are the types of all that is violent, and dark, and cunning. There are many examples in existence of the conjugal love among such birds, but they are so well known that reference to them is unnecessary. The case of the mandarin duck, already narrated, affords a strong instance of conjugal love wherein the lady was faithful and the husband avenged himself on the destruction of his domestic peace.

Figure 75. Mated For Life. Conjugal fidelity shown by a pair of doves.

So numerous as are the instances of love shown by parents among the lower animals towards their offspring, yet it is a very singular fact that few, if any, trustworthy accounts of Filial Love, or the love of children toward their parents, are to be found. But we must look to man if we would understand the lower animals. Even human nature must attain a high state of development before filial love can find any place in the affections. In savages it barely exists at all, and certainly does not survive into mature years. It is the glory of the North American Indian boy, at as early an age as possible, to despise his mother and defy his father. And the women are just as bad as the men. Rejoicing in the pride of youth and strength, they utterly despise the elder and feeble women, even though they be their own mothers, and will tear from their hands the food they are about to eat, on the plea that old women are of no use, and that the food would be much better employed in giving nourishment to the young and strong. The Fijians have not the least scruple in burying a father alive when he becomes infirm, and assist in strangling a mother that she may keep him company in the land of spirits. Both the Bosjesmen of South Africa and the Australian seem to have not the least idea that any duty is owing to a parent from a child, nor have they much notion of duty from a parent toward the child. If the father be angry with any one for any reason, he has a way of relieving his feelings by driving his spear through the body of his wife or child, whichever one of the two happens to be the nearer. Even the mother treats her child with less consideration than a cow does her calf, and leaves the little creature to shift for itself at an age when the children of civilized parents are hardly thought fit to be left alone for a few minutes. This being the case with parental love, it may be readily imagined that filial affection can have not the slightest chance for development, and it is very much to be questioned whether in the savage it can really be said to exist at all in the sense understood by enlightened peoples. Therefore, as in the lower human races, we find that filial love either is very trifling, or is absolutely non-existent, need we wonder that in the lower animals such few, if any, indications of its presence should be found?

Now, as to the subject of Parental Love, and the various ways in which it manifests itself. There are many writers who claim that parental love in the lower animals is not identical with that of man. They affirm that it is only a blind instinct, and, in order to mark more strongly the distinction between man and beast, call the parental love of the latter by the name of storgë. Speaking for myself, I must declare that I am unable to perceive any distinction between the two, save that in civilized man the parental love is better regulated than among the lower animals. But, as has been seen, it is not regulated at all among the uncivilized races, and, in truth, many of the beasts are far better parents than most savages. Nor can I understand why the word storgë should be applied to parental love among the lower animals and not to the same feeling in man. Among Greek writers the word, together with the verb from which it is derived, is applied to the love between human parents and children. It is so applied by Plato, and in the same sense by Sophocles and others. One argument adduced by those who deny the identity of the feeling in both cases is that parental love endures throughout life in man, while it expires with the adolescence of the young in the lower animals. This is doubtless true, as a rule, with civilized man, but in the case of the savage, as has previously been shown, it does not last longer than that of a bird, a cat or a dog, taking into consideration the relative duration of life. And the reason is identical in both cases. Were this love to exist through life in the savage, the beast or the bird, the race would become extinct, for neither race is able to support its children longer than their time of helplessness. The beast and the bird cannot, and the savage will not, provide for the future. It is therefore evident that if the young had to depend upon their parents for subsistence, they would soon perish from lack of food. Exceptions there are to this general rule, and always, as far as can be determined, in the case of domesticated animals whose means of subsistence are already insured.

Several of such cases have come to my notice. I shall instance but one. A friend of mine has two terriers, a mother and a daughter. The strongest bond of love and fellowship unites them. They always sit close together, and the mother playfully pinches her daughter all over. Should they by chance become separated, even for a very short time, the daughter comes up wagging her tail, and then licks her mother's nose and mouth. When hunting together, they always act in concert, each one taking a hole, and one keeping watch while the other scrapes away the earth. The meaning of each other's whine or bark is perfectly understood, and no two persons could understand their own language better than do these dogs theirs, nor be more comprehensible to each other.

Self-abnegation is perhaps one of the most beautiful characteristics which parental love can give. This is particularly shown when the young are in danger. A human mother in charge of her child will defy a danger before which she would shrink if alone, and in its defence would dare deeds of which most strong men would be incapable, for during the time her selfhood is extinguished, and her being is sunk into that of her child. Such abnegation becomes a true mother, for if she would not consent to do and dare for the sake of her offspring, she would degrade herself below the beasts and the birds, who hesitate not in performing that duty to their children, though *savants* do declare that they possess only storgë, whatever they may mean by it, and not parental love.

Figure 76. Evidence of Conjugal Affection. Male humming-bird feeding his partner, and ready to act in her defence.

Everyone who has paid even a passing attention to the habits of birds must have noticed the vigilance a pair of catbirds exercise over their nest when containing young birds. Neither parent, when the other is absent, relaxes this vigilance, for they consider no labor, no care, no watchfulness, too great or too exacting where their offspring are to be benefited. Let an enemy approach, even if it be man himself, and they are beside themselves with anger and resentment, flying into the very face of the audacious intruder, as though they would pluck his eyes out as a just punishment for his presumption and temerity. I have seen the nest of a catbird attacked by a black snake, and crushed within the folds of the hideous serpent the father-bird, but the disaster did not cause the mother-bird to desist from the attack, for, utterly oblivious of all else but her offspring and the snake, she fought on until the latter was forced to glide away into the bushes to escape her infuriated assaults. But no species of bird is more courageous in defence of its nest than the little ruby-throated humming-bird. It is really dangerous to visit the nest when with eggs or young. I would as soon attempt to assail the dome-shaped nest of our common hornet as that of this humming-bird. It is as much as one can do to protect his eyes from the lightning-like attacks of these birds, so swiftly and so unerringly do they direct their blows at these points.

So great is the affection and solicitude of the red-eyed vireo for her young, that she will scarcely leave the nest when the hand is stretched out a few inches over the mouth of the structure. And then when she does leave, it is not in a hurried, precipitate manner, but with a quiet, deliberate movement that excites one's admiration and makes one vow never to abuse such simple, childlike confidence. I have even placed my hand upon the sitting-bird without disturbing the current of her brooding thoughts, or the peaceful serenity of her soul. A rough dash at the nest tends to frighten her away *instanter*, but when the hand is reached out to it slowly and silently the bird seems to act as though it had nothing to fear, and remains calm and self-possessed.

Who is not familiar with the proverbial skill of the Carolina dove in feigning lameness when her nest is being approached? Without a cry, and with scarcely a rustle of her feathers, she slips out of her nest upon the ground, and by a series of manœuvres, as if desperately wounded, grovels along on her belly in the dust till she has led her enemy a long journey from the site of the nest, when she will take to wing and fly away into a coppice or a clump of brushwood.

That birds should manifest a love for the young which they hatch has always seemed a strange problem to me. I can see how that, in the case of a mammal, the mother should feel a love for the creature who is absolutely a part of herself—whose very life-blood is drawn from her veins. But this is not necessarily the case with birds. If, as often happens with poultry, the eggs of several hens are placed under one bird for hatching, the hen that hatches them knows no difference between the chickens that come from her own eggs and those which proceed from eggs laid by others. Even where the eggs belong to birds of different species, as to the common Muscovy-duck for example, the hen displays as much affection for the young ducklings, despite the disparity of instinct and habit, as she does had they proceeded from her own eggs. May it not be that parental love has different channels of transmission, and that in such a case as this the emanation from the sitting-hen may be the vehicle of parental love toward the young which are to be hatched? Certain it is that a sitting-hen, as many of us have observed, is altogether a changed being, both in attitude and expression. She is entirely absorbed in the eggs when she is incubating, and, though she may not have the intellect to distinguish a mere lump of chalk from one of her own eggs, yet love is altogether independent of intellect, and may exist in all its vigor, and yet may be wasted on an unworthy object.

Fishes, as is generally known, are not particularly emotional beings, and are not likely to entertain a lasting love for anything. Indeed, in some instances, parental love would be absolutely useless, as in the case of the cod-fish, which could be hardly expected to entertain a special love for each of the countless thousands of young it produces every year. The life of the mother would be an unenviable one, if her lot were to look after her young as soon as they are hatched, especially when the varied foes that beset her eggs as soon as they are produced, are considered. Just as there are fishes that possess conjugal love, so there are fishes that possess parental love, and prominent among these are the sticklebacks. But in the case of these fishes the most curious part is that parental love is shown by the father, and not by the mother, the latter having nothing to do but to lay the eggs, and leaving to the former the exclusive labor of providing for the young.

Enough of instances of true parental love among the lower animals could be given to fill this entire book, but a sufficient number have been adduced to show that the feeling is the same in man as in them, although, of course, the mode of manifesting it is different. We have shown the fallacy of the theory that parental love is life-enduring in man and very brief among the animals, and have seen that, in proportion to the duration of life, it is quite as brief among the savages as among the animals. And, again, we have seen where it has been lost and then restored, and also where it was never lost; where in animals, as in man, it has caused complete abnegation of self, the parents living for their children, and not for themselves, and where it has given strength to the weak and courage to the timid. Even the very fishes have been shown to be amenable to the same influences as man, and could we have carried our illustrations still lower down the scale we would have found the same influences existing among much humbler forms of animal existences. In conclusion, there is no resisting the fact that parental love, one of the highest and holiest feelings of which a loving and immortal soul can be capable, is shared equally by man and beast, according to their respective capacities.

Figure 77. Wood-Thrush Setting. Copyright 1900 by A. R. Dugmore.

Chapter 45

LIFE PROGRESSIVE

No one can doubt that the earth's crust, so far as it has been deciphered by man, presents us with a record, imperfect though it be, of the past. Whether, however, the known and admitted imperfections of its records, geological and palæontological, are sufficiently trustworthy to account satisfactorily for the lack of direct evidence recognizable in some modern hypotheses, may be a matter of individual opinion, but there can be little doubt that they are sufficiently extensive to throw the balance of evidence decisively in favor of some theory of continuity, as opposed to any theory of intermittent and occasional action, which some writers have strenuously and intelligently advocated. No marks of mighty and general convulsions of nature exist, as the seeming breaks which divide the grand series of stratified rocks into numerous isolated formations would indicate. They are simply indications of the imperfection of our knowledge. Science will never, in all probability, point to a complete series of deposits, or to a complete succession of life, which shall link one geological period to another. But that such deposits and such an unbroken succession must have existed at one time we may well feel sure, and stand ready to believe that nowhere in the long series of fossiliferous rocks has there been a total break, but that there has inevitably been a complete continuity of life, as well as a more or less complete continuity of sedimentation from the Laurentian period to the present day. One generation, speaking figuratively, hands on the lamp of life to the next, and each system of rocks is the direct offspring of its predecessor in time. Though it is apparent that there has not been continuity in any given area, still the geological chain could not have been snapped at one point and taken up again at a totally different one. Hence we arrive at the conviction that in geology, as in other sciences, continuity is the fundamental law, and that the lines of demarcation between the great formations are but gaps in our own knowledge.

Through the study of fossils, as is well known, geologists have been led to the all-important generalization that the vast series of fossiliferous or sedimentary rocks may be separated into a number of definite groups or formations, each of which being characterized by its own organic remains, but not properly and strictly, it must be understood, by the occurrence therein of any one particular fossil. However, a formation may contain some particular fossil or fossils not occurring outside of that formation, thus enabling an observer to identify a given group with tolerable certainty; or, as very often happens, some particular stratum or subgroup of a series, may contain peculiar fossils, whereby its existence may be determined with considerable readiness in divers localities. Each great formation, let it be said, is properly characterized by the association of certain fossils, the predominance of certain families or orders, or by an assemblage of fossil remains that represent the life of the period during which the formation was deposited.

Fossils, then, not only enable us to determine the age of the deposits in which they are found, but they also further enable us to arrive at some very important conclusions respecting the manner in which the fossiliferous bed was deposited, and, consequently, to the condition of the particular region occupied by the bed at the period of its formation. Beds that contain the remains of animals, such as now inhabit rivers, we know to be fluviatile in their origin, and that at one time they must have either constituted actual river-beds, or been deposited by the overflowing of ancient streams. But if the beds contain the remains of mollusks, minute crustaceans or fish, such as are found to-day in lakes, then we conclude that they are lacustrine, and were deposited beneath the waters of former lakes. And, lastly, if the remains of animals such as now people the oceans are to be met with in the beds, then we know that they are marine in origin, and that they are fragments of an old sea-bottom. On the whole, the conditions under which a bed was deposited, whether in a shallow sea, in the immediate vicinity of a coast-line, or in deep water, can often be determined with considerable accuracy from the nature of the relics of the organisms which they contain. But we have thus far been dealing with the remains of aquatic animals. When, however, we consider the remains of aerial and terrestrial animals, or of plants, the determination of the conditions of deposition is not made out with such an absolute certainty. Remains of land-animals would, of course, occur in sub-aerial deposits, that is, in beds, like blown sand, accumulated upon the land, but the most of such remains of such animals are found in deposits which have been laid down in water, and hence their present position is due to the fact that their former owners were either drowned in rivers or lakes, or borne out to sea by water-channels. Animals possessed of the power of flight might also similarly find their way into aqueous deposits, but, when it is remembered that many birds and mammals habitually spent a great part of their time in the water, it is not to be wondered at that they should present themselves as fossils in sedimentary rocks. Even plants, such as have undoubtedly grown upon land, do not prove that the bed in which they are found was formed on land, for many of their remains are extraneous to the bed in which they now occur, having reached their present site by falling into lakes or rivers, or by being carried out to sea by floods or gales of winds. Still, there are many cases which obviously show that plants have grown on the very spot where we now find them. The great coal-fields of the Carboniferous Age, it is now generally conceded, are the result of the growth *in situ* of the plants which compose coal, as well as that they grew on vast marshy or partially submerged tracts of level alluvial land.

While fossils enable us in many cases to arrive at important conclusions as to the climate of the period in which they lived, yet it is only in the case of marine fossils, which constitute the majority of such remains, that we acquire such knowledge, but it is mostly the temperature of the sea which can thus be determined. However, let it be remembered that, owing to the existence of heated currents, the marine climate of a designated area does not necessarily imply a correspondingly warm climate in the adjoining land, for land-climates can only be determined by the relics of land-animals or land-plants, and these are comparatively rare as fossils. But all conclusions on this head are really based upon the existing distribution of vegetable and animal life upon the globe, and are therefore liable to be vitiated by the considerations that no certainty exists that the habits and requirements of an extinct animal were exactly similar to those of its nearest living relative; that far back in time groups of organisms, so unlike anything we know at the present day, are met with, which render all conjectures of climate based upon their supposed habits more or less uncertain and unsafe; that in the case of marine animals we are as yet very far from knowing the precise limits of distribution of many species within our present seas as to render conclusions drawn from living forms in relation to extinct species unsatisfactory and, probably, incorrect; and, finally, that the distribution of animals to-day, is certainly dependent on other conditions than climate alone, the causes limiting the range of given animals being assuredly such as belong to the existing order of things, and are different from what they were in former times, not necessarily because the climate has changed, but because of the alteration of other conditions that are essential to the life of the species or conducive to its extension. But notwithstanding the difficulties in the way, we are able in many cases to deduce completely trustworthy conclusions concerning the climate of a given geological period by an examination of its fossil remains. In Eocene times, or at the beginning of the Tertiary Period, the climate of what is now Western Europe was of a tropical or sub-tropical character, the Eocene beds being found to contain the remains of cowries and volutes, such shells as now inhabit tropical seas, together with the fruits of palms and remains of other tropical plants. And further, it has been shown that in Miocene times, or about the middle of the same epoch, the central parts of Europe were peopled with a luxuriant flora resembling that of the warmer parts of the United States, and that Greenland, now buried for the most part beneath a vast ice-shroud, was warm enough to support a large number of trees, shrubs and other plants that are at present denizens of the temperate regions of the globe.

And lastly, from the study of fossils, geologists first learned to comprehend a fact, that is, that the crust of the earth is liable to local elevations and subsidences, which may be regarded as of cardinal importance in all modern geological theories and speculations. Long after the remains of shells and those of other marine animals were first observed in the solid rocks constituting the dry land, and at great elevations above the sea-level, attempts were made to explain this unintelligible phenomenon upon the hypothesis that these remains or fossils were mere *lusus naturæ*, due to some "plastic virtue latent in the earth." But the common-sense of science soon rejected this idea, and it was universally agreed that these bodies were really the relics of animals that once lived in the sea. When once this was admitted, further steps in the right way of thinking became comparatively easy, and at the present day no geological doctrine stands on a surer foundation than that which teaches that our existing continents and islands, fixed and immovable as they appear, have been repeatedly sunk beneath the ocean and just as repeatedly been lifted above its waters.

Not only have fossils an important bearing upon geology and physiography as has been seen, but they have relations, most complicated and weighty in character, with the science of biology, or the study of living beings. No adequate understanding of zoölogy and botany is possible without some acquaintance with the types of plants and animals that have passed away, for there are numerous speculative problems in the domain of vital science, which, if soluble at all, can only hope to find their key in researches carried out on extinct organisms.

No attempt will be made by the writer to discuss fully the biological relations of fossils. Such an undertaking would afford matter for a separate volume. All that I purpose in this chapter is to indicate very cursorily the principal points of palæontological teaching, so that my readers can acquire some idea of the progression from lower to higher types that life has made throughout the geological ages. Preliminary to the purpose held in view, let it be understood that the vast majority of fossil animals and plants are extinct, or, differently and perhaps more intelligently expressed, belong to species that no longer exist. So far from there being any truth in the old idea that there have been periodic destructions of all the living beings in existence upon the earth, followed by a corresponding number of new creations of plants and animals, the actual facts indicate that the extinction of old and introduction of new forms have been processes that have been continually going on throughout the whole of geologic time. Every species seems to come into existence at a definite point of time, and to disappear finally at another definite period, though there are few, if any, instances, in which the times of entrance and exit could be fixed with any degree of certainty or precision. Marked differences in the actual time during which different species have remained in existence are noticeable, and therefore corresponding differences in their vertical range, or in the actual amount and thickness of strata through which they present themselves as fossils, some species being found to extend through two or three formations, and even a few have had a more prolonged existence. More commonly, however, the species which begin in the commencement of a great formation die out at or before its close, while those which are introduced for the first time near its middle or end may either become extinct or pass into the next succeeding formation, animals of the lowest and simplest organization as a rule having the longest range in time. Microscopic or minute dimensions seem to favor longevity, for some of the Foraminifera appear to have survived, with little or no perceptible alteration, from the Silurian Period to the present day, whereas largely and highly-organized animals, though long-lived as individuals, rarely seem to live long specifically, and consequently have a restricted vertical range. Exceptions to this rule are, however, occasionally found in some persistent types, the Lampshells of the genus Lingula being little changed from the LinguIæ that swarmed in the Lower Silurian seas, while the existing Pearly Nautilus is the last descendant of a clan nearly as old. Some forms, on the other hand, the Ammonites, which are closely related to the Nautilus, and mostly restricted to certain zones of strata, seem to have enjoyed a comparatively brief lease of life.

Figure 78. Life In The Primordial Sea. Representing mollusks, sponges, crustaceans, worms and seaweeds.

But of the causes that have led to the extinction of plants and animals, little or nothing is known. All that can be affirmed, in our present knowledge, is that the attributes constituting a species do not seem to be intrinsically endowed with permanence, any more than those constituting an individual, though the former may endure whilst many successive generations of the latter have disappeared from the earth. Each species, it would seem, has its own life-period—its beginning, culmination and decay—the life-periods of different species being of very different duration. From all that has been said, it may be gathered that our existing plants and animals are for the most part of modern origin, using the term modern in its geological acceptation. Measured by human standards, many of our existing animals, those which are capable of being preserved as fossils, are known to have a high antiquity. Not a few of our shell-fish commenced their existence at some time in the Tertiary, while one species of Lampshell—*Terebratulina caput-serpentis*—is believed to have survived since the Chalk, and a number of the Foraminifera date from the Carboniferous Period. Thus, we learn the additional fact that our existing flora and fauna do not constitute an aggregation of organic forms which were introduced into the world collectively and simultaneously, but that they commenced their existence at very different times, some being extremely ancient, whilst others are of comparatively recent origin. And this introduction of existing plants and animals, as admirably shown by the study of the fossil shells of the Tertiary Period, was a slow and gradual process. Ninety-five per cent. of the known fossil shells in the earliest Tertiary are found to be species no longer in existence, the remaining 5 per cent. being forms that are known to live in our present seas. In the Middle Tertiary, the extinct types are much fewer in number, while at the close of the Period the proportion with which we started may be reversed, not more than 5 per cent. being extinct types.

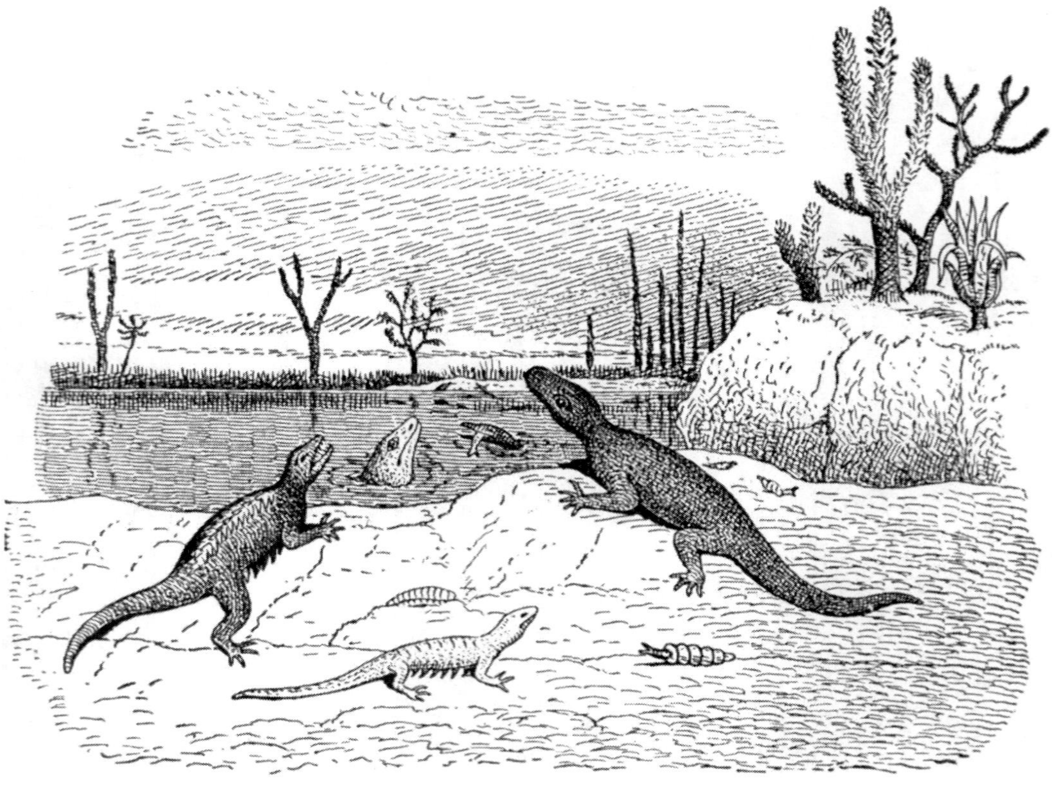

Figure 79. Carboniferous Times. Animals and plants that prevailed.

All existing animals belong to some five or six primary divisions, which are technically known as sub-kingdoms, each sub-kingdom to be regarded as representing a certain plan of structure, each and every animal embraced therein being merely a modified form of this common type. Not only are all known living animals reducible to these five or six fundamental plans, but also the vast series of fossil forms which have come to light in investigations of the earth's strata. While many fossil groups have no closely-related group now in existence, but in no case do we meet with a fossil animal whose peculiarities do not entitle it to be placed in one or other of the grand structural types already indicated. The old types differ in many respects from those now upon the earth, and the further we go back in time the more pronounced does the divergence become. A comparison of the animals that lived in the old Silurian seas with those now occupying our oceans, would indicate differences so great in many instances as almost to place us in another world, this divergence being most marked in the Palæozoic forms of life, less so in those of the Mesozoic, and still less so in the Tertiary. Each successive formation has therefore presented us with animals becoming gradually more and more like those now in existence. Though there is, however, an immense and striking difference between the Silurian animals and those of the present day, yet this difference is considerably lessened when a comparison is instituted between the Silurian and the Devonian, and this with the Carboniferous, and so on down to the present period.

Thus it follows that the animals of any given formation, and the plants as well, where the records are preserved, are more like those of the next formation below and of the next formation above, than they are like any others. This fact of itself is an inexplicable one. But if we believe that the animals and plants of any given formation are, in part at any rate, the lineal descendants of those of the preceding, and the progenitors, also in part at least, of those of the succeeding formation, then the fact is readily comprehensible. So frequently confronted is the palæontologist with the phenomenon of closely-related forms, especially of animals, succeeding one another in point of time, that he is compelled to believe that such forms have been developed from some common ancestral type by some process of evolution. Upon no other theory can we comprehend why the Post-Tertiary mammals of South America should consist of edentates, llamas, tapirs, peccaries, platyrhine monkeys and other forms now characterizing this continent, while those of Australia should be exclusively referable to the order of marsupials; and on no other view can we explain the common occurrence of transitional forms of life, filling in the gaps between groups now widely distinct. But, on the other hand, there are facts which point clearly to the presence of some other law than that of evolution, and probably of a deeper and more far-reaching character. No theory of evolution can offer a satisfactory explanation for the constant introduction throughout geological time of new forms of life, which do not appear to be preceded by pre-existent allied types. The graptolites and trilobites have no known predecessors, and leave no known successors. Insects appear suddenly in the Devonian, and spiders and myriopods in the Carboniferous, but all under well-differentiated and highly-specialized forms. With equal apparent suddenness the Dibranchiate Cephalopods show themselves in the older Mesozoic deposits, and no known type of the Palæozoic period can be pointed to as a possible ancestor. And so does the wonderful dicotyledonous flora of the Upper Cretaceous similarly surprise us without any prophetic annunciation from the older Jurassic. Many other instances might be cited, but enough has been said to show that the problem is one environed with profound difficulties.

Figure 80. Mesozoic Flora and Fauna. Cycads, Pandanus, Deinosaurs, Birds and Pterodactyl.

As we pass from the older rocks into the newer, we not only find that the animals of each successive formation become gradually more and more like existing species upon the globe, but we also find that there has been a gradual progression and development in the types of animal life which characterize the geological ages. Taking the earliest-known and oldest examples of any given group, it can sometimes be shown that these primitive forms, even though they are highly organized themselves, possessed certain characters such as are now only to be met with in the young of their existing representatives. Such characters, which are technically called embryonic characters, do not prevent the frequent attainment by their possessors of sizes much more gigantic than those of their nearest living relatives. Moreover, these ancient forms of life represent what are called comprehensive types, or types that possess characters in combination such as are nowadays found separately developed in different groups of animals. Such permanent retention of embryonic characters and comprehensiveness of structural type are signs of what zoölogists consider to be comparatively low grades of organization, and their prevalence in the earlier forms of animals is a very astonishing phenomenon, though they are none the less perfectly organized so far as their peculiar type is concerned. As we ascend the geological scale, these features will be found to gradually disappear, higher and even higher forms will be introduced, and specialization of type take the place of the former comprehensiveness. That there has been in the past a general progression of organic types, and that the appearance of the lower forms of life has in the main preceded that of the higher forms in point of time, is a widely-accepted generalization of palæontology.

Now that it has been seen that there has been a gradual progression and development of animal types all through the ages up to the era of man, the question naturally occurs whether or not the changes are still going on which will result in a higher development. Man coexisted in Western Europe with several remarkable mammals in the later portion of the Post-Pliocene Period. While we do not know the causes which led to the extinction of the mammoth, woolly rhinoceros, cave-lion and others, yet we do know that scarcely any mammalian species have become extinct during the historical period. The species with which man coexisted are such that presumably required a very different climate to that now prevailing in Western Europe. Some of the deposits in which man's remains have been found in association with the bones of extinct mammals incontestably show that great changes in the physiography and surface-configuration of the country had taken place since the period of their accumulation, the human implements themselves bearing evidence of an exceedingly barbarous condition of the human species. Post-Pliocene, or Palæolithic man, was clearly unacquainted with the use of the metals. Not only was this the case, but the workmanship of these ancient races was much inferior to that of the later tribes, who were also ignorant of the metals, and who also used nothing but weapons and tools of stone, bone, etc., in war, chase and domestic affairs. When first man spread over the earth, he had no domestic animals, perhaps not even the dog, and had no knowledge of agriculture. His weapons were of the rudest character, and his houses scarcely worthy of the name. No doubt can exist that his food, habits and entire manner of living have varied as he has passed from country to country, for he must then have been far more subject to the influence of external circumstances, and in all probability more susceptible of change. Moreover, his form, which is now stereotyped by long ages of repetition, may reasonably be presumed to have been more plastic than is now the case. As long as man led a mere animal existence, he would be subject to the same laws, and would vary in the same manner as the rest of his fellow-creatures. But when at last he had acquired the capacity of clothing himself, and of making weapons or tools, he has taken away from nature, in a great measure, that power of changing the external form and structure which she exercises over all other animals. From the time, then, when his social and sympathetic feelings came into active operation, and his intellectual and moral faculties became fairly developed, man's physical form and structure would not be so much influenced by natural laws, and, therefore, as an animal, he would become almost stationary, his environment ceasing to have upon him that powerful modifying effect which it exercises over other parts of the organic world. But from the moment that his body became less subject to the changes of the surrounding universe, his mind would become acted upon by the influences which the body had escaped. Every slight variation in his mental and moral nature, which would consequently be brought about, and which would enable him better to guard against adverse circumstances, and league together for mutual comfort and protection, would be preserved and accumulated. The better and higher specimens of our race would therefore increase and diffuse themselves, while the lower and more brutal would succumb and successively die out, and that rapid advancement of mental organization would occur, which has raised the very lowest races of men, whose mentality was scarcely superior to the animal, to that high position which it has attained in the Germanic races. It would be too bold an assertion to say that man's body has become stationary. Slow and gradual changes still take place, although his mere bodily structure long ago became of less importance to him than that subtle energy, which is termed mind. No one can doubt that *this* gave his naked and unprotected body clothing against the varying inclemencies of the seasons and enabled him to compete with the deer in swiftness and the wild bull in strength by giving him weapons wherewith to capture or subdue them both. Though less capable than most other animals of subsisting on the herbs and the fruits of unaided nature, it was this wonderful faculty that taught him to govern and direct nature to his own benefit, and compel her to produce food for him when and where he pleased. From the moment, then, when the first skin was used as a covering, the first rude spear fashioned to aid in the chase, and the first seed sown or shoot planted, a grand revolution was effected in nature, a revolution which had had no parallel in all the previous cycles of the world's history, for a being had arisen who was no longer necessarily subject to a changing universe, a being who was in some degree superior to nature, inasmuch as he knew how to control and regulate her action, and could maintain himself in unison with her, not by a change brought about in the body, but by a growth and advance in mind. Therein are shadowed forth the true grandeur and dignity of man. Not only has he achieved for himself a great victory in this rising by the power of mind superior to nature in a sense, but he has also gained a directing influence over other existences, in that he has been able to grasp from nature some of that power which, before his appearance, she universally exercised. From all that man has accomplished in the past, it is easy to anticipate the time when only cultivated plants and domestic animals will be produced by the earth, and when the ocean, which, for countless cycles of ages ruled supreme over the globe, will be the only domain in which that power can be exercised.

That man has improved under civilization there can be no question. Statistics show that, since the introduction of civilization, the population of the earth in general has increased. No one can fail to observe that under its influence the means of subsistence have increased even more rapidly than the population. Far from suffering for lack of food, the most densely peopled countries are those in which it is, not only absolutely but even relatively most abundant. A thousand men live to-day in plenty upon an area of ground that would scarcely afford a scanty and precarious subsistence to a single savage. There is no denying the fact that happiness is increased by civilization. To talk of the free and noble savage is folly. The true savage is neither free nor noble. He is a slave to his own wants, his own passions. Imperfectly protected as he is from the weather, he suffers at night from the cold and by day from the heat of the sun. Ignorant of agriculture, living by the chase, and improvident in success, hunger ever stares him in the face, and often drives him to the dreadful alternative of cannibalism or death. The life of all beasts in their wild state is certainly an exceedingly anxious one. So it is with the savage. He is always suspicious, always in danger, always on the watch. He can depend on no one, and no one can depend upon him, for he expects nothing from his neighbor, and does unto others as he believes that they would do unto him. His life is one prolonged scene of selfishness and fear. Even in his religion, if he has any, he creates for himself a new source of terror, and peoples the world with invisible enemies. More wretched is the position of the female savage than that of her master, for she not only shares his sufferings, but has also to bear his ill-humor and ill-usage, being little better than his dog, little dearer than his horse. Few of them, it is believed, are so fortunate as to die a natural death, being despatched ere they become old and emaciated, that so much good food shall not be lost. Indeed, so little importance is attached to women, either before or after death, that it may be doubted whether the man does not esteem his dog, when alive, quite as much as he does his woman, and think of both quite as often and as lovingly after he has made a meal of them. Not content, moreover, with the pleasures incident to their mode of life, savages appear to take a melancholy delight in self-inflicted sufferings. They not only tattoo their bodies, but practise the most extraordinary methods of disfigurement and self-torture, some amputating the little finger, while others drill immense holes in the under-lip, or pierce the cartilage of the nose. These and many other curious practices, none the less painful because they are voluntary, are in vogue among savage people. Turning now to the bright side of the question, we cannot but conclude that the pleasures of civilized man are greater than those of the savage. While man will never be able to improve the organization of the eye or the ear, yet, on the other hand, the invention of the telescope and the microscope is equivalent in its results to an immense improvement of the eyes, thus opening up to us new worlds, fresh sources of interest and happiness, while the training of the ear will enable us to invent new musical instruments and compose new melodies. The savage, like a child, sees and hears only that which is brought directly before him, but the civilized man questions nature, and by the various processes of chemistry, electricity and magnetism, and a thousand ingenious contrivances, forces nature to reveal herself, thereby discovering hidden uses and unsuspected beauties, quite as marvellously as though he were endowed with some entirely new organ of sense. Through the discovery of printing, we are brought into communion with the greatest minds, and thus the thoughts of a Shakespeare or a Tennyson, or the discoveries of a Newton or a Darwin, become the common property of mankind. Already the results of this all-important, though simple, process have vastly improved our mental faculties, and day by day, as books become cheaper, schools are established and education more general, a greater and greater effect will be produced.

Nor are all these new sources of happiness accompanied by any new liability to suffering. On the contrary, while our pleasures are increased, our pains are lessened. In a thousand ways we can avoid or diminish evils which to our ancestors were great and unavoidable. No one can estimate the misery which, for instance, the simple discovery of chloroform has spared the human race. The capacity for pain, so far as it can serve as a warning, remains all the same, but the necessity for endurance has been greatly diminished. With increased knowledge of the laws of health, and attention thereto, disease will become less and less frequent, and those tendencies to disease which we have inherited from our ancestors will gradually die out, and, if fresh seeds are not sown, the race will one day enjoy the inestimable advantages of a more vigorous and healthy existence. Thus, then, with the increasing influence of science we may confidently look forward to a great improvement in the condition of man. But it may be alleged that our present sufferings and sorrows arise chiefly from sin, and that any moral improvement must come from religion and not from science. This separation of the two mighty agents of improvement, the great misfortune of humanity, has done more than anything else to retard the progress of civilization. But even if we admit for the nonce that science will not render us more virtuous, it must certainly make us more innocent, for in fact the most of our criminal population are mere savages, persons who can rarely read and write, and whose crimes are but injudicious and desperate attempts to live a savage life in the midst, and at the expense, of a civilized community. Men do wrong either from ignorance or in the hope, unexpressed perhaps even to themselves, that they may enjoy the pleasure and yet avoid the penalty of sin. All that they have to do they think, when they have committed sin, is to repent. The religious teaching of the day has much to do with this misapprehension. Repentance is too frequently regarded as a substitute for punishment. Sin it is thought is followed either by the one or the other. So far, therefore, as this world is concerned, this is not the case; repentance may enable a man to avoid sin in future, but has no effect on the consequences of the past. The laws of nature are not only just and salutary, but they are also inexorable. While all men admit that "the wages of sin is death," yet they seem to think that this is a general rule to which there may be many exceptions, that some sins may possibly tend to happiness. That suffering is the inevitable consequence of sin, as surely as an effect follows a cause, is the stern yet salutary teaching of science. And certainly if this lesson were thoroughly impressed upon our minds, that punishment and not happiness is the consequence of sin, then temptation, which is the very root of crime, would be cut away, and mankind must therefore necessarily become more innocent. May we not go still further and say that science will also render us more virtuous? He who studies philosophy can only obtain a just idea of the great things for which Providence has fitted his understanding. Such a study not only makes our lives more agreeable, but it also makes them better, and every motive of interest and duty should constrain a rational being to direct his mind towards pursuits which all experience has shown to be the sure path of virtue and happiness.

Man is in reality but on the threshold of civilization. Far from showing any indication of having reached the end, the tendency to improvement seems laterally to have proceeded with augmented impetus and accelerated rapidity. There is no reason to suppose that it must now cease. Man has not attained the limits of intellectual development, nor exhausted the infinite capabilities of nature. There are many things not yet dreamt of in our philosophy which science must reveal, many discoveries yet to be made which will confer upon the human race advantages which as yet, perhaps, we are not in a condition to grasp and appreciate. We seem, when we compare our present knowledge with the great ocean of truth that lies all undiscovered before us, like little children playing on the sea-shore, and picking up a smoother pebble and prettier shell than any they had met with before. Thus, it is obvious, that our most sanguine hopes for the future are justified by the entire experience of the past. It is surely unreasonable to presume that a process which has been going on for so many thousand years should have now suddenly ceased; and he must indeed be blind who thinks that our civilization is unsusceptible of improvement, or that we ourselves are in the highest state possible for man to attain. Theory, as well as experience, forces the same conclusion upon us. That principle of Natural Selection, which in animals affects the body and seems to have little influence on the mind, in man affects the mind and has little influence on the body. In the former it leads mainly to the preservation of life, and in the latter to the improvement of the mind, and consequently to the increase of happiness. It ensures, in the words of Spencer, "a constant progress towards a higher skill, intelligence, and self-regulation—a better coördination of actions—a more complete life." Nearly all the evils under which we suffer, it will be conceded, may be attributed either to ignorance or sin. That ignorance will be diminished by the progress of science is, of course, self-evident; and that the same will be the case with sin, seems little less so. Thus, then, do both science and theory point to the same conclusion. That which poets hardly dared to hope for, the future happiness of our race, science boldly predicts. Even in our own time we trust to see some wonderful improvement. But the unselfish mind, however, will find its highest gratification in the belief that, whatever may be the case with ourselves, our descendants will understand many things which are mysterious to us now, will better appreciate the beautiful world in which we live, avoid much of the suffering to which we are subject, enjoy many blessings of which we are not yet worthy, and escape many of those temptations which we deplore but cannot wholly resist.

We have thus seen that all life has been progressive. There has been through the ages a steadily growing upward tendency to higher life. But the changes have mainly been in the line of physical form and structure. And such, too, had been the case with man, until his social, intellectual and moral faculties had begun to assert themselves, when his body ceased in a great measure to be acted upon by physical laws, and development began to manifest itself in a higher type of mental organization. From the low, simple, childlike mind of palæolithic man has come that wonderful intellect which now characterizes the Germanic races, and which is destined to make itself felt in its contact with all the earth.

Those peoples that are able to embrace the new civilization brought to their doors, so to speak, will survive, while the others, unable to adapt themselves thereto, like the Tasmanian, will succumb in the struggle with a superior being and go to the wall. Animals and plants will be brought into new relations and new conditions, and such as can meet the new requirements will, as certain species have done before, endure. They will, in other words, have partaken of an enlightened civilization. Thus things will go on until all life, vegetal and animal, will be brought under the controlling and elevating influence of man, and then will be inaugurated on earth that condition when the lion and the kid shall lie down together, and a little child shall be found in their midst. Nothing harmful will anywhere exist. Heaven will then have been brought down to earth, and peace and harmony will universally prevail. Then will have come the complete triumph of mind over body. All growth and development of the reformed and regenerated earth-man will be in the direction of mind, and his accomplishments will he share with the inferior subjects of his peaceful and happy domain. Progression, however, will not cease, but will go on steadily advancing as the years increase. And if there is a life beyond the earth-life, then the intellect or mind, or soul if you please, shall, in some form or other, exist therein, and reach up into higher and yet higher growth and development.

Chapter 46

SURVIVAL OF THE FITTEST

Among organic beings in a state of nature there is some individual variability. This is an admission about which there can be no dispute. But the mere existence of individual variability and of a few well-marked varieties, though necessary as the foundation for the work, assists us but little in understanding how species originate in nature. Those exquisite adaptations of one part of the organization to another part, and to the conditions of life, and of one organic being to another being, which we know to exist, seem as mysteries. We see them in the humblest parasite that clings to the hairs of a quadruped or the feathers of a bird, in the structure of the beetle that dives through the water, and in the plumed seed that is wafted by the gentlest breeze. In short, we see beautiful adaptations everywhere and in every part of the organic world. And yet, how few have paused while admiring these beautiful and wonderful co-adaptations to ask themselves the question: How have these been perfected?

If the existence of any well-marked varieties be admitted, how is it that these varieties, which may be denominated incipient species, become ultimately converted into good and distinct species, which in the generality of cases obviously differ from each in a greater degree than do the varieties of the same species? How do these groups of species, which constitute what are authoritatively called genera, and which differ from each other more than do the species of the same genus, arise? All these results, as will presently be seen, follow from the Struggle for Existence. Owing to this struggle, all variations, no matter how slight they may be, or from what cause soever they may proceed, will, if they be in any degree profitable to the individuals of a species in their infinitely complex relations to other organic beings and their physical conditions of life, unavoidably conduce to the preservation of such individuals, and generally be inherited by the offspring. The offspring, too, will thus have a better chance of surviving, for, of the many individuals of a species that are periodically born, but a very small number can survive. That principle, by which each slight variation, if useful to the individual, is preserved, has been termed Natural

Selection by Darwin, in order to distinguish it from the selection which is exercised by man over the plants and animals which he has brought under subjection for his own wants. But the expression—Survival of the Fittest—so frequently used by Spencer, is more accurate, and sometimes equally convenient. Man can certainly produce great results by this power, and can adapt, through the accumulation of slight but useful variations given to him by the hand of nature, organic beings to his own uses. But Natural Selection, as is well known, is a power incessantly ready for action, and is as infinitely superior to man's feeble efforts as the works of nature are to those of art.

All organic beings are exposed to severe competition. Nothing is easier than to admit in words the truth of the universal struggle for life, or more difficult than constantly to bear this conclusion, which has been reached through the investigations and researches of De Candolle, Lyell, Herbert, Darwin and others, in mind. Unless, however, it be thoroughly ingrained in the mind, the whole economy of nature, with every fact on distribution, rarity, abundance, extinction and variation, will be but dimly perceived or quite misunderstood. We behold the face of nature radiant with gladness, and food everywhere in excessive abundance, but we do not see that the birds which are happily singing round us mostly live on insects or seeds, and are thus constantly destroying life, or we fail to remember how largely these songsters, or their eggs, or their nestlings, are destroyed by birds and beasts of prey. Yes, we do not always bear in mind that, though food may now be superabundant, it is not so at all seasons of each recurring year. The term, Struggle for Existence, must be used in a large and metaphorical sense. It must be construed to include the dependence of one being on another, and also not only the life of the individual but also its success in leaving offspring. Two carnivores, in a time of scarcity of food, may be truly said to struggle with each other for maintenance of life. But a plant on the edge of a desert is said to struggle for life against the drought, though, properly speaking, it is dependent for its existence upon the moisture. A plant, however, that annually produces many thousand seeds of which on an average only one comes to maturity, may in a much truer sense be said to struggle with the plants of the same and other kinds which already invest the ground. While the mistletoe is dependent on the apple and some other trees, yet it cannot be said, unless in a far-fetched sense, to struggle with these trees, for, if too many of these parasites are found upon the same tree, it will certainly languish and die. Several seedling mistletoes, however, growing close together upon the same branch, may more truly be said to struggle with each other.

From the high rate at which all organic beings tend to increase, there must inevitably follow a Struggle for Existence. Every being which, during its natural lifetime, produces several eggs or seeds, must necessarily suffer destruction during some part of that period, and during some season or occasional year, otherwise, on the principle of Geometrical Increase, its numbers would become so inordinately excessive that no country would be able to support its product. Therefore, as more individuals are produced than can possibly survive, there must be in every case a Struggle for Existence, either one individual struggling with another of the same kind, or with individuals of distinct kinds or species, or with the conditions of the environment. This is the doctrine of Malthus applied with manifold force to the entire vegetable and animal kingdoms. Although some species may be now increasing at a very high rate in numbers, yet all cannot do so, for the earth would not be able to contain them. Slow-breeding man has doubled in twenty-five years, and should he go on at this rate for a few thousand years, there would literally not be standing room for his progeny. It has been calculated that, if an annual plant produced only two seeds, and their seedlings next year produced two, and the same rate of increase was kept up for twenty years, there would be a million of plants as the result. Even the elephant, which is reckoned the slowest breeder of all known animals, would after a period of from seven hundred and forty to seven hundred and fifty years leave nearly nineteen million elephants as descendants from the first pair.

Much better evidence than mere theoretical calculations are not wanting on this subject. Instances are recorded of the astonishingly rapid increase of various animals in a state of nature, when conditions have been favorable to them, during two or three succeeding seasons. More striking, however, is the evidence from domestic animals that have run wild in several parts of the world. Were not the statements of the rate of increase of cattle and horses in South America, and latterly in Australia, where millions now abound, well authenticated, they would have been incredible. Cases could be mentioned of introduced plants that have become quite common throughout entire islands in a period of less than twelve years. Several of these plants, the cardoon and a rare thistle, which were introduced from Europe, clothe square leagues of the surface of the wide plains of the La Plata almost to the exclusion of all other plants; and there are plants which now range in India, from Cape Comorin to the Himalaya, which have been imported from America since its discovery. In all such cases, and endless instances could be adduced, no intelligent person supposes that their fertility has been increased in any sensible degree by change of habitat, the obvious explanation being that the conditions of environment have been very favorable, and that there has consequently been less destruction of old and young, and that nearly all the latter have been enabled to breed. The extraordinarily rapid increase and wide diffusion of naturalized productions in new homes, a result which never fails to evoke surprise, is only to be explained on the principle of the Geometrical Ratio of Increase. As in nature almost every plant produces seed, and there are very few animals that do not annually pair, therefore we can confidently assert that all plants and animals are tending to increase in a geometrical ratio; that all would most rapidly stock every station in which they could in any way exist, and that the tendency to increase must be checked by destruction at some period of life. Among our larger domestic animals we see no great destruction falling on them. We forget that thousands are annually slaughtered for food, and that in a natural state an equal number would have to be disposed of in some way or other. Between organisms which annually produce seeds or eggs by the thousands, and those which produce extremely few, the only difference is that the slow breeders would require a few more years to people, under favorable conditions, a whole district, let it be ever so large. But a couple of eggs are laid by the condor, while the ostrich lays a score. Yet in the same country the condor may be the more abundant of the two. The Fulmer petrel lays but a single egg, yet it is believed to be the most numerous bird in the world. A large number of eggs is of some importance to those species which depend upon a rapidly-fluctuating quantity of food, for it permits them to increase rapidly in number; but the real importance of a large number of eggs or seeds is to make up for the great destruction that goes on at some period of life, and this period in the vast majority of cases is an early one. If an animal can in any way protect its own eggs or young, a small number may be produced, and the average stock be kept up; but if many eggs or young are destroyed, then many must be produced or the species will become extinct. Therefore, the average number of any animal or plant depends, though only indirectly, upon the number of its eggs or seeds. We should never forget, in taking a survey of nature, that every single organic being around us may be said to be striving to the utmost to augment its members; that each lives by a struggle at some period of its existence, and that heavy destruction falls either on the young or old during each generation or at recurrent intervals. Let any check be lightened, or the destruction be mitigated ever so little, and the number of the species will almost instantaneously increase to any extent.

But of the nature of the checks to increase we know little, although this subject has been very ably treated by writers of eminence. Eggs or very young animals seem generally to suffer the most, but this is not invariably the case. While there is a vast destruction of the seeds of plants, but it is the seedlings which are believed to suffer the greatest, from germinating in ground already thickly stocked with other plants, and from being destroyed in large numbers by various enemies. The amount of food for each species of course determines the extreme limit to which each can increase, but very often it is not the obtaining of food, but the serving as prey to other animals which fixes the average number of a species. Thus there seems to be little doubt that the stock of partridges, grouse and hares on any large estate depends mainly on the destruction of vermin. Were not a single head of game shot during the next twenty years in England, says Darwin in substance, and no vermin were at the same time destroyed, there would in all probability be less game than at present exists, although hundreds of thousands of game animals are now annually killed for the market. In some cases, on the other hand, as in the case of the elephant, none are destroyed by beasts of prey, for even the tiger in India, bold and venturesome as he is known to be, rarely dares to attack a young elephant protected by its mother. Climate, also, plays an important part in determining the average number of a species, and periodical seasons of extreme cold or drought are seemingly the most effective checks of all. The action of climate appears at first sight to be altogether independent of the Struggle for Existence; but in so far as it chiefly acts in the reduction of food, it brings on the most severe struggle between the individuals, whether of the same or different species, which subsist on the same kind of fare. Even when climate, extreme cold for example, acts directly, it will be the least vigorous animals, or those which have been the poorest fed through the advancing winter, that will suffer the greatest. This will be most readily seen from what we shall now relate. When we travel from south to north, or from a damp region to a dry, we invariably see some species getting rarer and rarer by degrees, and finally disappearing. Change of climate being conspicuous, we are inclined to ascribe the entire effect to its direct action, but this is a false interpretation of the phenomenon, for we fail to remember that each species, even where it most prevails, is constantly suffering enormous destruction at some period of its existence, from enemies or competitors for the same station and food; and if these enemies or competitors be the least favored by any slight change of climate, they will necessarily increase in numbers, while the other species, each area being already stocked with inhabitants, will correspondingly decrease. And when we travel southward and see a species decreasing in numbers, we may feel reasonably sure that the cause lies quite as much in other species being favored as in this being hurt. So it is when we travel northward, though in a less degree. When we go northward, or when we ascend a mountain, we far oftener meet with stunted forms, due to the directly injurious action of climate, than we do when we go southward or descend a mountain. When, however, we reach the Arctic regions, or explore snow-capped summits, or absolute deserts, we perceive the struggle for life to be almost exclusively with the elements.

That climate operates mainly, but indirectly, in favoring other species, may be clearly seen in the prodigious numbers of garden plants that can thoroughly well endure our climate, but which can never become naturalized, inasmuch as they cannot compete with native vegetation nor resist destruction by native animals.

When a species, owing to highly favorable conditions, increases inordinately in numbers in a small tract of country, epidemics, especially in game animals, often occur, and here we have a limiting check independent of the Struggle for Existence. But some of these so-called epidemics appear to be due to parasitic worms, which have from some cause, possibly in part through ease of diffusion among the crowded animals, been disproportionately favored, and here comes in a sort of struggle between the parasite and its more illustrious prey.

But, on the other hand, as is frequently the case, a large stock of individuals of the same species, relatively to the number of its enemies, is absolutely essential to its preservation. We thus see how it is possible to raise with ease a plentiful supply of corn in our fields, because the seeds are greatly in excess of the number of birds which feed thereon. Nor can the birds, though blessed with a superabundance of food at this one season, increase in number in proportion to the supply of seed, as their numbers are checked during the winter. Any one, however, who has made the experiment, knows how troublesome it is to get seed from a few wheat or other such plants sown broad-cast in a garden. Some singular facts in nature, such as that of very rare plants being sometimes extremely abundant in the few spots where they do occur, and that of some social plants being social, or abounding in individuals, even on the extreme confines of their range, are readily explainable by this view of the necessity of a large stock of the same species for its preservation, for in such cases we may believe that a plant could only exist where the conditions of its life were so favorable that many could exist together and thus save the species from extinction.

Complex and varied are the checks and relations between organic beings which have to struggle together in the same country. In the case of every species, many different checks, some very complicated and unintelligible to man at present, acting at different periods of life, and during different seasons or years, come into play, some one check or some few being generally the most powerful, but all concurring in determining the average number or even the existence of the species. Widely-different checks sometimes act on the same species in different districts. Looking at the plants and bushes that clothe an entangled bank, we are tempted to ascribe their proportional numbers and kinds to what we call chance. But this is a very false view to take of the matter. Chance has no part in such things. They follow in obedience to laws of which we know comparatively little. When an American forest is cut down a very different vegetation springs up. Ancient Indian ruins have been observed in the southern parts of the United States, which must in former times have been cleared of trees, but which now display the same beautiful diversity and proportion of kinds as are now found in the surrounding virgin forest. What a struggle must have gone on during long centuries between the several kinds of trees, each annually scattering its seeds by the thousand, and what a war between insect and insect, and between insects, snails and other animals with birds and beasts of prey, all striving to increase, all feeding on each other, or on the trees, their seeds and their seedlings, or on the other plants which once clothed the soil, and thus checked the growth of the trees! It is easier to account for the fall of an apple from a tree, or the descent of a stone to the earth when hurled into the air, than to account for the action and reaction of the innumerable plants and animals that have determined in the course of untold centuries the proportional numbers and kinds of trees that are now found growing on these old Indian ruins. But the struggle will almost invariably be the severest between individuals of the same species, for they frequent the same districts, require the same food and are exposed to the same dangers. In the case of varieties of the same species, the struggle will generally be almost equally severe. If several varieties of wheat be sown together, and the mixed seed be re-sown, some of the varieties which best suit the soil or climate, or are naturally the most fertile, will beat the others and so yield more seed, and will consequently in a few years supplant the others. Such extremely-close varieties as the variously-colored sweet-peas must be separately harvested each year, and the seed mixed in due proportion, or the weaker kinds will steadily decrease in number and disappear. So, again, with the varieties of sheep. Certain mountain-varieties will starve out other mountain-varieties, so that they cannot be kept together. Similar results have followed from keeping together different varieties of the medicinal leech. In view of all that has been said, it is questionable whether the varieties of any of our domestic plants and animals have so exactly the same vigor, constitution and habits that the original proportions of a mixed stock could be kept up for a half-dozen generations if they were permitted to struggle together like beings in a state of nature, if the seed or young were not annually assorted.

Species of the same genus having usually, though not invariably, much similarity in habits and constitution, and always in structure, the struggle will be more severe between species of the same genus, where they come into competition with each other, than between species of distinct genera. One species of swallow has caused in certain parts of the United States the decrease of another species, just as the missel-thrush in parts of Scotland has caused the decrease of the song-thrush. The small Asiatic cockroach has everywhere in Russia driven before it its great congener, and the imported European hive-bee is rapidly exterminating in Australia the small, stingless bee, indigenous to the country. Hundreds of such cases might be cited, but we forbear. We can clearly see why the competition should be most severe between allied forms, which fill nearly the same place in the economy of nature; but it is perhaps not possible to individualize a case and say with preciseness why such species has been victorious over another in the battle of life. That the structure of every organic being is related, in the most essential yet often hidden manner to that of all the other organisms with which it comes into competition for food or residence, or from which it has to escape, or on which it preys, is a corollary of the highest importance deducible from the foregoing remarks. Very obvious is this in the structure of the teeth and talons of the tiger, and in that of the legs and claws of the parasite which clings to the hair on the tiger's body. But in the beautifully-plumed seed of the dandelion and the flattened and fringed legs of the water-beetle the relation seems at first restricted to the elements of air and water, yet the advantage of plumed seeds undoubtedly stands in the most intimate relation to the land, being already densely clothed with other plants, so that the seeds may be widely diffused and fall on unoccupied ground, while in the water-beetle, the structure of its legs, so admirably adapted for diving, allows it to compete with other aquatic insects, to hunt for its own prey and to escape destruction by other predaceous animals. All organic beings, it will thus be seen, are not only striving to increase in numbers, but are called upon some time in their lives to struggle for existence or to suffer serious if not utter destruction. When we reflect on this struggle, we can console ourselves with the full belief that this war of nature is not incessant, that no fear is felt, that death is generally sudden, and that the vigorous, healthy and happy survive and multiply.

Seeing what a potent influence the principle of Selection has in the hands of man, in regard to variation, can it be applied in nature? We can see that it can act most effectually. But in our domestic productions the variability is not directly produced by man, for he can neither originate varieties nor prevent their occurrence. All he can do is to preserve and accumulate such as do occur. Unintentionally he exposes organic beings to new and changing conditions of life, for under domestication, plant and animal organizations become in some degree plastic, and variability ensues. Similar changes, however, do occur in nature. When it is borne in mind how infinitely complex and close-fitting are the mutual relations of all organic beings to each other, and to their environment, and consequently what infinitely-varied diversities of structure may be of advantage to each being under altered conditions, can it then be thought improbable, seeing that variations useful to man have undoubtedly occurred, that other variations useful in some way to each being in the great and complex battle of life should sometimes occur in the course of tens of thousands of generations? If such do occur, can we doubt, when it is remembered that many more individuals are born than can possibly survive, that individuals possessing any advantage, no matter how slight, over their fellows would have the best chance of surviving and of procreating their kind? Any variation, on the other hand, we may feel sure if in the least degree injurious would be rigidly destroyed. This preservation of useful and favorable variations, and the destruction of those that are injurious, is called Natural Selection, or the Survival of the Fittest. Variations neither advantageous nor deleterious would not be affected by Natural Selection, and would be left either a fluctuating element, as seen in certain polymorphic species, or would alternately become fixed, owing to the nature both of the organism and its conditions.

We shall best understand the probable cause of Natural Selection by taking a country undergoing some physical change, as of climate for example. The proportional number of its inhabitants would almost immediately undergo a change, and some of its species might become extinct. From the complex and very intimate manner in which the inhabitants of each country are bound together, we may conclude that any change in the numerical proportion of some of its inhabitants, independently of the change of climate itself, would seriously affect the others. Were the country open on its borders, new forms would certainly immigrate, and this, too, would often seriously disturb the relations of some of its former inhabitants. In the case, however, of an island, or a country hemmed in by barriers, into which new and better-adapted forms could not readily enter, we would then meet with places in the economy of nature which would assuredly be better filled up, if some of the original occupants were in some manner modified, for had the area been open to immigration, these same places would have been seized by intruders. Thus, slight modifications, which any way favored the individuals of a species, would by better adapting them to changed conditions tend to become preserved, and Natural Selection would there have free scope for the work of improvement. Changes in the conditions of life cause or excite a tendency to vary. In the foregoing case the conditions are supposed to have changed, and this would manifestly be favorable, by giving a better chance of profitable variations occurring, to Natural Selection, for unless such do occur, Natural Selection can do nothing. As man, by adding up in any given direction individual differences, can certainly produce a great result with his domestic animals and plants, so could Natural Selection, but far more easily from having an incomparably longer time for its action. No great physical change, as of climate, nor any unusual degree of isolation to check immigration, is actually necessary, it would seem, to produce new and unoccupied places for Natural Selection to fill up by modifying and improving some of the varying inhabitants, for as all the inhabitants of a country are struggling together with nicely-balanced forces, extremely-slight modifications in the structure or habits of one species would often give it an advantage over others; and still further modifications, so long as the species continued under the same conditions of life and profited by similar means of subsistence and defence, would often still further augment the advantage. No country can be mentioned whose native inhabitants are now so perfectly adapted to each other and to their environment that none could be better adapted and improved, for in all countries the natives have been so far conquered by naturalized productions as to have allowed them to take firm possession of the land. And as foreigners have thus in every country beaten some of the natives, it may be safely concluded that the latter might have been modified with profit so as to have better resisted the intruders.

A man by his methodical and unconscious means of selection can produce and has produced great results. What may not Natural Selection effect? Man can only operate on external and visible characters, but nature cares nothing for appearances, except in so far as they are beneficial to any being. She can act on every internal organ, on every shade of constitutional difference and, in fine, on the entire machinery of life. Man selects exclusively for his own advantage, but nature solely for that of the being she tends, and under her judicious selection the slightest difference of structure or constitution may well turn the nicely-balanced scale in the Struggle for Existence, and thus be preserved. As fleeting as are the wishes and efforts of man, and as short as is his earthly career, so poor, therefore, must be the results which he accomplishes when compared with those accumulated by nature during whole geological periods. Is it a wonder, then, that her productions should be far *truer* in character than man's, and that they should be infinitely better adapted to the most complex conditions of life and should bear the stamp of far higher workmanship? Metaphorically speaking, Natural Selection may be said to be daily and hourly scrutinizing, throughout the world, the slightest variations, rejecting the bad, preserving and adding up the good, and silently and insensibly working, whenever and wherever opportunities occur, at the betterment of each organic being in relation to its organic and inorganic conditions of life. So slow is her work that we see nothing of the changes in progress, and only when the hand of time has marked the lapse of ages do we perceive that changes have been produced; but then so imperfect is our view into long-past geological periods, that we see only that the forms of life are now different from what they formerly were. That any great amount of modification in any point should be effected, a variety once formed must again, perhaps after a long interval of time, present individual differences of the same favorable character, and these must again be preserved, and so onward step by step. As individual differences of all kinds perpetually recur, this can hardly be considered as an unwarrantable assumption. Judged by the extent the hypothesis accords with and explains the general phenomena of nature, notwithstanding the ordinary belief that the amount of possible variation is a strictly-limited quantity, we are justified, it seems to us, in assuming that all this has actually taken place. But in looking at many small points of difference between species, which in our ignorance seem quite unimportant, we must not lose sight of the facts that climate, food and modes of life may have produced some direct effect, and also of the truth that, owing to the Law of Correlation, when one part varies, and the variations are accumulated through the Survival of the Fittest, other modifications often of the most unlooked-for nature will ensue.

As under domestication these variations are known to appear at a particular period of life, and tend to reappear in the offspring at the same period, so, in a state of nature, it is reasonable to infer that Natural Selection will be enabled to act on and modify organic beings at any age, by the accumulation of variations useful at that age, and by their inheritance at a corresponding age. Thus, if it be profitable to a plant to have its seeds more and more widely disseminated by the wind, there can be no greater difficulty in conceiving this to be effected through Natural Selection than in conceiving the increasing and improving of the down in the pods on his cotton-trees by a wise selection upon the part of a cotton-planter. Natural Selection may modify and adapt the larva of an insect to a score of contingencies, wholly different from those which affect the mature insect, and these modifications through Correlation may work changes in the structure of the adult. On the other hand, modifications of the adult may affect the structure of the larva, but in all such cases Natural Selection will insure that these changes shall not be injurious, for, if they were so, the extinction of the species would be the inevitable result. Thousands of instances might be given to show the influence which Natural Selection, or Sexual Selection, which is only a less vigorous phase of the former, has had all through the ages in the adaptation of life to the places in nature which it was intended to occupy in pursuance of the plan formulated by the Great Originator and Designer of the Universe.

Despite the imperfection of the geological record, which has been urged as a serious objection to the theory of descent with modification, sensible, intelligent, educated men no longer doubt that species have all changed, and that they have changed in the way required, for they have changed slowly and in a graduated manner. This is clearly seen in the fossil remains from consecutive formations being invariably much more closely allied to each other than are those from widely-separated formations. It is true geological research does not yield those infinitely fine gradations between past and present species which the theory of Natural Selection requires, but when it is remembered that only a small portion of the world has been geologically explored; that only organic beings of certain classes, at least in any great number, can be preserved in a fossil condition; that many species when once formed never undergo any further change, but become extinct without leaving any modified descendants; that dominant and widely-ranging species vary the most and the most frequently, and that varieties are often at first only local, it is not at all surprising that the discovery of intermediate links to any considerable extent should not have been made. Local varieties, as is well known, will not diffuse themselves into other and distant localities until they have become very much modified and improved, and when they have thus diffused themselves, and are discovered in a geological formation, they will appear as if suddenly created there, and will simply be ranked as new species. Besides, formations have often been intermittent in their accumulation, and their duration has probably been shorter than the average duration of specific forms. And as successive formations in most cases are separated from each other by blank intervals of time of considerable length, and as fossiliferous formations thick enough to withstand future degradation can as a general rule be accumulated only where much sediment is laid down in the subsiding bed of the ocean, it follows that during the alternate periods of elevation and of stationary level the record will generally be blank or devoid of fossil remains. During these latter periods there will doubtless be more variability in the forms of life, and during the periods of subsidence a greater amount of extinction. Now, as geology plainly declares that each land has undergone great physical changes, we have a right to expect that organic beings have varied under nature in the same manner as they have varied under domestication, and such have scientific study and research found to be the case. And if there has been any variability under nature, such a fact would seem unaccountable unless Natural Selection, or the Survival of the Fittest, did not come into play. Upon the view that variations have occurred in nature and have been preserved and accumulated by Natural Selection, and not in the ordinary view of independent creation, we can understand why the specific characters, or those by which the species of the same genus differ from each other, should be more variable than the generic characters in which they all agree. Inexplicable as is the occasional appearance of stripes on the shoulders and legs of the different equine species and their hybrids on the theory of creation, yet how simply is the fact explained if we believe that they are all descended from a striped progenitor just as the different domestic breeds of pigeons are descended from the blue and barred rock-pigeons. Why, for example, should the color of a flower be more likely to vary in any one species of a genus, if the other species, supposed to have been created independently, have differently-colored flowers, than if all the species of the genus have the same colored flowers? On the theory that species are only well-marked varieties, of which the characters have become in a high degree permanent, the fact is intelligible, for they have already varied in certain characters since they branched off from a common progenitor, and by these characters they have come to be specifically distinct from each other. Therefore, these same characters would be more likely again to vary than the generic characters which have been inherited without change for an enormous period of time.

Upon the theory of Natural Selection, or the Survival of the Fittest, with its contingencies of extinction and divergence of character, we can see how it is that all past and present organic beings can be arranged within a few classes, in groups subordinate to groups, and with the extinct groups often falling in between the recent groups. We can see how it is that the mutual affinities of the forms within each class are so complex and diversified, and only adaptive characters, though of superior importance to the beings, are of scarcely any significance in classification, while those derived from rudimentary parts, though of no recognized service, are often of high classificatory value, and only embryological characters are frequently the most valuable of all. The real affinities of all organisms, in contradistinction to their adaptive likenesses, are due to inheritance or community of descent. Hence, a natural system of classification is a genealogical arrangement, with the acquired grades of difference, denoted by varieties, species, genera, families, etc., and their lines of descent have to be discovered by the most permanent characters, whatever they may be and how little of vital importance they may possess.

That species are immutable productions, which was until quite recently the current belief by laymen and naturalists, was almost unavoidable so long as the world was considered to be of short duration. But now that some idea has been acquired of the time that has elapsed since the beginning of earth-life, we are too apt to assume, without proof, that the geologic record is so complete, that it would have afforded us some plain evidence of the mutation of species, if they had undergone mutation. But the principal cause of our unwillingness to admit that one species has given birth to other and distinct species, is that we are always slow in admitting any great change of which we do not discern the intermediate steps. Just such a difficulty was felt by many geologists when Lyell first insisted that long lines of inland cliffs had been produced, and great valleys excavated, by the agencies which are still at work in the earth. No effort of mind can adequately grasp the meaning of even ten million of years, nor add up and perceive the full effects of the many slight variations to which species have been subjected during an almost infinite number of generations. The day, however, is not distant, when mankind will have become just as thoroughly convinced that species have been modified during a long course of descent, mainly through the Natural Selection of innumerous successive, slight and favorable variations as they are that the attraction of gravitation is an important element in the maintenance of the harmony that exists among the planetary spheres. That the law of the attraction of gravity, which is perhaps the greatest discovery ever made by man, is subversive of natural and revealed religion, which was at one time maintained by a no more distinguished person than Leibnitz, is now no longer objected to, even though its discoverer was unable to explain what is the essence of the principle he had discovered. No nobler conception of Deity could be entertained than that which attributes to Him the creation of a few original forms capable of self-development into other and needful forms, or the origination *de novo* of these simple forms from inorganic nature. It places a higher estimate upon His Omnipotence than the belief that He required a fresh act of creation to supply the voids caused by the action of His laws. That science is as yet unable to throw any light on the far higher problem of the essence or origin of life, should constitute no valid objection to the theory of descent.

When all beings are looked upon not as special creations, but as the lineal descendants of some beings that existed long before the first bed of ancient Siluria was deposited, they seem to become ennobled. Judging from the past, we think it safe to conclude that no existing species will transmit its unaltered likeness to a distant futurity. Few, very few living species will transmit progeny of any kind, for the manner in which all organisms are grouped shows that the majority of species in each genus, and all the species in many genera, have left no descendants, but have become utterly extinct. It will only be the common and widely-spread species, belonging to the larger and dominant groups within each class, that will ultimately prevail and procreate new and dominant species. Since all the living forms of life are the lineal descendants of forms that lived long anterior to the Silurian epoch, it is reasonably certain that the ordinary succession by generation has never once been broken, and that no cataclysmic disaster has laid waste the entire world. Therefore, we may look into the future with some confidence of an equally secure and inappreciably enduring earth-life. And as Natural Selection operates solely by and for the good of each being, all corporeal and mental endowments will tend to progress toward perfection.

Figure 81. Palæolithic Men Attacking Cave Bear. Mammoth, Woolly Rhinoceros, Musk-Sheep and Irish Stag in Background.

When we contemplate a tangled bank, with innumerable plants of diverse kinds, and many-voiced birds singing in concert, or waging destruction on manifold insects that are flitting about, or the long, slimy worm that has come up from its underground retreat, we are lost in wonder and admiration, and can only reflect that these elaborately constructed forms, so different from each other, and so strangely and intricately dependent on each other, have all been evolved by laws that act all around us. These are the laws of Growth with Reproduction; Inheritance, which is almost implied by reproduction; Variability from the action, direct and indirect, of the conditions of life, and from use and disuse; a Ratio of Increase so high as to lead to a Struggle for Existence, and as a consequence to Natural Selection, or Survival of the Fittest, entailing thereby Divergence of Character and the Extinction of less-improved forms. And thus, from the war of nature, and from famine and death, have arisen the higher mammalia, in which man, the *summa summarum* of life, is included. He occupies the summit, toward which the efforts of millions of buried ages seem to have been tending. There is a grandeur in this view of life, with its several powers, originally breathed, by the operation of the natural laws, into one or a few forms of life, and that, while the earth, in obedience to the fixed principle of gravitation, has gone cycling on, endless forms, most beautiful and most wonderful, have been, and are being, evolved from so simple a beginning.

While thus it has been shown that life has been progressive, successive forms of life being the result of modification through descent, those faring the best in the Struggle for Existence surviving, by reason of some advantage, physical or otherwise, gained over their competitors, yet little, bearing specially upon man, has been expressed in this chapter. After he had acquired those intellectual and moral faculties which largely distinguish him from the lower animals in a state of nature, he would have been but little liable to have his bodily structure modified through Natural Selection or any other means, for man is enabled, through his mental faculties, "to keep with an unchanged body in harmony with the changing universe." He has a most wonderful power of adapting his habits to altered conditions of life. Tools, weapons and various devices are invented by him for the procurement of food and bodily defence. And when he migrates into a colder climate, he uses clothes, builds sheds and makes fire, and by its aid cooks food that would otherwise be indigestible. The lower animals, however, must have their bodily structure modified in order to survive under greatly changed conditions. They must be rendered stronger, or acquire more effective teeth or claws, or both, if they would successfully defend themselves from new enemies, or they must be reduced in proportions, so as to escape detection and danger. When they remove into colder climates they must become clothed in thicker fur, or have their constitutions altered, for failure to be thus modified must ultimately result in their ceasing to exist. But in the case of man's intellectual and moral faculties, as has been shown by Wallace, it is widely different. These faculties are quite variable, and there is reason to believe that the variations tend to be inherited. Therefore, if they were formerly of high importance to palæolithic man and his ape-like progenitors, they would have been perfected or advanced through Natural Selection. But of the high importance of the intellectual faculties there can be no question, for man owes to them in a great measure his preëminent position in the world. It can be seen that, in the rudest state of society, the individuals who were the most sagacious, and who were the most skilful in the invention of weapons or traps, and who were the best able to defend themselves, would rear the greatest number of offspring, and that the tribes which included the largest number of men possessed of such superior endowments would increase in number and eventually supplant the other tribes. Numbers depend primarily on the means of subsistence, and this on the physical nature of the country, but in a much higher degree upon the arts therein practised. As a tribe increases and is victorious, it is often still further increased by the absorption of other tribes, and after a time the tribes which are thus absorbed into another tribe assume, as has been remarked by Mr. Maine in his "Ancient Law," that they are the co-descendants of the same ancestors. Stature and strength in the men of a tribe are also of importance in its success, and these are dependent in part upon the character and the quantity of food that can be obtained. Men of the Bronze Period in Europe were supplanted by a larger-handed and more powerful race, but their success was probably due in a much higher degree to their superiority in the arts. All that is known by savages, as inferred from their traditions and from old monuments, shows that from the most remote times successful tribes have supplanted others. Relics of extinct tribes have been found on the wild plains of America and on the isolated islands in the Pacific Ocean. Civilized nations are everywhere at the present time supplanting barbarous peoples, excepting where climate opposes a fatal barrier, and they thus succeed in a great measure, though not exclusively, through the arts, which are the products of the intellect. With mankind, then, it is highly probable that the intellectual faculties have been gradually perfected through Natural Selection. Undoubtedly it would have been interesting to have traced the development of each separate faculty from the state in which it exists in the lower animals to that in which it exists in man, but this would have been a task of no easy accomplishment. As soon, however, as the progenitors of man became social, and this probably occurred at a very early period, the advancement of the intellectual faculties would have been aided and modified in an important manner, for if one man in a tribe, more sagacious than his fellows, had invented a new snare or a weapon, or other means of attack or defence, the plainest self-interest, with no great help of reasoning power, would have prompted the other members to have imitated him, and thus all would have been profited. Habitual practice of each new art must likewise in some slight degree strengthen the intellect. If the new invention were an important one, the tribe would increase in numbers, spread and supplant other tribes, and thus rendered stronger numerically there would be a better chance of the birth of other superior and inventive members. Should these last be so fortunate as to leave children to inherit their mental superiority, the chance of the birth of still more ingenious members would be somewhat better, and in a very small tribe would be decidedly better.

That primeval man, or his ape-like progenitors, should have become social, they must have acquired the same instinctive feelings which impel other animals to live in a body, and they doubtless exhibited the same general disposition. When separated from their companions, for whom they would have felt some degree of love, they would have experienced a feeling of uneasiness. They would have warned each other of danger, and have given mutual aid in attack or defence. All this implies some degree of sympathy, fidelity and courage. Such social qualities, whose paramount importance to the lower animals is undisputed, were doubtless acquired by the progenitors of men in a similar manner, namely, through Natural Selection, aided by inherited habit. In the never-ceasing wars of savages, fidelity and courage are all-important, and certainly when two tribes of primeval man, living in the same country, came into competition, the one that contained the greatest number of courageous, sympathetic and faithful members, who were ever ready to warn each other of danger, and to assist and defend each other, would without doubt succeed the best and conquer the other. The advantage which disciplined soldiers have over undisciplined hordes follows mainly from the confidence which each soldier has in his comrades. Obedience is of the highest importance, for any form of government is better than none. Selfish and contentious people will not cohere, and without coherence nothing can be effected. Thus, a tribe possessing these qualities in an eminent degree would spread and be victorious over other tribes. But, in the course of events, or all past history is a myth, this successful tribe would in its turn be overcome by some other more highly-endowed tribe; and thus would the social and moral qualities tend slowly to advance and be diffused throughout the world.

Praise and the blame of our fellow-men are much more powerful stimuli to the development of the social qualities. These virtues are primarily due to the instinct of sympathy, and this instinct, like all other social instincts, was doubtlessly acquired through Natural Selection. How early man's progenitors, in the course of their development, became capable of feeling and being impelled by the praise or blame of their fellow-men, we are unable to say. Even dogs appreciate encouragement, praise and blame, and it would be strange if such could not be predicated of beings higher in the scale. The wildest savages feel the sentiment of glory. This is clearly shown by their preservation of the trophies of their bravery, by their habit of excessive boasting, and even by the extreme care they take of their personal appearance and adornments. Unless, however, they regarded the opinion of their comrades, such habits would be without meaning and senseless. How far the savage experiences remorse, is doubtful. He certainly feels shame and contrition for the breach of some of the lesser rules of his tribe. It is true that remorse is a deeply-hidden feeling, but it is hardly credible that a being who will sacrifice his life rather than betray his tribe, or give himself up as a prisoner rather than violate his parole, would not feel remorse, though he might, if he failed in a duty which he held sacred, hide it from view.

Primeval man must have been, at a very remote time, influenced by the praise and blame of his fellows. That the members of the same tribe would approve of conduct that appeared for the general good, and reprobate such as seemed to carry with it evil, there can be no question. To do good unto others, or to do unto others as you would that they should do unto you, is the foundation-stone of morality. It is, therefore, hardly possible to place too high an estimate upon the importance of the love of praise and fear of blame during rude, barbaric times, for a man, who was not impelled by any profound instinctive feeling to sacrifice his life for the good of others, but who was raised to such a noble action by a sense of glory, would by his example excite a similar wish for glory in the bosoms of other men, and would thereby engender and strengthen by exercise the laudable feeling of admiration. With increased experience and reason, those more remote consequences of his actions, such as temperance, chastity, etc., which during his very early times were utterly disregarded, would come to be highly esteemed or even held sacred. And ultimately there would have been developed from the social instincts a highly-complex sentiment which, largely guided by the approbation of his fellow-men, and ruled by reason, self-interest, and latterly by deep religious feelings, confirmed by teaching and habit, would constitute his moral sense or conscience. Although a high standard of morality gives but little if any advantage to each individual man and his children over the other men of the same tribe, yet it must be borne in mind that it is an advancement in the standard of morality and an increase in the number of well-endowed men that certainly give a telling advantage to a tribe over another, for the tribe that includes many members who, from possessing in an eminent degree the spirit of patriotism, fidelity, obedience, courage and sympathy, and who were always prepared to give aid to each other, and to sacrifice themselves for the common weal, would be victorious over most other tribes. And this would be Natural Selection. Tribes at all times throughout the world have supplanted other tribes. Now, as morality is one element in their success, the standard of morality and the number of well-endowed men will thus everywhere tend to rise and increase.

Very difficult it is to form any judgment why one particular tribe and not another has been successful in the Struggle for Existence and has risen in the scale of civilization. Many savages are still in the same condition of degradation as when first discovered. The greatest part of mankind has never evinced the slightest desire that their civil institutions should be improved. Progress is not, as we are apt to consider, the normal rule in human society. Many concurrent favorable conditions, far too complex to be followed out, seem to determine human progress. A cool climate, it has been remarked, by leading to industry and the various arts, has been indispensable thereto, but if the climate has been too severe, as in the Arctic regions, there is a check to continual progress. Pressed by hard necessity, the Esquimaux have succeeded in many ingenious inventions, but they can never attain, for the reason already assigned, to any very great success. Nomadic habits, whether along the shores of the sea, or over wide plains, or through dense tropical forests, have in all cases proved detrimental. Perhaps, the possession of some property, a fixed abode, and the union of many families under a leader or chief, are indispensable requisites for civilization, as such habits almost necessitate the cultivation of the ground. From some such accident as the falling of the seeds of a fruit-tree on a heap of refuse and producing an unusually fine variety may probably have resulted the first steps in cultivation, for if the fruit were profitable and good for food, it would be a very dull intellect that could not readily perceive, especially among a people that had given up a roving habit of life, the advantage which would accrue from the planting of some more trees of a similar kind. They would undoubtedly be led to cultivation for themselves by a simple observation of the plan by which nature contrives in keeping up a continuation of her many kinds of plants. Instead of dropping the seeds upon the ground as nature is prone to do, and trusting to their burial by accident or otherwise, seeing the advantage to be gained by burying them out of the reach of noxious influences, whether of climate or animal life, they would soon learn to take the matter of planting under their own watchful care rather than leave it to the seemingly thoughtless provision of nature. But the problem of the first advance of palæolithic man toward civilization, is at present much too difficult to be solved, for it involves the consideration of certain elements which we know too little about, and their disentanglement from others whose value is of recognized significance in the domain of biological science.

While it has been shown how it has been possible for primeval man to have acquired a moral sense or conscience, yet it must not be forgotten that the lower animals, at least such as have come under the civilizing influence of man, have also come into possession of the same highly complex sentiment which has been of such inestimable service to man for his progressive advancement. Other faculties, such as the powers of imagination, wonder, curiosity, an undefined sense of beauty, a tendency to imitation, and the love of excitement or novelty, have also been of immense importance in this direction, for they could not fail to have led to the most capricious changes of customs and fashions. Caprice, it has been rather oddly claimed by a recent writer, is "one of the most remarkable and typical differences between savages and brutes." It is not only possible to perceive how it is that man is capricious, but the lower animals, as has been previously shown, are capricious in their affections, aversions and sense of beauty. And there is good reason to suspect that they love novelty for its own sake. Self-consciousness, individuality, abstraction, general ideas, etc., which have been held by several recent writers as making the sole and complete distinction between man and the brutes, seem useless subjects for discussion, since hardly any two authors agree in their definitions of these high faculties. In man, such faculties could not have been fully developed until his mental powers had advanced to a high state of perfection, and this implies the use of a highly-developed language. No one supposes that one of the lower animals reflects whence he comes or whither he goes, or what is death or what is life, but can one feel sure that an old dog with an excellent memory, and some power of imagination as shown by his dreams, never reflects on his past pleasures in the chase? And this would be a form of self-consciousness. On the contrary, as Büchner ably remarks, how little can the hard-worked wife of an Australian savage who scarcely uses any abstract words and whose ability to count does not extend beyond four, exert her self-consciousness, or reflect on the origin, nature and aim of her own existence. That animals retain their mental individuality is unquestioned, for when any voice awakens a train of old associations in the mind of some favorite dog, as in the case of my dog Frisky, already referred to, he must have retained his mental individuality, although every atom of his brain had probably undergone change more than once during the five or six years he lived in my family. Animals have some ideas of numbers. The crow has been known to count as far as the number six, and a dog I once had knew as well as I did when Saturday came. The sense of beauty, which has been declared peculiar to man, is innate in birds. Certain bright colors and certain sounds, when in harmony, excite in them pleasure as they do in man. The taste for the beautiful, at least so far as female beauty is concerned, is not of a special nature in the human mind, for it differs widely in the different races of man, and is not quite the same even in the different nations of the same race. If we are to judge from the hideous ornaments and the equally hideous music admired by most savages, it might be urged that their æsthetic faculty was less highly developed than it is in some species of birds. No animal, it is obvious, would be capable of admiring the nocturnal heaven, a beautiful landscape, or refined music. And this should not be wondered at, for such high tastes, dependent as they are upon culture and complex associations, are not even enjoyed by barbarous or by uneducated persons.

Seeing that man in a state of nature has no preëminence above the lower animals so far as his mental and moral qualities are concerned, and in many instances ranks far below the so-called brute, let us examine fora short time his religious nature. No evidence exists to show that man was aboriginally endowed with the ennobling belief in the existence of an Omnipotent God. On the contrary, ample evidence, not from hasty travellers, but from men who have long resided with savages, can be adduced to show that numerous races have existed, and still exist, who have no idea of one or more gods, and who have no words in their languages to express such an idea. If under the term religion is included the belief in unseen or spiritual agencies, the case is entirely different, for this belief seems to be almost universal with the less civilized races. Nor is it difficult to understand how it originated. With the development of the imagination, wonder and curiosity, and of a moderate power of reasoning, man would naturally have craved to understand what was going on around him, and even have vaguely speculated on his own existence. According to McLennan man must, in his efforts to arrive at some explanation of the phenomena of life, feign for himself. Judging from the universality of this life, the same author remarks that "the simplest hypothesis, and the first to occur to men, seems to have been that natural phenomena are ascribable to the presence in animals, plants and things, and in the forces of nature, of such spirits prompting to action as men are conscious they themselves possess." Probably, as has been clearly shown by Tyler, dreams may have first given rise to the notion of spirits. Savages do not readily discriminate between subjective and objective phenomena. When a savage dreams, the figures which appear in his vision are believed to have come from a distance and to stand over him, or the soul of the dreamer goes out on a journey and returns with a remembrance of what has been seen. That tendency in savages to imagine that natural objects and agencies are animated by living or spiritual beings may be illustrated by a little fact which I have frequently noticed. Standing on the corner of a street, waiting for a closed snow-sweeper, which was driven by electricity, to pass, my attention was directed to a young horse that was geared to a hansom. The horse was at rest, and its driver, evidently awaiting some one, sat upon the box. Upon the appearance of the sweeper the horse reared, turned his face directly toward the object of his fear, pawed the pavement in the most impatient manner possible, and then looked wistfully and pleadingly at his master, as though imploring protection from some fearful and gigantic monster. Another sweeper passed while I was still in waiting, and the poor animal went through the same trying and fearful ordeal as before. He must, I think, have reasoned in a rapid and unconscious manner, that movement without any apparent cause indicated the presence of some strange living agent, which was about to do him some serious physical harm. Belief in spiritual agencies would thus easily pass into a belief in the existence of one or more gods, for savages would naturally ascribe to spirits the same passions, the same line of vengeance or simple form of justice, and the same affections which they themselves experienced.

Religious devotion is a highly complex feeling. Love, complete submission to an exalted and mysterious superior, a strong sense of dependence, fear, reverence, gratitude, hope for the future and other elements enter into its composition. No being could experience so complex an emotion unless his intellectual and moral faculties had attained a moderately high level. Some approach to this high state of mind is visible in the profound love of a dog for his master, for it is associated with complete submission, some fear, reverence, gratitude and perhaps other feelings. A dog's behavior towards his master, after a long absence, is widely different from that which he shows towards his fellows, for his transports of joy in the latter case are less intense, and his every action savors of a mere sense of equality. But upon his master, as Prof. Braubach goes so far as to maintain, he looks as on a god.

These high mental faculties, which first led man to believe in unseen spiritual agencies, and subsequently in fetishism, polytheism and monotheism, would infallibly lead him, as long as his reasoning powers remained at a very low level, to various strange superstitions and customs, many of which, such as the sacrifice of human beings to a blood-loving god and the trial of innocent persons by the ordeal of poison or fire, are too terrible to contemplate. It is well, however, to reflect occasionally on these superstitions, for they show us what an infinite debt of gratitude we owe to improved reason, science and accumulated knowledge. How much better is the life of civilized man than that of the savage, for as Lubbock has well remarked, "it is not too much to say that the horrible dread of unknown evil hangs like a thick cloud over savage life, and embitters every pleasure."

From the opinions advanced, it is evident that the belief in God has been the ultimate outcome of belief in unseen spiritual agencies. There has been a gradual leading up through fetishism and polytheism to monotheism. If religion implies belief in unseen agencies, as well as belief in a personal agency in the universe strong enough to influence conduct in any degree, then it is obvious that there has been a progressive advancement in religious thought, each succeeding form of religion by its superior advantages over its predecessor tending to supplant it wherever and whenever its beneficent influences are felt. It is true that fetishism and polytheism still prevail among rude, uncultured peoples, as well as the worship of false deities and prophets, but with the spread of the civilizing and elevating influence of Christianity these religions in the fitness of time will disappear. Christianity, from its foundation in Judaism, has throughout been a religion of sacrifice and sorrow. It has been a religion of blood and tears, and yet one of profoundest happiness to its votaries. While fakirs hang on hooks, and pagans cut themselves and even their children, for the sake of propitiating diabolical deities, yet Christianity, which has its roots in Judaism, has no need for such practices. It is *par excellence* the religion of sorrow, because it reaches to truer and deeper levels of our spiritual nature, and therefore has capabilities both of sorrow and joy which are presumably non-existent except in civilized man. They are the sorrows and joys which arise from the fully-developed consciousness of sin against a God of Love, as distinguished from propitiation of malignant spirits. These joys and sorrows are wholly spiritual, not merely physical. "Thou desirest no sacrifice." God's only sacrifice at the hands of sinful man is a troubled spirit.

Estimated by the influence which He has exerted on mankind, there can be no question, even from a secular point of view, that Christ is much the greatest man who has ever lived. That the revolution which His teachings have effected in human life is immeasurable and unparalleled by any other movement in history is unquestioned. Though most nearly approached by the religion of the Jews, of which it is a development, so that it may be regarded as of a piece with it, it is evident that this whole system of religion is so immeasurably in advance of all others that it may be truthfully said, if it had not been for the Jews, the human race would have had no religion worthy of serious consideration. Had it not been for this religion man's spiritual side would not have been developed in civilized life. And although there are numberless individuals who are all unconscious of its development in themselves, yet these have been influenced to an enormous extent by the religious atmosphere by which they are surrounded.

Not only is Christianity so immeasurably in advance of all other religions, but it is no less of every other system of thought that has ever been promulgated in regard to what is moral and spiritual. Neither philosophy, science nor poetry has ever produced results in thought, conduct or beauty in any degree comparable with it. What has science or philosophy done for the thought of mankind compared with what has been done by the single doctrine, "God is love?" The Story of the Cross, from its commencement in prophetic aspiration to its culmination in the Gospel, is preëminently the most magnificent presentation in literature. Only to a man wholly destitute of religious perception can Christianity fail to appear the greatest exhibition of the beautiful, the sublime, and of all else that appeals to our spiritual nature, which has ever been known upon the earth. It is not only adapted to men of the highest culture, but the most remarkable thing about it is its perfect adaptation to all sorts and conditions of men. Its problems, historical and philosophical, open up to you worlds of material, over which you may spend your life with the same interminable interest as the student meets in the fields of natural science.

Whatever our theory of the origin of man, there can be no doubt that we all feel that his intellectual part is higher than the animal; and that the moral is higher than the intellectual, whatever our theory of either may be; and that the spiritual is higher than the moral, whatever our theory of religion may be. It is what is understood by his moral, and still more by his spiritual qualities, that make up what is called his character, and, astonishing to say, it is character that tells in the long run. Morality and spirituality are two different things, for a man may be highly moral in conduct without being in any degree spiritual in nature, and the reverse, though to a less extent. Objectively, the same distinction subsists between morals and religion. Intellectual pleasures are more satisfying and enduring than sensual, or even sensuous; and spiritual, to those who have experienced them, than intellectual, an objective fact, abundantly testified to by those who have had experience, which seems to indicate that the spiritual nature of man is the highest part of man—the culminating point of his being. That there will always be materialists and spiritualists, as Renan says, is probably true, inasmuch as it will always be observable on the one hand that there is no thought without brain, while, on the other hand, the instincts of man will always aspire to higher beliefs. If religion is true, and life is a state of probation, this is just what ought to be. It is not probable that the materialistic position, which is discredited even by philosophy, is due simply to custom and a want of imagination. Else why the inextinguishable instincts which we have thus shown to exist?

Figure 82. Era of Mind and Heart. Things as they will exist in a future earth-life.

Evolution, not only of the earth, but of its organic machinery, by natural causes, is now no longer doubted. That this has taken place by degrees is equally unquestioned. Now, if there is a Deity, the fact is certainly of the nature of a first principle, and it must be first of all first principles. No one can dispute this, nor can any one dispute the necessary conclusion that, if there be a Deity, he is knowable, if knowable at all, by intuition and not by reason. From its very nature, as a little thought is sufficient to show, reason is utterly incapable of adjudicating on the subject, for it is a process of inferring from the known to the unknown. It would be against reason itself to suppose that Deity, even if He exists, can be known by reason. He must be known, if knowable at all, by intuition. If there is a Deity, then it seems to be in some indefinite degree more probable that He should impart a Revelation than that He should not have done so. As a mere matter of evidence, a sudden revelation might be much more convincing than a gradual one, but it would be quite out of analogy with causation in nature. Besides, a gradual one might be given easily, and of demonstrative value, as by making prophecies of historical events, scientific discoveries and other things so clear as to be unmistakable. But a demonstrative revelation has not been made, and there may well be good reasons why it should not have been made. If there are such reasons, as, for example, our state of probation, we can well see "that the gradual unfolding of a plan of revelation, from earliest dawn of history to the end of the world, is much preferable to a sudden manifestation sufficiently late in the world's history to be historically attested for all subsequent time." Gradual evolution, as has been said before, is in analogy with God's other work. If Revelation has been of a progressive character, then it follows that it must have been so not only historically, but intellectually, morally and spiritually, for in such sequence could it be always adapted to the advancing conditions of the human race.

Thus it will be seen that all through the ages some mighty influence has been at work, directly or indirectly, in preparing this earth by slow and gradual changes for a steadily progressive succession of vegetable and animal life. That life best fitted to meet new and changing conditions of environment being preserved by a process of natural selection. And from a few primordial types, far simpler than the lowest of existing structureless moners, or from some living protoplasmic mass, elaborated by some form of energy acting upon inorganic nature, there have been evolved in the millions of years of earth-life our existing flora and fauna. Man, the pinnacle of animal life, has come up through the life that preceded him, and bears in the history of his development from the ovum to the adult state the line of his descent. Not only has his physical nature been evolved through the action of natural laws impressed upon living matter by Deity, but that subtle principle, termed mind, which has attained such a wonderful growth in his civilized condition, is but the outcome of the mind of a long line of life antecedent to his appearance on the globe. His moral nature was similarly acquired, and most probably in the manner already explained. Palæolithic man, like the Australian of to-day, was, as has been shown, but little superior in intelligence to some of the animals with whom he was contemporaneous. He lived the life of the mere animal, and as an animal could be said to have had no preëminence above a beast. Like the latter, he was a living, breathing frame, or body of life; a *living*, but not an *everliving*, soul. In time, as conditions became favorable, he passed *into* the moral stage of his being, but not without increased intellectuality, and would thus have continued, but going on and adding to his mental and moral possessions, had not Deity, in the fitness of time, prepared the way through Christ, whereby his corruptible nature should be made incorruptible and immortal. Unless man is "born of the spirit" he cannot inherit the kingdom of God. He must be "changed into spirit," put on incorruptibility and immortality of body, or he will be physically incapable of retaining the honor, glory and power of the kingdom forever, or even during Christ's reign of a thousand years upon earth.

That there is a distinction between *a living soul* and a *spiritual body* cannot be questioned. Speaking about *body*, the apostle Paul says, "there is *a natural body*, and there is *a spiritual body*, but he does not content himself with simply declaring this truth, but goes further and proves it by quoting the language of Moses, saying, "for so it is written, the first man Adam was made into *a living soul*;" and then adding, "the last Adam *into a spirit* giving life." And in another place, speaking of the latter, he says of Him, "now the Lord is the spirit. And we all, with unveiled face, beholding as in a mirror, the glory of the Lord, are changed into His image from glory into glory, as by *the Lord the Spirit*." Therefore, the proof of the apostle's proposition, that there is a *natural body* as distinct from a *spiritual body*, lies in the testimony that "Adam was made into *a living soul*," showing that he considered a natural, or animal body, and a living soul, as one and the same thing. If he did not, then there was no proof in the quotation of what he affirmed. Mortality, then, is life manifested through a corruptible body, and immortality is life manifested through an incorruptible body. Hence, the necessity laid down in the saying of the apostle, "this corruptible body must put on incorruption, and this mortal put on immortality," before death can be "swallowed up in victory,"—a doctrine of "life and incorruptibility" that was new to the Greeks and Romans, and brought to light only through the gospel of the kingdom and name of Jesus Christ. To them it was foolishness, and to many at the present day incredible, because they do not understand the glad tidings of the age to come. God could have created all things upon a spiritual or incorruptible basis at once, but in that case the globe would have been filled with men and women equal to the angels in nature, power and intellect, and hence would have been without a history, and its population characterless. And this would not have been according to His plan, for in it the animal must precede the spiritual just as surely as the acorn must precede the oak. The Bible has to do with things and not with imaginations; with bodies and not phantasms; with *living souls* of every species; with *corporeal* beings of other worlds, and with incorruptible and undying men, but is as silent as the grave about such *souls* as men pretend to cure. For the sons of Adam to become sons of God, they must be the subjects of an adoption, which is attainable only by a divinely appointed means. It must be by a process of selection. "Since by a man came death, by a man also came a resurrection of dead persons. For as in the Adam they all die, so also in the Christ shall they all be made alive. But every one in his order. Christ the first fruits; afterward *they that are Christ's* at His coming." Here it is obvious that the apostle is not writing of all the individuals of the human race, but only such that become the subject of *a pardon of life*. It is true that all men do die, but it is not true that they are all the subject of pardon. Those who are pardoned are "the many," who are sentenced to live forever. The sentence to pardon of life is through Jesus Christ who in pouring out His blood upon the cross, was made a sacrifice for sin. "He was delivered for our offences, and raised again for our justification," that is, for the pardon of those *who believe the gospel*. As it is written, "he that believeth the gospel, and is baptized, shall be saved." Hence, "*the obedience of faith*" is made the condition of righteousness, and this obedience implies the existence of a "*law of faith*," as attested by that of Moses, which is "*the law of works*." Having believed the gospel and been baptized, such a person is required to "walk worthy of the vocation," or calling, "wherewith he has been called," that by so doing he may be "accounted worthy" of being "born of spirit," that he may become "spirit," or a spiritual body, and so enter the kingdom of God, crowned with "glory, honor, incorruptibility and life." From all the above, it must be obvious to the unbiassed mind, that all will not arise to newness of life, "for as many of you, as have been *baptized into Christ* have put on Christ, and if ye be Christ's, *then* are ye the seed of Abraham, and *heirs* according to the promise." When they have been thus baptized, then they have received the spirit of adoption, or have been elected into God's family, and then they can address God as their Father who is in heaven.

Thus adopted into God's family through faith in Jesus Christ, it must not be supposed that they have attained to that perfect condition of knowing all that is to be known. New glories will continually open up to their admiring vision, and new facts be revealed through the eternity of futurity. Man will carry his earth-acquired knowledge into the other world, and little by little will he add to his fund. Those who have made the best of their time in their probationary existence, will rank as much above their fellows in the heaven-life as they did in the earth-life, and like the others will reach up to higher acquirements. There will be no equalization of talents, capacities and possessions, but each will be satisfied with his own, and all will endeavor to be as like unto Christ as the conditions of their heavenly environment will permit. There will be grades of ability and character in the new life, but all of the very highest standard when measured by what prevails in the earth-life. This is the teaching of the Scriptures. "*There is* one glory of the sun, and another glory of the moon, and another glory of the stars; for *one* star differeth from *another* star in glory. So also *is* the resurrection of the dead."

Now as to the part that animals and plants shall figure in the new existence. Revelation, as has been seen, was given to man. This does not imply that the lower forms of life were not made "partakers of the divine nature." When man was placed upon this earth, or rather when in the sequence of events, which was brought about by the prescribed scheme of Divinity, he appeared upon the earth, he was given the control of all the creatures of God's hands, to rule them as his judgment seemed best. They were a necessary part of the plan of creation. God gave the man directions concerning them, and what they are, and we refer to the domesticated species especially, they have thus been made through man's wise, intelligent and thoughtful selection. This has been the instrument through which God has worked in building up a history and a character for the humbler works of His hands. That they shall pass into the future life with him, at least such as have shown their fitness to endure, there can be no doubt in the mind of any one who pauses a few brief moments in the rush and turmoil of everyday life and considers the matter with all due seriousness. All existence, as we have elsewhere claimed, is a unit. All life, like all love, is divine. There can nothing exist that does not contain some sort of development of soul. There is no escape from this assertion. Instead of isolating ourselves then from the humbler creatures of God's workmanship, let us recognize them as our kin and include them in the grand scheme of redemption, and as partakers with us in the future state of Divine Love and in higher and endlessly higher development and progress.

Chapter 47

MAN'S PREEMINENCE

There is a popular tradition that somewhere in the Scriptures we are taught that of all living denizens of the earth, man alone possesses a spirit, and that he alone survives in spirit after the death of the material body. Were this the truth, no room would exist for argument to those who profess belief in a literal rendering of the Scriptures, and who base their faith upon that literal belief. However much such a statement might seem to controvert all ideas of benevolence, justice and common-sense, such believers would feel bound to accept it on trust, and to wait a future time for its full comprehension.

Even the possession of reason is denied by many persons to animals, their several actions being ascribed to the power of instinct, and it is therefore not the least bit strange that all but a comparatively few should believe that when an animal dies, its life-principle dies too. The animating power, they claim, is annihilated, while the body is resolved into its constituent elements so as to take form in other bodies.

Two passages of Scripture, one in the Psalms and the other in Ecclesiastes, are almost entirely, if not wholly, responsible for this belief. The former, which runs in the authorized version, "Nevertheless, man being in honor, abideth not; he is like the beasts that perish," is that which is generally quoted as decisive of the whole question. "Man, being in honor, hath no understanding, but is compared to the beasts that perish" is another translation, but differs not materially from the other. The second passage referred to from Ecclesiastes, reads: "Who knoweth the spirit of man that goeth upward, and the spirit of the beast that goeth downward to the earth?" Now, it is upon the strength of these two passages that we are called upon to believe that when a beast dies its life, like that of an expired lamp, goes out forever. Nothing is more dangerous in the exposition of Scripture than attempting to explain a passage, however simple it may seem to be, without reference to the original text, for the translator may have mistaken the true sense of the words, or he may have inadequately expressed their signification, or, owing to a change in meaning, the words of a passage may now bear an exactly contrary sense to that conveyed when they were first written.

But laying aside this point for the present, and accepting the passage as it stands, as well as the literal meaning of the words as generally understood, there can be no doubt that we must believe that beasts are not possessed of immortal life. If, however, we are to take the literal sense of the Bible, and no other, we are equally forced to believe that man has no life after death. The book of Psalms is full of examples. Let us take a few from the many that might be given: "In death there is no remembrance of thee: in the grave, who shall give thee thanks?" "The dead praise not the Lord, neither any that go down into silence." "His breath goeth forth, he returneth to his earth; in that very day his thoughts perish." Taken solely in their literal sense, there can be no doubt of their meaning. Nothing more gloomy, dreary or more despondent can be found in the entire range of heathen literature than these passages, and others that might be quoted from the inspired Psalmist, in the contemplation of death. In the very book from which the single passage was taken, which is claimed to deny immortality to the lower animals, there are five times as many passages that proclaim the same sad end to the life of man. We are distinctly and definitely told therein that those who have died have no remembrance of God, and cannot praise Him. Death has been spoken of as the "land of forgetfulness"—the place of darkness, where all man's thoughts perish. Certainly no more than this can be said of the "beasts that perish."

Other holy writers make similar affirmations. Speaking of mankind in general, who "dwell in houses of clay," Job says: "They are destroyed from morning to evening; they perish forever, without any regarding it." Again he says, and the passage is more definite than the preceding: "As the cloud is consumed and vanisheth away, so he that goeth down to the grave shall come up no more." And still again: "Man dieth, and wasteth away: yea, man giveth up the ghost, and where is he? As the waters fail from the sea, and the flood decayeth and drieth up: so man lieth down, and riseth not." Chapters III and X tell of the piteous lamentations of Job over his life, wherein he complains that he ever was born, that existence was ever given to him, that he was ever taken from a state of absolute nonentity, and that even death itself can bring no relief to his miseries except extinction.

Turning to Ecclesiastes, in which book occurs the solitary passage which is held to disprove a future existence to the lower animals, there are passages which are even more emphatic as to the immortality of man. Read what is declared: "I said in my heart concerning the estate of the sons of men, that God might manifest them, and that they might see that they themselves are beasts. For that which befalleth the sons of men befalleth beasts; even one thing befalleth them. As the one dieth, so dieth the other; yea, they have all one breath, so that a man has no preëminence over a beast: for all is vanity. All go unto one place; all are of the dust, and all turn to dust again." Further it is said: "For the living know that they shall die, but the dead know not anything, neither have they any more a reward, for the memory of them is forgotten." "Whatsoever thy hand findeth to do, do it with thy might; for there is no work, nor device, nor knowledge, nor wisdom in the grave whither thou goest." Literally interpreted, no one can doubt the import of these words from Ecclesiastes, for they definitely state that, as regards a future life, there is no distinction between man and beast, and that when they die they all go to the same place. It is also distinctly stated that after death man can do no work, know nothing, nor receive any reward. Were we to deduce our ideas of the condition of man after death from the irrepressibly sad and gloomy passages from Job and Ecclesiastes, most deplorable and hopeless would be the very thought of dissolution. But we do not accept them in this light. They are written symbolically, and there underlies them a spiritual sense. It is not, however, the latter sense that concerns us at present, but the literal meaning of the translation, and, according to that literal meaning, if we take two texts to prove that beasts have no future life, we are compelled by no less than fourteen passages to believe that man, in common with beasts, has no better prospect. We have no right to say which passages are to be taken literally, and which parabolically, but must apply the same test to all alike, and treat all in a similar manner.

All classical readers are familiar with that wonderful eleventh book of Homer's Odyssey, called the Necyomanteia, or Invocation of the Dead, in which Ulysses is depicted as descending into the regions of departed spirits for the purpose of invoking them and obtaining advice as to his future adventures. Dreary, and horrible indeed, are the revelations which the whole of the strange history makes of the condition of the future life. All is wild and dark, and hunger, thirst and discontent prevail. Nothing is heard of elysian fields, where piety, wisdom and virtue abound. Gloom, misery and vain regrets for earth pervade the entire episode. When is considered this heathen poet's ideas concerning the future state of man, it is no wonder that sensual pleasures should be held as the principal object of his life when he is to look forward to such a future, a future from which neither wisdom, nor virtue, nor piety could save him, and where there is nothing but an eternity of gloom, remorse and hopeless despondency. Sad as this picture is, yet it is far brighter than that of the Psalmist, the Preacher, or Job. Those who have passed into the world of spirits still retain their individuality after death, being distinguished in the spirit as they had been in the flesh. Memory survives the body's death. Naught of their earthly career is forgotten. They still have an interest in their friends that remain in the body whom they love, and over whose well-being they unceasingly watch. No such consolation, as has been described, exists in the future state of man if the passages of Scripture that have been quoted are taken in a literal sense. Man, in that event, passes at death into a place of darkness, forgetfulness and silence, where there is no work, nor device, nor knowledge, nor wisdom, and where even his very thoughts perish. No other interpretation, if taken literally, can be put upon them, for the statements are too explicit to be explained away or softened.

In the outward sense of their writings the Psalmist, Job and the Preacher are on an equality with Horace in their absolute unbelief in a future existence, and in a consequent desire to snatch what fleeting pleasures they can from earth before the inexorable law of fate consigns them to dark oblivion. Startling as it may seem to compare the teachings of a

Greek idolater and of a Latin Epicurean heathen with those of sacred writers, yet it is still more startling to show that the teachings of the Epicurean sensualist are not a whit wiser than those of the Scriptural writer, while those of the Greek poet are very much better. Such, however, is the fact, and, if we are to be bound by the literal interpretation of the Scriptures, there is no possibility of denying it without doing violence to reason and common-sense.

We are now brought face to face with the point previously mentioned. Does the authorized version give a full and correct interpretation of the original? It is claimed that it does not. The word "perish," it is said, does not occur at all in the Hebrew text, nor is even the idea expressed. No such translation as "beasts that perish," which appears twice in our version, is justified by the Hebrew, the words of the original implying "dumb beasts." The idea of perishing, in the sense of annihilation, does not seem to be implied. Let us take the Jewish Bible, which is acknowledged to be the best and closest translation in the English language, and examine it. Both in verses 12 and 20 of Psalm XLIX, where the passage occurs, the rendering reads: "Man that is in honor, and understandeth this not, is like the beasts that are irrational." As an alternative reading for "irrational," the word "dumb" is given in a footnote. A somewhat similar reading is found in the Septuagint, which, according to Brunton, runs as follows: "Man that is in honor understands not; he is compared to the senseless cattle, and is like them." In Wycliffe's Bible, which is a translation from the Vulgate, the passage is rendered: "A man whanne he was in honour understood not; he is comparisoned to unwise beestis, and is maad lijk to tho." The "Douay" Bible, made by the English Roman Catholic College of Douay, and which is the version accepted by that branch of the Church in England, renders the passage: "Man, when he was in honor, did not understand; he hath been compared to senseless beasts and made like to them." Numerous other translations might be adduced, and it is safe to say that scarcely any of them imply the idea of perishing in the sense of being reduced to nothing. Even supposing that the word "perish" is translated correctly, it does not therefore follow that annihilation is meant. Take the tenth verse of the same Psalm in our authorized version: "For he seeth that wise men die, and likewise the fool and the brutish person perish, and leave their wealth to others." Surely no sensible, intelligent person would construe this passage into a declaration that the wise and fool and brutish had no existence after the death of the body.

That the last verse of the Psalm is a summary of the whole poem, seems not improbable. A vivid picture of the true object of man's life in this world is drawn by the Psalmist, and also of his tendency to lose sight thereof. In it he sets forth the shortness of human existence, and shows that neither riches, station in life, nor fame, which appertain to the mere earthly career of man, can endure after his death. He, therefore, reasonably concludes that men who fix their hearts upon these earthly vanities ignore the honor of their manhood, and degrade themselves to the plane of the dumb beasts, whose operations are, as far as we know, restricted to this present world.

From what has been adduced it will at once be evident that the idea that beasts are said by the Psalmist to have no future life may be dismissed from our minds, and that the passage may be rejected as totally irrelevant to the subject. This is of the greatest importance, as the passage in question is the only one which even appears to make any definite statement as to the condition of the lower animals after death. Every reasonable person will now see how essential it is that the true meaning of the Hebrew text should be known, and that the Psalmist should not be charged with the introduction of a doctrine to which, whether true or false, he makes not the slightest reference.

Having settled beyond the possibility of refutation the true meaning implied by the "beasts that perish," we will now turn to the passage in Ecclesiastes, which, as has been seen, is the only one which contains any direct reference to the future of the lower orders of animal existence: "Who knoweth the spirit of man that goeth upward, and the spirit of the beast that goeth downward to the earth?"—exclaimeth the Preacher. Here we have an admission that, whether the spirit ascend or descend, both man and beasts do have spirits, and these are undoubtedly the same in essence, for the Hebrew word is identical is both cases. In the Jewish Bible the rendering is verbatim the same as that of our authorized version. Read, instead of an isolated verse, the entire passage:—

"I said in mine heart concerning the estate of the sons of men, that God might manifest them, and that they might see that they themselves are beasts.

"For that which befalleth the sons of men befalleth beasts; even the one thing befalleth them: as the one dieth, so dieth the other; yea, they have all one breath; so that a man hath no preëminence above a beast: for all is vanity.

"All go to one place; all are of the same dust, and all turn to dust again.

"Who knoweth the spirit of man that goeth upward, and the spirit of the beast that goeth downward to the earth?

"Wherefore I perceive that there is nothing better than that a man should rejoice in his own works; for that is his portion; for who shall bring him to see what shall be after him?"

Every page of Ecclesiastes breathes of the self-reproach of the Preacher for a wasted life. Speaking from his own sad, bitter experience, he shows that riches, glory, pleasure and even wisdom are nothing but utter emptiness. The same theme pervades the forty-ninth Psalm, but the Psalmist treats it with grave solemnity, admonishing his hearers of the shortness of human life, and showing that if a man forgets the glory of his manhood, made in the image of God, he puts himself on the level of the dumb brutes. Though reaching the same conclusion, yet the Preacher views the subject from a different standpoint. Employing biting sarcasm rather than solemn warning, he exposes the vanity of all worldly and selfish pleasures, and the miserable fate that awaits the voluptuary, and then ironically advises his readers to place in such their entire happiness.

So palpable is the bitter irony of the author throughout the book, and even in the twenty-first verse of the third chapter, yet by no manner of interpretation can this specialized text be made to mean that beasts are annihilated after death, while men rise again and soar above earthly things to honor and glory. Ironically the writer assumes in it that his readers do not know the difference between the spirit of man and that of beast, and, reasoning from that position, advises them that "there is nothing better for a man than that he should eat and drink, and that he should make his soul enjoy good in his labor."

From what has been shown, it is evident that the passage from Psalms does not even contain the idea of annihilation as regards beasts, and that the one from Ecclesiastes is entirely misapprehended. That they have no bearing upon the subject must now be manifest. We cannot, therefore, resist the conclusion that the Scriptures do not deny future life to the inferior animals.

This admission gives courage for a step still further forward. Man's latest achievement is to conceive that all existence is a unit. One spirit pervades the whole natural world, an emanation from the Spirit of Him who sitteth enthroned in the Eternal Heavens, and who not only is, as Moses declares, "God of the spirits of all flesh," but God of the spirits of all animate nature. We cannot divorce the two great kingdoms of nature. If there is a futurity of existence for man, whom we are told was "made a little lower than the angels," but who in these latter days seems to have deteriorated, and who in thousands of instances displays a character far less noble and honorable than that of the dog which he kennels and feeds, then there must be for the so-called brute, the companion of his joys and his sorrows. If for beast, bird, reptile, fish and insect, and none can be so foolish in the face of the most indubitable evidence to deny it, then there must be for tree, shrub and flower, for God, who is infinite in love, mercy and charity, would not be God if solely concerned with the future of the smallest fractional part of His children. Man is psychically related to all life. There is soul, in some sort of development, in everything; and certainly God meant in His grand scheme of redemption to lift the world, not a portion of it, but the entire world, out of its lower ideas into its higher beauties and realities.

Chapter 48

FUTURE LIFE

That the Scriptures, contrary to popular tradition, do not deny a future life to the lower animals has already been conclusively shown. But do they declare anything in favor of another world for beast as well as for man? This is a question which we shall now endeavor to answer. As to man's immortality, the Old Testament Scriptures teach the doctrine by inference rather than by direct assertion, for the reason, as has been presumed, that the writers of the several books, which were selected at a comparatively late period from among many others and formed into the volume popularly designated the Bible, assumed as a matter of course that man was immortal, and therefore did not concern themselves about a matter which they supposed everybody knew. But as far as the Old Testament goes, inference tells more strongly in favor of the beast's immortality than that of man. Although in either case there does not appear to be any definite assertion of a futurity of existence, yet there is no such denial of the immortality of the beast as has already been shown in the case of the man.

Beasts, as readers of the Old Testament only too well know, were included in the merciful provision of the Sabbath, which, in its essence, was a spiritual and not simply a physical ordinance. And, again, we find many provisions in the ancient Scriptures against maltreating the lower animals, or giving them unnecessary pain, and these provisions stand side by side in the Divine Law with those which apply to man. All are familiar with the prohibition of "seething a kid in its mother's milk," and the non-muzzling of the ox in treading out the corn lest he should suffer the pangs of hunger in the presence of the food which he may not eat. Even bird's nesting was regulated by Divine Law. "If a bird's nest chance to be before thee in the way in any tree, or on the ground, *whether they* be young ones, or eggs, and the dam sitting upon the young, or upon the eggs, thou shalt not take the dam with the young: *But* thou shalt in any wise let the dam go, and take the young to thee; that it may be well with thee, and *that* thou mayest prolong *thy* days." Moreover, as many animals must be killed daily, some for sacrifice and others solely for food, the strictest regulations were enjoined that their death should be sharp and quick, and that the whole of their blood should be poured out upon the ground lest they suffer lingering pain.

In keeping with the same consideration felt by Deity towards the kid and ox and bird, as expressed in the Law, we would refer to the few concluding sentences of the Book of Jonah:—

> "Thou hast had pity on the gourd, for the which thou hast not labored, neither madest it grow; which came up in a night, and perished in a night.
>
> "And should I not spare Nineveh, that great city, wherein are more than six score thousand persons that cannot discern between their right hand and their left hand; *and also much cattle*?"
>
> "Every beast of the forest is mine," saith the Lord, "and the cattle upon a thousand hills." And again, "I know all the fowls of the mountains: and the wild beasts of the field are mine." Similar passages, in which God announces himself as the protector of the beast as well as of man, could be given, for the Scriptures are full of them. Who does not recall the well-known saying of our Lord respecting the lives of the sparrows: "Are not two sparrows sold for a farthing? and one of them shall not fall on the ground without the notice of your Father."

Cowper in his "Task," makes allusion to this branch of our subject in the following lines:—

> "Man may dismiss compassion from his heart,
> But God will never. When He charged the Jew
> To assist his foe's down-fallen beast to rise;
> And when the bush-exploring boy, that seized
> The young, to let the parent-bird go free;
> Proved He not plainly that His meaner works,
> Are yet His care, and have an interest all—
> All in the universal Father's love?"

One passage there is which certainly does point to a future for the beast as well as for man, and which places them both on the very same plane. It is found in Genesis, ninth chapter and fifth verse, and constitutes a part of the law which was delivered to Noah, and which was subsequently incorporated in the fuller law given through Moses. "And surely your blood of your lives will I require," said God to Noah and his sons, "at the hand of every beast will I require it, and at the hand of every man; at the hand of every man's brother will I require the life of man." In Exodus, chapter twenty-one and twenty-eighth verse, we read, "If an ox gore a man or a woman, that they die: then the ox shall be surely stoned, and his flesh shall not be eaten; but the owner of the ox *shall be* quit."

While there are no passages of Scripture, as has been seen, which deny immortality of life to the lower animals, yet there are certainly some which tend to show it by inference.

But the Scriptures were written for human beings, and not for the lower animals, and therefore it could hardly be expected that any information could be gained therefrom on the subject. As we find so few direct references to the future state of man, it is not at all to be expected that we should receive direct instruction upon the after-life of the beast.

But just as man has had within himself for untold ages an intuitive witness to his own immortality, yet there are those, lovers and friends of the so-called brute, who have an instinctive sense that animals, some of whom surpass in love, unselfishness, generosity, conscience and self-sacrifice many of their human brethren, must share with him in addition to these virtues an immortal spirit in which they take their rise. No more eminent personage than Bishop Butler was a believer in this idea. Substantially he asserts that the Scriptures give no reasons why the lower animals should not possess immortal souls. Similar sentiments have been voiced by equally distinguished writers.

Southey, writing of the death of a favorite spaniel that had been the companion of his boyhood, says:—

"Ah, poor companion! when thou followedst last
Thy master's parting footsteps to the gate
Which closed forever on him, thou didst lose
Thy best friend, and none was left to plead
For the old age of brute fidelity.
But fare thee well. Mine is no narrowed creed;
And He who gave thee being did not frame
The mystery of Life to be the sport
Of merciless man. There is another world
For all that live and move—a better one!
Where the proud bipeds, who would fain confine
Infinite Goodness to the little bounds
Of their own charity, may envy thee."

Thus does Lamartine, in "Jocelyn's Episode," beautifully express himself in addressing a faithful and affectionate canine by the name of Fido:—

"I cannot, will not, deem thee a deceiving,
Illusive mockery of human feeling,
A body organized, by fond caress
Warmed into seeming tenderness;
A mere automaton, on which our love
Plays, as on puppets, when their wires we move.
No! when that feeling quits thy glazing eye,
'Twill live in some blest world beyond the sky."

Not by man alone have these higher qualities been accorded to the brute. Women have praised the good within the lower animals, and been quite as willing to share with them the benefits of an immortal life. Eugenie de Guérin, a woman distinguished for her devotional piety, and an author of no mean repute, was, like the most of her sex, quite passionately fond of pets. Hers was a turtle-dove. Its voice was the first to greet her in the morning. There was a pleasure in its soft, gentle cooings, as they fell upon her ear, that sent a sweet consolation to her busy, thinking soul. But the time came at last when she must part with her treasure. The morn dawned bright, an August morning, and the bird was well and happy, but, with the falling of the shadows at even-tide, its little life went out. A bitter trial it was for the mistress, who loved with a perfect love her feathered friend. While wrestling with her intense sorrow, and after she had sincerely placed its mortal remains in a dainty cavity beneath the roses, it was that she wrote: "I have a tolerably strong belief in the souls of animals, and I should even like there to be a little paradise for the good and gentle, like turtle-doves, dogs and lambs. But what to do with wolves and other wicked animals? To damn them?—that embarrasses me."

Less devotional, perhaps, and looking rather to logic than to intuition, was the mind of Mrs. Somerville. With such a difference in constitution between the two women, we would naturally look for the greatest divergence of opinion upon a matter of this kind, but, astonishing to relate, there is noticeable a marked unanimity. Speaking of death, and the accompanying change of environing objects, this gifted writer, in her eighty-ninth year, says in her "Memoirs":—

"I shall regret the sky, the sea, with all their beautiful coloring; the earth, with its verdure and flowers; but far more shall I grieve to leave animals that have followed our steps affectionately for years, without knowing for certainty their ultimate fate, though I firmly believe that the living principle is never extinguished. Since the atoms of matter are indestructible, as far as we know, it is difficult to believe that the span which gives to their union life, memory, affection, intelligence and fidelity is evanescent.

"Every atom in the human frame, as well as in that of animals, undergoes a periodical change by continual waste and renovation: the abode is changed, not its inhabitant. If animals have no future, the existence of many is most wretched. Multitudes are starved, cruelly beaten, and loaded during life; many die under a barbarous vivisection.

"I cannot believe that any creature was created for uncompensated misery: it would be contrary to the attributes of God's mercy and justice. I am sincerely happy to find that I am not the only believer in the immortality of the lower animals."

To have given the many opinions that have been expressed by the good and wise of the past in favor of the belief that animals received, in common with man, a particle of the divine essence, and hence became immortal, would have extended this chapter beyond intended limits. We have room for just another witness. No one is better known for his convictions upon this subject than the late Dr. Wood, whose contributions to natural history are known the world over. Speaking of the death of his dog Rory, a creature that manifested in the flesh the strongest affection for his keeper, the Doctor says:—

> "I could not believe that an animal which would die of grief, as he died, for the absence of his master, would have his existence limited to this present world, and that such intensity of love should terminate at the same moment that the material heart ceased to beat."

When we think of the apparent inequality that is everywhere to be seen in the lives both of man and beast, we cannot believe, as Mrs. Somerville has remarked, that any being was "created for uncompensated misery." Some human beings are endowed with everything that a man can desire—health, strength, riches, accomplishments and capacity for enjoyment—while others are destitute of all these accessories to happiness. Putting aside the fact that those whose lots seem to be the most enviable are the least to be envied, we cannot help acknowledging that this disparity does exist, and that the earthly lot of some is very hard, while that of others is very easy. But we must remember that there is taught in the New Testament the grand doctrine of Compensation. Paul alludes to this when he remarks that the sufferings of this world are not to be compared with the glories of the world to come, and that the troubles, trials and tribulations of this life are but the precursor of that glorified existence where all these things will be utterly unknown. That some such arrangement would be nothing more than justice there can be no question, and that some principle of Divine Justice must exist was instinctively known long before it was explicitly declared by the inspired apostle, for references to such compensation are found throughout the Psalms. Even Job himself, sunk as he was in the very depth of afflictions, could say: "Though He slay me, yet will I trust in Him; but I will maintain my own ways before Him. He also *shall be* my salvation; for an hypocrite shall not come before Him." So far, then, as man is concerned, this problem of apparent inequality is not so difficult of solution, for he knows only too well that in spite of his hard and bitter earth-life that Divine Justice will be more than vindicated in the life beyond the grave to which he aspires. But in the case of the lower animals, granting that they have no future existence, what, I ask, becomes of Divine Justice? In this land of enlightenment we meet with many animals that are treated with the greatest kindness by their masters, and others, endowed with capacities that are not a whit inferior to their more fortunate brethren, that are treated with the utmost cruelty. While one is petted and pampered, another is abused and given over to the pangs of hunger and starvation. If there is a future life for these animals, it is simply impossible to recognize in their Maker that justice which sensible, reasoning man should expect. Such an injustice, as shown by the lives which we have contrasted, would be too flagrant for any human being to perpetrate, unless such a being was wholly deficient in the ideas of right and wrong. But on the supposition that these animals possess immortal souls, and that there is for them a future life in which these souls shall be developed to their fullest capacities, then these apparent discrepancies can be reconciled with Absolute Justice and Perfect Love. In His dealings with the lower animals, as with ourselves, God looks to the spiritual rather than the material world, and by the means of the one instructs and prepares his pupils for the other. With Paul I firmly believe that suffering in the present world has for its object a preparation for and an introduction to a future life, and therefore am thoroughly convinced that any creature capable of suffering has in that capacity its passport to an eternal world.

Another step, that is, the possession of Individuality, as connected with Immortality, now presses forward for consideration. As for man, did he not possess Individuality, no diverseness of management would be needed, for all would be treated in a similar manner. No two faces in man are precisely alike, for the very simple reason that no two souls, of which the countenance is an indication, are alike. The same rule, no matter what may be affirmed to the contrary, holds good among the lower animals. To the casual observer no apparent difference can be detected between any two individuals of a flock of sheep, a portrait of one equally resembling that of any other. But a shepherd, who understands his business, will readily distinguish every sheep of his flock, as well as describe the mental peculiarities of each individual. One ordinary yellow canary looks just like another yellow canary to the ordinary vision, while in reality the mental character of each bird is impressed just as strongly upon its countenance as are human qualities upon that of man. This quality it is, both in man and beast, that implies a separate treatment for each individual, and becomes a plea for an immortality of life. I am not alone in this idea. It is simply astounding how Individuality in the lower animals is ignored by man. The generality of grooms treat all horses as though they were just so many machines turned out of the same mould, and to be treated just like machines. There is in every species a double kind of Individuality. One kind there is that is common to the entire species, and then there is in addition to this common characteristic another that distinguishes each separate being from its fellows. It is the former that makes a species what it is, and there can be no doubt that each will exist in the future life, and that both may be capable of development. The dog, the horse, the lion and the elephant, and in truth all animals that may be fitted to survive, will be in the other world what they are in this. They will be better animals in that world, just as we hope to be better men, but they will not approach us any nearer than they do in the earth-life.

Man does not, as some are foolish enough to claim, lower the condition of humanity the least by granting immortality to the lower animals. If they be immortal, as the evidence adduces most strongly shows, there is not the slightest use of denial. We cannot shirk a fact, and even if we could, we ought not to do it. Such an argument, which seeks to elevate man by depreciating his lower fellow-creatures, is not very creditable to humanity. In announcing the belief that the lower animals share immortality with man in the higher world, as they share mortality in this, does not claim for them the slightest equality. Man will be man and beast will be beast, and insect will be insect, in the next world as they are in this. They are living exponents of Divine Ideas, as is evident from the Scriptures and the teachings of science, and will be wanted to continue in the world of spirit the work which they have begun in the world of matter. True it is, as has been asserted, that because a man can transmit his ideas to the lower animals, there is evidence that they possess a spirit which is able to communicate with the spirit of man. When a man gives an order to his dog, and is obeyed, there is proof that both possess spirits, similar in quality, though differing in degree. We know that to give an order to a plant would be useless and absurd, because the plant has not the spirit that can respond to the spirit of the man in the same manner that a dog's or a horse's spirit can, but the inability so to respond does not prove that the plant is devoid of a spirit. That the spirit of the plant does respond to the spirit of the man, when it adapts itself to the conditions which the spirit of the man has imposed upon it, there can be no question, or the many hundred plants which have been reclaimed from a state of wildness by a judicious and careful management upon the part of man would have been among the impossibilities of modern civilization. The spirit of man must have entered into the spirit of the plant, and held communion therewith, or the world to-day would not have been blessed with its manifold cereals, fruits and vegetables, all of which have been rendered possible for use by the spirit of man entering into an understanding with the nature, wants and peculiar dispositions of the plants about him. No less are plants living exponents of Divine Ideas than worms, insects, beasts and men are, and as such living exponents, they are as much needed in the future existence, at least such as are fitted to continue in the spirit-world the work begun in the world of matter, as are the higher forms of animal beings. As plants go a great ways towards making this earth-life a paradise of beauty and delight, and have ever been associated through the ages with animal life, each of the two great kingdoms of life from simple beginnings attaining to higher and still higher development up to the present period—the Era of Mind—it cannot be possible that the two will have become suddenly divorced when the temporal or earth-life is about to pass into the eternal or spirit-life. Heaven would not be Heaven without the plants that we have cultured, and tended, and admired.

Concluding, then, let me say, I claim not for the lower animals the slightest equality with man. What I claim for them is a higher *status* in creation than is generally attributed to them. I claim for them a future life, where they will receive a just compensation for the sufferings which so many of them have to undergo in this world. Most of the cruelties which are perpetrated upon animals are due to the habit which man has, in his exalted opinion of self, of considering them as mere automata, without susceptibilities, without reason and without the capacity of a future. That I have achieved the purpose, with which I set out, of proving that all life is immortal, or that soul exists in plants and animals, I think must be admitted. If this doctrine of immortality shall have the effect of bringing about a more humane treatment of the animals over which man has been given dominion, and thus contribute, be it ever so little, to their well-being and happiness, even in this life, then the object attained will be felt to be a just and worthy recompense for the thought and labor which have been expended in its support and defence. Not alone are we of the upper walks of being made the possessors of the inner life, but all nature shares it in common with us, and love is its expression and the method of its action.

Index

I

L

M

N

O

P

R

S

T

V

W

Y

Z